Antiviral Drug Development

A Multidisciplinary Approach

NATO ASI Series

Advanced Science Institutes Series

A series presenting the results of activities sponsored by the NATO Science Committee, which aims at the dissemination of advanced scientific and technological knowledge, with a view to strengthening links between scientific communities.

The series is published by an international board of publishers in conjunction with the NATO Scientific Affairs Division

A	**Life Sciences**	Plenum Publishing Corporation
B	**Physics**	New York and London
C	**Mathematical and Physical Sciences**	D. Reidel Publishing Company Dordrecht, Boston, and Lancaster
D	**Behavioral and Social Sciences**	Martinus Nijhoff Publishers
E	**Engineering and Materials Sciences**	The Hague, Boston, and Lancaster
F	**Computer and Systems Sciences**	Springer-Verlag
G	**Ecological Sciences**	Berlin, Heidelberg, New York, and Tokyo

Recent Volumes in this Series

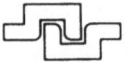

Series A: Life Sciences

Antiviral Drug Development

A Multidisciplinary Approach

Edited by

Erik De Clercq

Katholieke Universiteit Leuven
Leuven, Belgium

and

Richard T. Walker

The University of Birmingham
Birmingham, United Kingdom

Plenum Press
New York and London
Published in cooperation with NATO Scientific Affairs Division

Proceedings of a NATO Advanced Study Institute on
Antiviral Drug Development: A Multidisciplinary Approach,
held May 10–23, 1987,
in Il Ciocco, Lucca, Italy

Library of Congress Cataloging in Publication Data

NATO Advanced Study Institute on Antiviral Drug Development: a Multidisciplinary Approach (1987: Lucca, Italy)
   Antiviral drug development.

   (NATO ASI series. Series A, Life sciences; v. 143)
   "Proceedings of a NATO Advanced Study Institute on Antiviral Drug Devlop-
ment: A Multidisciplinary Approach, held May 10–23, 1987, in Il Ciocco, Lucca,
Italy"—T.p. verso.
   "Published in cooperation with NATO Scientific Affairs Division."
   Bibliography: p.
   Includes index.
   1. Antiviral agents—Congresses. I. De Clercq, Erik. II. Walker, Richard T. III.
North Atlantic Treaty Organization. Scientific Affairs Division. IV. Title. V. Series.
RM411.N36 1987   616.8'7   87-35747

ISBN-13: 978-1-4684-7277-6   e-ISBN-13: 978-1-4684-7275-2
DOI: 10.1007/978-1-4684-7275-2

© 1988 Plenum Press, New York
Softcover reprint of the hardcover 1st edition 1988
A Division of Plenum Publishing Corporation
233 Spring Street, New York, N.Y. 10013

ACKNOWLEDGEMENTS

This Advanced Study Institute was sponsored by NATO (ASI no 237/86) and co-sponsored by FEBS (Advanced Course no 87/02). The Course Directors also acknowledge the financial assistance of the following companies :

- Astra Alab AB, Södertälje, Sweden
- Bayer AG, Wuppertal, W.-Germany
- Beecham Pharmaceuticals, Epsom, Surrey, Great Britain
- Behringwerke AG, Marburg, W.-Germany
- Boehringer Mannheim GmbH, Mannheim, W.-Germany
- Bristol-Myers Company, New York, New York, USA
- Farma-Biagini s.p.a., Castelvecchio Pascoli (Lucca), Italy
- Farmitalia Carlo Erba, Milano, Italy
- Glaxo, Greenford, Middlesex, Great Britain
- Hoechst AG, Frankfurt am Main, W.-Germany
- Hoffmann-La Roche Inc., Nutley, New Jersey, USA
- ICN Pharmaceutical Inc., Costa Mesa, California, USA
- Janssen Pharmaceutica, Beerse, Belgium
- Merz + Co GmbH & Co., Frankfurt am Main, W.-Germany
- Ortho Pharmaceutical (Canada) Ltd., Don Mills, Ontario, Canada
- Pfizer Central Research, Sandwich, Kent, Great Britain
- Praxis Pharmaceuticals Inc., Los Angeles, California, USA
- Roche Products Ltd., Welwyn Garden City, Hertfordshire, Great Britain
- Sandoz Forschungsinstitut GmbH, Wien, Austria
- Schering Corporation, Bloomfield, New Jersey, USA
- Sclavo s.p.a., Siena, Italy
- SeaPharm, Fort Pierce, Florida, USA
- Syntex Research, Palo Alto, California, USA
- Toyo Jozo Co., Ltd., Ohito, Shizuoka, Japan
- Upjohn Company, Kalamazoo, Michigan, USA
- Wellcome Research Laboratories, Beckenham, Kent, Great Britain

The Editors are very grateful to Christiane Callebaut for her skilful editorial assistance.

CONTENTS

ANTIVIRAL CHEMOTHERAPY: AN INTRODUCTION AND REASONS FOR THE SLOW PROGRESS,

PARTICULARLY TOWARDS RATIONAL DESIGN

Richard T. Walker

Chemistry Department
Birmingham University
Birmingham B15 2TT, Great Britain

INTRODUCTION

Six years ago, Sir Charles Stuart-Harris opened a Beecham Colloquium on Problems of Antiviral Therapy[1] with a lecture entitled "Antiviral Chemotherapy: an Introduction and Apology for the Slow Progress". Today, I think we no longer have to apologise; progress is still very slow but as I hope to show, there are many reasons for this, some of which are outside the control of scientists and given the constraints, no one can reasonably expect progress to be any faster.

In this introduction, I wish to examine the current status of antiviral chemotherapy and set the scene for the review lectures which are to follow and in which specific areas will be discussed in more detail.

This is the third Course in this series, the first being held eight years ago.[2] Already at that time, even in this pre-AIDS period, it was becoming clear that the old-fashioned view concerning antiviral chemotherapy, which was a safe and specific antiviral agent would be unlikely to exist, was not correct. Acyclovir[3] had been synthesised and shown to be effective and there were rumours of other compounds being discovered in many laboratories.

Four years later,[4] the initial enthusiasm had waned somewhat. The progress of acyclovir continued unabated but most of the other compounds were running into difficulties and few if any new leads had been discovered. However, we were learning more about how the so-called second generation of antivirals might work and in an elegant introduction to our meeting four years ago, Bill Prusoff[5] showed us some of the specific viral targets which could be attacked so that efficacious and safe antiviral drugs could be developed. Now, four years later, acyclovir is on the market but apart from the area of anti-AIDS compounds, progress has been extremely disappointing. Why is this ?

WHY HAS PROGRESS BEEN SLOW ?

In order to generate discussion, I will be deliberately provocative and I would suggest that one of the main reasons for lack of progress is that in general, industry and government are not interested in preventing or alleviating the effects of viral infection by means of chemotherapy. The lawyers have got such a domination of this area, that what was already a high-risk area in terms of money invested when balanced against the possible return, is now so unattractive that it is unlikely that there are many areas where the risk is worth taking.

That governments and industry are not interested has been highlighted in the last few years by their attitude to the spread of HIV. Suddenly they have realised that viral diseases could not be cured with penicillin and now that a lethal viral disease is at large among the populations of the industrialised nations, it has had the effect of focussing the minds of politicians on the necessity to find a cure. Thus, in a remarkably short space of time,[6] a drug (AZT) which has been around for years, has been tested, synthesised in very large quantities and licensed for use. This alternative of no treatment is not acceptable. However, outside the industrialised countries there are many other viral infections which cause death and misery and in which little interest has been shown.

Until recently, it was not possible to develop a vaccine against hepatitis B virus. This problem has now been solved[7] but no successful antiviral compound had been found to prevent the disease, despite the fact that currently there are 200 million carriers of hepatitis B virus in the world of whom 40 million will die of cirrhosis and 10 million of hepatocarcinoma. It will be several years, if ever, that deaths from AIDS will reach this level.

Thus I have to conclude that with such a high-risk future, it is unlikely that many antiviral compounds will ever reach the market. It has been shown that when public demand insists, compounds can be found very quickly, most of the safety and testing standards can be bent and a compound with a very low therapeutic index can be licensed, but most viral diseases can be allowed to run their natural course.

With that as introduction, I would now like to detail some of the major reasons why advance in antiviral chemotherapy has been so slow. These reasons follow in no particular order, but taken together they all contribute to a general atmosphere which is not encouraging.

Firstly, there is a general impression that the most reasonable way to cure viral diseases is by their prevention by using vaccines. Vaccines in many areas have had outstanding success and apart from total victory over smallpox, many other of the common childhood viral diseases are almost unknown in countries like the USA where vaccination is taken seriously by the health authorities. Until recently there have been some problem areas but we have already referred to one, hepatitis B virus, which has recently succumbed to the molecular virologist. However, there are still some outstanding examples of economically important viral diseases where no ideal vaccine yet exists such as influenza virus, rhinovirus, rotavirus, herpesvirus and HIV. Herpesvirus may have its own problems because of latency, the others cause problems because of the diverse spectrum of epitopes present and because mutants continuously evolve. Thus, there are few virus diseases which present a commercially viable target once vaccines have been discussed.

However, consideration of vaccines raises two other problems which are equally applicable to antiviral chemotherapy and these are costs of development and use, and safety. Although presumably someone undergoing chemotherapy already has the disease, whereas by definition someone being vaccinated does not have it, it is rare in a developed country (with the exception of HIV) for people to die or suffer permanent damage from viral infections (at least on a scale which governments notice). This means that the safety regulations for licensing are correctly very stringent, but then the product is going to be very expensive, can only therefore be used in developed countries where it has to be supremely safe because, left to itself, the virus will probably not do too much damage and the lawyers are every-ready to take advantage of any possible side-effects which might be the result of treatment.

Another problem which retarded research in the early days into antiviral chemotherapy was the competing research on interferon. Many large pharmaceutical companies invested large amounts of money into research in this area in the hope of finding an "antiviral penicillin". This resulted in lack of funds for antiviral chemotherapy and yet despite all this investment in interferon, very little in terms of clinically useful products has resulted, certainly as far as the antiviral field is concerned; and the future does not look bright.

As I have previously mentioned, the existence of specific viral targets for attack has been recognised for some time and yet with amantadine as an exception, all the licensed antivirals are nucleosides and are targeted against virally coded enzymes. None of the other potential targets seem to have been tackled and no non-nucleosides have been found to be specific enzyme inhibitors. I suggest that the main reason for this is that because of the limited resources that are available, we do not have sufficient information for any designed attack to be made on targets other than the virally coded enzymes. Even in the HIV field, all the initial success has been with nucleoside analogues with attempts to produce reverse transcriptase inhibitors or chain terminators. Yet, almost inevitably, nucleosides are likely to have some side effects as the compounds or their metabolites may be recognised by other enzymes in the cell. Even when the HIV target is not the reverse transcriptase,[8] compounds designed to prevent translation of the tat-III gene are still nucleoside derivatives in the form of antisense DNA. However, with our current lack of knowledge, when it is still difficult, if not impossible, to design rationally even a nucleoside analogue, it is clearly too much to expect a rational design of non-nucleoside molecules.

Before I discuss further the prospects for the design of antivirals, there are more problems which contribute to the lack of success in antiviral chemotherapy, which will no doubt be covered in some detail during this Course. Among these are the relevance of test systems and toxicity assays and the problems of early viral disease diagnosis. The latter, with the help of molecular biology is reasonably easily overcome nowadays if the money is available. However, it requires a change in attitude of many clinicians, which is often that "if it is a viral disease, it really doesn't matter too much which one it is because little treatment is available for any of them".

The relevance and use of testing and screening assays is altogether another problem. Antiviral chemotherapy is (usually) aimed at preventing or curing a viral disease in a human patient; yet, with the exception of anti-HIV compounds, that is the last thing which one is allowed to do. The close association of a virus with its host, not only brings problems in the design of safe antivirals but it also makes the design of relevant test systems very difficult. Because of cost and the quantity of material available, one is usually initially restricted to in vitro assays in cell culture. Some viruses cannot so be cultured, other systems are so artificial that their relevance to the clinical disease is tenuous at best and yet this is the sort of system we often have to use to find leads. Sometimes it is relatively easy to stop viral growth in infected cells but it proves to be much more difficult when the compound is used in vivo. This can be because either transport or cell targeting or uptake is very different or that the drug is metabolised before it reaches the target site. In vitro cell lines usually do not contain a full complement of functional enzymes and thus a compound like E-5-(2-bromovinyl)-2'-deoxyuridine (BVDU) works even better in vitro than it does in vivo where it is susceptible to nucleoside phosphorylase which is absent or present at only low levels in many cell lines. However, this doesn't have to be an all or none effect. Should a compound require activation by one or more enzymes before the active metabolite is produced, then small changes in concentrations of intermediates could mean that the K_m for subsequent enzyme-mediated transformations may not be achieved and so the active metabolite is not produced or some intermediate is side-tracked down another pathway.

Animal models often have precisely the same problems. Usually one is trying to cure or prevent a human disease and the animal model, apart from being very costly, may or may not be relevant. Uptake, distribution and metabolism of the drug may be very different, as is the nature of the viral infection.

Toxicity assays also may not be relevant. Recently it was reported that 1-(2'-deoxy-2'-fluoro-β-D-arabinofuranosyl)-5-iodocytosine (FIAC)[9] is poorly deaminated by dogs used for toxicity studies, whereas in humans it is rapidly deaminated to the corresponding uracil derivative, FIAU. What then is the purpose of doing toxicity studies of this compound in dogs when one is concerned with a different compound in the human situation ? It is a very naive assumption to make, that all metabolism of compounds in test animals and humans is likely to be the same or even to be similar, particularly when huge doses of compounds are used.

Thus all these factors add up to give the present situation where antiviral chemotherapy is still regarded as a high risk business. Although targets are clearly presented, we do not have sufficient information for rational design. The chance of getting anything clinically-useful from an in vitro screen which will then show promise in animal models, pass the toxicity tests and yet still work against the clinical disease is very small and the number of viral diseases worth commercial consideration is also small.

Let me finally turn to the problem of design and I will restrict myself to an area where I have had some experience and which is probably one of the areas in which the most information is available; the design of an antiherpesvirus nucleoside.

We have several lead compounds: 5-iodo-2'-deoxyuridine (IDU), acyclovir, FIAC and BVDU. How do they work, do they have anything in common and ought we to be able to design something better ? All these compounds require phosphorylation by a kinase.[10] The latter three compounds are only phosphorylated by the virally coded thymidine kinase, whereas IDU shows little specificity, is also phosphorylated by the normal cellular kinase and is therefore much more toxic.

It is known that acyclovir monophosphate is further taken to the diphosphate by a cellular guanylate kinase and presumably the triphosphate is a substrate for the viral polymerase and acts as a chain terminator.

This at least is a plausible mechanism because it is not clear to me how the other nucleosides exert their antiviral effect. They too apparently require further phosphorylation (BVDU is relatively ineffective against HSV-2 and here the second phosphorylation step is less efficient although this may not be cause and effect).

For many years it was a tenet of faith that incorporation of these analogues into viral DNA was responsible for the cessation of viral replication. For instance, DNA containing BVDU has been claimed to be more susceptible to single-stranded breaks than normal DNA[11] although in vitro it is known that DNA containing quite large quantities of 5-substituted pyrimidine nucleosides can be replicated and transcribed. Such incorporation may be mutagenic (although BVDU is not mutagenic in the Ames test and other test systems[12]) but should not necessarily be responsible for the very efficient way in which viral DNA synthesis is stopped. It is also possible that the nucleoside analogue triphosphates are viral DNA polymerase inhibitors rather than alternative substrates but again one has to question whether the effects seen in vivo are likely to be caused by such inhibition at the concentrations of analogue likely to be achieved.

Recently[13,14] it has become possible to study the effect of the activation of BVDU to the monophosphate in a mammalian carcinoma cell line. Murine mammary carcinoma cells have been transformed with the thymidine kinase gene from HSV-1[13] or HSV-2[14] and the effect of BVDU investigated. If one assumes that what one now sees is due exclusively to the production of BVDUMP in the cells, the result is indeed dramatic as BVDU is now toxic with an ID_{50} of 500 pg/ml for the HSV-1 transformed line and 50 pg/ml for the HSV-2 transformed line. Thus, BVDU is 5 orders of magnitude more toxic than it is for the wild-type cell line and yet acyclovir is only 100-fold more toxic for these transformed cells. Why is BVDU now so toxic and does this have any relevance to its antiviral activity ? At the moment, we do not have the answers to these questions but at least it should begin to raise doubts in our mind as to whether a compound which is toxic at 50 pg/ml can conceivably exert this effect, whether against a virus or a transformed cell line, by either being incorporated into DNA or by inhibiting a viral polymerase.

This discussion is relevant to the problem of design because we now have to bear in mind that we do not definitely know how some of the antiviral agents available now, actually work. However, for the purpose of further discussion let us make the most simple and naive assumption that we require a pyrimidine analogue 5'-triphosphate which will inhibit a viral polymerase and let us examine the metabolic pathway open to the nucleosides as it enters a cell.

Here, even in our present ignorant state, we can see that at least eight enzymes are involved; four of which have to be avoided and for four of which the analogue or its metabolites need to be substrates. The enzymes in the first category are : 1) nucleoside phosphorylase; 2) cellular thymidine kinase; 3) thymidylate synthetase (for the analogue 5'-monophosphate); and 4) cellular DNA polymerase. The first enzyme inactivates the analogue; reaction with the others would almost certainly cause toxicity. The enzymes in the second category are : 1) viral thymidine kinase; 2) viral (or cellular ?) thymidylate kinase; 3) nucleoside diphosphate kinase; 4) viral DNA polymerase. We know almost nothing about the substrate specificity of any of these enzymes and some of the information we do have seems to lack any chemical logic (why does the viral thymidine kinase phosphorylate specifically both acyclovir and BVDU ?). Thus I would submit that, at present, it is hopelessly optimistic to expect to achieve a rational pathway as those compounds which already show activity. This, however, would be a minor task when considering what needs to be done if the design of a non-nucleoside is required or that of a product which is to be aimed at one of the other possible targets.

To highlight how little interest has been taken in the design of antiviral drugs, let me suggest an example where design could have been started years ago but presumably it has been thought to be not worthwhile, even though the product would be aimed at herpesvirus infections which are one of the few commercially viable targets for antiviral chemotherapy.

It is known (and there are many examples available), that the specificity of the thymidine kinases of normal human cells, HSV-1 and HSV-2 are different. It is likely that a thymidine kinase inhibitor might show antiviral properties and it should be possible to use the difference in specificity to design HSV-1 thymidine kinase - or HSV-2 thymidine kinase - inhibitors (not nucleosides) which are not inhibitors of the normal cellular enzymes. As a start to this project, if any rationality is to be brought into the design, we need to know how these three key enzymes work. The genes for these enzymes can all be cloned, expressed, probably crystallised and their 3-dimensional structures determined and active site located. Then perhaps an inhibitor could be designed but as far as I am aware, little has been done in this direction even though the chance of success would seem to be reasonably high. So far, with very few exceptions, most nucleosides investigated are either not recognised by the kinase or are substrates for it, but it should surely be possible to design a non-nucleoside inhibitor if the information discussed above were available ?

However, I suspect that no one takes the problem seriously enough to take up the challenge. Hopefully during this meeting it will be shown that this pessimistic view I have taken of the current status of antiviral chemotherapy is wrong and that in the post-AIDS era there will be a resurgent interest in the rational design of antiviral compounds.

REFERENCES

1. Sir Charles H. Stuart-Harris, Antiviral chemotherapy: an introduction and apology for the slow progress, in: "Problems of Antiviral Therapy", Sir Charles H. Stuart-Harris and J. Oxford, eds., Academic Press, New York, p. 1 (1983).
2. R.T. Walker, E. De Clercq, and F. Eckstein, "Nucleoside Analogues. Chemistry, Biology, and Medical Applications", NATO Advanced Study Institutes Series. Series A: Life Sciences, Vol. 26, Plenum Press, New York (1979).
3. G.B. Elion, P.A. Furman, J.A. Fyfe, P. de Miranda, L. Beauchamp, and H.J. Schaeffer, Selectivity of action of an antiherpetic agent, 9-(2-hydroxyethoxymethyl)guanine, Proc. Natl. Acad. Sci. USA 74:5716 (1977).
4. E. De Clercq and R.T. Walker, "Targets for the Design of Antiviral Agents", NATO Advanced Study Institutes Series. Series A: Life Sciences, Vol. 73, Plenum Press, New York (1983).

5. W.H. Prusoff, T.-S. Lin, W.R. Mancini, M.J. Otto, S.A. Siegel, and J.J. Lee, Overview of the possible targets for viral chemotherapy, in "Targets for the Design of Antiviral Agents", NATO Advanced Study Institutes Series. Series A: Life Sciences, Vol. 73, E. De Clercq and R.T. Walker, eds., Plenum Press, New York, p. 1 (1983).

6. H. Mitsuya and S. Broder, Strategies for antiviral therapy in AIDS, Nature 325:773 (1987).

7. M.R. Hilleman, Perspectives in disease prevention by vaccines, in: "Frontiers in Microbiology. From Antibiotics to AIDS", E. De Clercq, ed., Martinus Nijhoff Publ., Dordrecht, p. 179 (1987).

8. P.C. Zamecnik, J. Goodchild, Y. Taguchi, and P.S. Sarin, Inhibition of replication and expression of human T-cell lymphotropic virus type III in cultured cells by exogenous synthetic oligonucleotides complementary to viral RNA, Proc. Natl. Acad. Sci. USA 83:4143 (1986).

9. K.A. Watanabe, U. Reichman, K. Hirota, C. Lopez, and J.J. Fox, Nucleosides. 110. Synthesis and antiherpes virus activity of some 2'-fluoro-2'-deoxyarabinofuranosylpyrimidine nucleosides, J. Med. Chem. 22:21 (1979).

10. E. De Clercq and R.T. Walker, Chemotherapeutic agents for herpesvirus infections, Progress Med. Chem. 23:187 (1986).

11. W.R. Mancini, E. De Clercq, and W.H. Prusoff, The relationship between incorporation of E-5-(2-bromovinyl)-2'-deoxyuridine into herpes simplex virus type 1 DNA with virus infectivity and DNA integrity, J. Biol. Chem. 258:792 (1983).

12. H. Marquardt, J. Westendorf, E. De Clercq, and H. Marquardt, Potent anti-viral 5-(2-bromovinyl)-uracil nucleosides are inactive at inducing gene mutations in Salmonella typhimurium and V79 Chinese hamster cells and unscheduled DNA synthesis in primary rat hepatocytes, Carcinogenesis 6:1207 (1985).

13. J. Balzarini, E. De Clercq, D. Ayusawa, and T. Seno, Murine mammary FM3A carcinoma cells transformed with the herpes simplex virus type 1 thymidine kinase gene are highly sensitive to the growth-inhibitory properties of (E)-5-(2-bromovinyl)-2'-deoxyuridine and related compounds, FEBS Lett. 185:95 (1985).

14. K. Shimizu, L. Ren, D. Ayusawa, T. Seno, J. Balzarini, and E. De Clercq, Establishment of mutant murine mammary carcinoma FM3A cell strains transformed with the herpes simplex virus type 2 thymidine kinase gene, Cell Structure Functions 11:295 (1986).

DESIGN OF NUCLEOSIDE ANALOGS AS POTENTIAL ANTIVIRAL AGENTS

Roland K. Robins and Ganapathi R. Revankar

Nucleic Acid Research Institute
3300 Hyland Avenue
Costa Mesa, California 92626

INTRODUCTION

Viruses are composed of either RNA or DNA, encased in protein shells and often further wrapped in lipid-containing envelopes. Viruses multiply only within living cells, commandeering the host cell to synthesize viral proteins and viral nucleic acids followed by final assembly into new virion particles. Viral infection may vary in severity from mild and transitory infections to illnesses that terminate in death. Persistent viral infection, which for years may not be accompanied by symptoms, can eventually cause chronic degenerative diseases with fatal outcome. It is known that viruses cause certain cancers in animals and man. Successful treatment of viral diseases with antiviral agents requires interruption of virus multiplication by specific inhibition of viral enzymes.

Nucleosides and nucleoside analogs constitute the major class of compounds which to date exhibit significant in vitro and in vivo antiviral activity. Since viruses possess considerably fewer viral associated and viral encoded enzymes than bacteria, the discovery of significant antiviral agents which exhibit selective inhibition has lagged considerably behind the isolation of antibiotics and synthetic antibacterial agents. Since viruses might be considered small parasites which are reduced to the barest of essential features for replication within the cell, viral RNA and DNA synthesis is vital for viral reproduction. Certain viruses contain viral specific RNA or DNA polymerases which are contained within the virion or are encoded in the viral genome(1). In general it would appear that the nucleotide substrate requirements for the viral enzyme may be less stringent than for comparable corresponding cellular enzymes. One of the reasons that nucleosides have been studied as potent antiviral agents is that these compounds rapidly cross the plasma membrane of the cell by a facilitated transport mechanism (2,3), thus gaining rapid entry into the cell. This overcomes one of the major barriers for an antiviral agent which must reach the viral infection within the cell to be effective. Most nucleoside derivatives are phosphorylated within the cell by a viral or cellular kinase to the 5'-phosphate or nucleotide analog which is further converted to the 5'-triphosphate form, the active form of the drug. Once the nucleoside is phosphorylated it does not readily leave the cell and should then stay within the cell long enough to successfully inhibit viral proliferation. Such a nucleoside 5'-triphosphate should <u>not</u> be incorporated into host cellular nucleic acid in order to avoid <u>host</u> toxicity and should tightly bind to the viral specific enzyme for maximum inhibitory activity. The antiviral nucleoside should be orally active for widest use and should possess a significant half life so that it is not too readily excreted. In order to illustrate these principles in more detail, several examples of nucleosides and nucleoside analogs exhibiting significant antiviral activity against viral infections in animal model systems will be reviewed.

NUCLEOSIDE ANALOGS ACTIVE AGAINST DNA VIRUSES <u>IN VIVO</u>

I. Pyrimidine Nucleoside Analogs

5-Iodo-2'-deoxyuridine (Idoxuridine, IDU, <u>1</u>). The first antiviral nucleoside, 5-iodo-2'-deoxyuridine, was synthesized from 2'-deoxyuridine by Prusoff(4) in 1959. Idoxuridine, the prototype of pyrimidine nucleoside antiviral agents, was shown to be active <u>in vitro</u> against various DNA viruses by Herrmann(5) in 1961. Kaufman(6,7) successfully employed IDU for the treatment of herpes simplex keratitis in rabbits and man in 1962. This topical treatment was the first successful use of an antiviral drug for the treatment of herpes infections in man and unequivocally established its use in preventing blindness due to herpes infections of the eye. FDA approval of IDU for this use in the USA quickly followed and it is interesting to note that 5-iodo-2'-deoxyuridine was the first antiviral chemotherapeutic agent to be licensed for human use. It became a prototype for further nucleoside syntheses in the search for other antiviral agents.

IDU   BVDU   FIAC   Ara-T

<u>1</u>   <u>2</u>   <u>3</u>   <u>4</u>

E-5(2-Bromovinyl)-2'-deoxyuridine (BVDU, <u>2</u>). BVDU was reported in 1979 synthesized by Jones, Verhelst and Walker(8) and shown by De Clercq and coworkers(9) to be a potent and selective inhibitory agent for herpes simplex virus type-1 (HSV-1). BVDU, <u>2</u>, differs from IDU, <u>1</u>, in mechanism of action in that it is specifically phosphorylated at the 5'-position by HSV-1 induced thymidine kinase. This restricts the action of BVDU to the virally infected cell. BVDU is also a very potent inhibitor of varicella zoster virus (VZV) and pseudorabies virus and other herpes viruses in various animal model systems and in humans(10). BVDU may be incorporated into DNA, but since it is phosphorylated specifically in the virally infected cell, such incorporation into DNA is largely confined to the virus-infected cell(10). Wildiers and De Clercq(11) have reported that BVDU has been used orally to successfully treat patients with disseminated herpes zoster. Studies by Datema and colleagues(12) of BVDU in HSV-1 infected cells suggest that BVDU appears to be responsible for significantly inhibiting protein glycosylation in these cells.

2'-Fluoro-5-iodo-1-β-D-arabinofuranosylcytosine (FIAC, 3). The synthesis of FIAC was reported in 1979 by Watanabe, Fox and coworkers (13). Animal experiments(14) reported for FIAC show this pyrimidine nucleoside to be capable of protecting mice against a lethal infection of HSV-1. When given to mice, FIAC is substantially deaminated to 1-(2'-fluoro-2'-deoxy-β-D-arabinofuranosyl)-5-iodouracil (FIAU), which is also the major metabolite in man(15). In cancer patients evidence has been presented(16) that the nucleoside 3 can abort cutaneous varicella zoster. Indeed FIAC was found to be superior to Ara-A in this regard in a double-blind study(16). FIAC in this study caused few toxic reactions and in immunosuppressed patients with herpes virus infection showed similar results(17).

1-β-D-Arabinofuranosylthymine (Ara-T, 4) was described in 1950 by Bergman and Feeney(18), and named spongothymidine since it was isolated from certain Caribbean sponges. 1-β-D-Arabinofuranosylthymine, 4, was first synthesized by Fox and coworkers(19) in 1957 and was shown by Underwood et al.(20) in 1964 to have activity against HSV-1 in the rabbit cornea. Gentry and Aswell(21) showed in 1975 that 4 inhibited herpes DNA replication in vitro and was selectively toxic to cells infected by herpes virus. Saneyoshi(22) in 1979 improved the synthesis of 4 and prepared a series of 1-β-D-arabinofuranosyluracils, the most active of these nucleosides(23) against HSV-1 was 4. Ara-T inhibited viral DNA synthesis in herpes infected cells but did not inhibit cellular DNA synthesis in uninfected cells. In vivo experiments with 4 against experimental HSV-1 induced encephalitis in mice showed remarkable efficacy with Ara-T being as potent as Ara-AMP. Ara-T is relatively nontoxic to mice (LD_{50} i.p. 10 g/kg, oral 15 g/kg).

II. Purine Type Nucleoside and Nucleotide Analogs

A. Adenosine Analogs. 9-β-D-Arabinofuranosyladenine (Ara-A, 5) was first synthesized by Baker, Goodman and coworkers(24) at Stanford Research Institute in 1960. Ara-A was later discovered as a nucleoside antibiotic by Miller and coworkers(25). Ara-A was first shown to have significant antiviral activity in vitro against herpes virus and vaccinia virus by de Garilhe and de Rudder(26) in 1964. The in vivo antiviral activity of Ara-A against intracerebral vaccinia infection in mice and herpes simplex viral keratitis in hamsters and intracerebral HSV-1 infections in mice were reported by Sidwell, Schabel and coworkers(27) in 1968.

Ara-A   Cyclaradine   (S)-HPMPA   8
  5   6   7

Ara-A, because of its insolubility, must be administered by slow infusion requiring large volumes of fluid. Ara-AMP, the 5'-phosphate derivative of 5, is considerably more soluble and can be administered by infusion with good tissue distribution. Treatment with Ara-AMP of varicella and disseminated zoster in immunosuppressed patients in comparison with Ara-A has recently been made(28). Good results were obtained with both drugs, and the clinical effectiveness of both drugs were comparable. The greater solubility of Ara-AMP permitted a higher dosage during the first hours of treatment. Ara-A is rapidly deaminated by adenosine deaminase in vivo to give 9-β-D-arabinofuranosylhypoxanthine (Ara-Hx), which is approximately ten times less active than 5 in cell culture experiments. An excellent review of the antiviral effects of Ara-A is available(29). Ara-A is phosphorylated in mammalian cells to Ara-AMP by adenosine kinase and deoxycytidine kinase. Further phosphorylation to Ara-ADP and Ara-ATP also occurs. Ara-A is also converted to Ara-ATP in HSV-1 infected cells. There is a direct correlation between levels of Ara-ATP and HSV replication. The selective action of Ara-A against HSV appears to be a consequence of the preferential action of Ara-ATP against HSV-1 and HSV-2 polymerases. Ara-ATP also inhibits cellular DNA polymerases, which is believed to be responsible for observed toxicity. Shipman, Drach and coworkers(30) have shown that herpes DNA synthesis was considerably more sensitive to Ara-A than was cellular DNA synthesis. Muller et al.(31), using purified preparations of herpes-induced DNA polymerase, showed that Ara-ATP was significantly more inhibitory to herpes-induced DNA polymerase than to the x- or β-DNA polymerase of uninfected cells. Muller et al.(32) have also shown that Ara-A treatment of HSV-infected cells results in accumulation of short DNA fragments. Drach and coworkers(33) have observed that Ara-A is incorporated uniformly throughout the HSV-1 genome, which results in defective viral DNA.

Cyclaradine, 6, another adenosine analog possessing significant activity against certain DNA viruses, is the carbocyclic analog of 9-β-D-arabinofuranosyladenine, prepared in 1977 by Vince and Daluge(34). This nucleoside has the furanose ring oxygen of Ara-A replaced by a methylene group. This derivative is resistant to adenosine deaminase and in contrast to Ara-A is highly active against HSV-2. Topical (intravaginal) treatment of genital (HSV-2) herpes infection in guinea pigs with the 5'-methoxyacetyl derivative of cyclaradine beginning three days after infection largely prevented the development of viral lesions(35). This 5'-ester of 6 was superior to acyclovir in this study. The low systemic toxicity of 6 and its stability toward enzymatic degradation suggest that this antiviral agent deserves further study.

(S)-9-(3-Hydroxy-2-phosphonylmethoxypropyl)adenine, (S)-HPMPA, 7. The nucleotide analog 7 has been prepared by Holý(36) and reported by De Clercq, Holý et al.(37) to be active against a number of DNA viruses in vitro and is active against a cutaneous HSV-1 model infection in hairless mice, and a lethal systemic HSV-1 infection in mice when given orally (37). (S)-HPMPA is also active in the herpetic eye infection model in rabbits. This nucleotide analog 7 is also active in vitro against retroviruses. (S)-HPMPA is not enzymatically dephosphorylated and penetrates cells directly as the phosphonate(37). It appears to be incorporated as a chain terminator. It is active against herpes thymidine kinase lacking mutants resistant to acyclovir and BVDU (37).

4-Amino-7-β-D-xylofuranosylpyrrolo[2,3-d]pyrimidine, 8. Of considerable interest is the xylofuranosyl analog of tubercidin, 8, prepared by Robins and coworkers(38) in 1977. De Clercq, Robins and coworkers(39) have recently shown that, unlike tubercidin which is generally cytotoxic at very low concentrations, xylotubercidin (8) exhibited decreased cytotoxicity with potent activity against HSV-1 and HSV-2 in vitro.

When **8** was administered i.p. at 50 mg/kg/day, a significant reduction of the mortality rate of mice infected with HSV-2 was achieved, (85% survival compared to 12% for the controls)(40). Acyclovir was not effective in prolonging survival rate in this model, even up to 250 mg/kg/day(40). Further studies against intracutaneous HSV-2 in hairless mice xylotubercidin applied topically in 1% DMSO gave 100% survivors without observable toxicity. Acyclovir at 1% concentration was ineffective, however, at 10% concentration 90% survivors were obtained(40). This nucleoside would appear to have considerable potential for topical treatment of HSV-2 infections. It is of interest to speculate regarding the possible mechanism of antiviral action of **8**. 9-β-D-Xylofuranosyladenine was shown to possess in vitro antiviral activity against herpes viruses(26) as early as 1964. Glazer and Peale(41) have shown that 9-β-D-xylofuranosyladenine inhibits nuclear RNA methylation by interference with the synthesis of S-adenosylmethionine. Perhaps the xylofuranosyl analog of tubercidin (**8**) acts selectively to interfere with methylation or processing of herpes viral mRNA. The nucleoside **8** is also highly active against vaccinia virus in vitro(40).

B. Guanosine Analogs. Acyclovir (**9**). Schaeffer and coworkers(42) described the antiviral activity of acyclovir in 1977. The first synthesis of acyclovir, 9-(2-hydroxyethoxymethyl)guanine was that reported by Schaeffer and coworkers(43) from 2,6-dichloropurine and was later greatly improved by direct alkylation of guanine in DMSO in the presence of sodium hydride(44). Acyclovir has proved to be active against HSV-1 and HSV-2 in many animal models and in man(45). Acyclovir is an effective agent for the treatment and prophylaxis of HSV infections in man and is approved for this use in many countries of the world. The drug is well tolerated in humans and may also be employed successfully to treat immunocompromised patients(46). Acyclovir therapy appears to be superior to Ara-A in the treatment of herpes simplex encephalitis in man(47). A double-blind, placebo controlled study on infectious mononucleosis in man with acyclovir suggests acyclovir may have some effect in the treatment of Epstein-Barr viral infections(48). Acyclovir had little or no effect on vaccinia virus, adenovirus type 5, and a range of RNA viruses including rhinovirus 1B, Mengo, Semliki Forest, Sindbis, Bunyamwera, Yellow fever, measles, respiratory syncytial virus, and the NWS strain of influenza virus(43).

2'-Deoxyguanosine   Acyclovir **9**   DHPG **10**   DHBG **11**

The antiviral mechanism of action of acyclovir has been reviewed by Elion(49). Elion and coworkers(42) showed in 1977 that in herpes infected cells acyclovir is converted to the monophosphate by a viral specified thymidine kinase. In the uninfected host cell phosphorylation occurs only to a very limited extent.

Enzymatic conversion of acyclovir to its mono-, di-, and triphosphates

Acyclovir, **9**, was found to be phosphorylated 30-120 times faster with extracts of Vero cells infected with HSV-1 than with similar extracts with uninfected cells. Evidence that this phosphorylation is due to viral specified thymidine kinase has been presented(50). HSV-1 and HSV-2 virally infected cells convert acyclovir to the 5'-monophosphate, which is further phosphorylated by host cell GMP kinase to acyclovir diphosphate, **12**, which is in turn further phosphorylated to acyclovir triphosphate, **13**, by unidentified cellular kinases. Acyclovir triphosphate has been shown to inhibit HSV-1 viral DNA polymerase whereas cellular α-DNA polymerase was found insensitive to acyclovir triphosphate at similar concentrations. The incorporation of **13** into calf thymus DNA primer-template has been shown to be much more rapid and extensive with HSV-1 DNA polymerase than with Vero cell DNA polymerase α. This incorporation of acyclovir ceased after 15 minutes since the template is chain terminated by acyclovir incorporation, since there is no 3'-hydroxyl group on which to continue elongation. The viral DNA polymerase was also inactivated by tight binding to the terminated template. McGuirt and coworkers (51) have identified small DNA fragments synthesized in HSV-1 infected cells in the presence of acyclovir. These DNA fragments were shown to be viral in origin by hybridization to purified HSV-1 DNA. Acyclovir monophosphate terminated DNA thus prepared was found to strongly inhibit HSV DNA polymerase-catalyzed DNA synthesis. Further supportive evidence has been obtained using highly purified HSV-1 and HSV-2 induced DNA polymerase and acyclovir triphosphate (**13**), which was found to be a competitive inhibitor of dGTP entering viral DNA(52). In contrast, HeLa DNA polymerase is insensitive to inhibition by **13**. The result of these selective enzymatic differences in the HSV infected cell suggests a 300 to 3,000-fold difference between the toxicity of acyclovir to the herpes viruses and the human cell.

9-(1,3-Dihydroxy-2-propoxymethyl)guanine (DHPG, **10**). Variations in the carbohydrate portion of acyclovir has resulted in a number of active analogs of acyclovir. The 2'-deoxyguanosine analog DHPG, **10**, was prepared by several research groups simultaneously. DHPG is more soluble than acyclovir and has greater bioavailability. The synthesis and antiviral activity of **10** has been reported by Martin et al.(53,54), Field et

al.(55) and Ogilvie et al.(56). Orally administered DHPG is reported to
be 50-fold more efficacious than acyclovir in the treatment of systemic
or local HSV-1 or HSV-2 intravaginal infections in mice. Smee and co-
workers(54) report that in mice infected with HSV-2 mortality reduction
to 50 percent could be achieved by 7-10 mg/kg/day of DHPG orally, whereas
an equally effective dose of acyclovir required 500 mg/kg/day oral administration. The mechanism of antiviral action of DHPG appears to be generally similar to that for acyclovir. DHPG is reported to be a 30-fold
more efficient substrate for HSV-1 thymidine kinase than acyclovir and
cellular GMP kinase phosphorylates DHPG monophosphate more readily than
acyclovir monophosphate. These differences result in seven times more
DHPG triphosphate in HSV-1 infected rabbit kidney cells than acyclovir
triphosphate six hours after infection. DHPG triphosphate is reported to
be a more selective inhibitor of the viral DNA polymerase compared to the
cellular DNA polymerase α. The Wellcome research group(57) has reported
that DHPG triphosphate was active against two acyclovir triphosphate
resistant DNA polymerases isolated from acyclovir resistant strains of
HSV-1. Cheng and coworkers(58) have shown that DHPG triphosphate competitively inhibits incorporation of dGTP into DNA catalyzed by DNA
polymerase specified by both HSV-1 and HSV-2. DHPG acted as an alternate
substrate for dGTP for the viral specified DNA polymerase. Incorporation
of DHPG into viral DNA resulted in a gradual slowing of viral DNA
synthesis. It was found that DHPG was incorporated into viral DNA in
both terminal and internal linkages(58), due to the presence of the
additional hydroxyl group of DHPG.

DHPG-cP   (R-)DHPG-MP   (S-)DHPG-MP   15
14

Tolman and coworkers(59) prepared the cyclic monophosphate of DHPG,
(60,61) 14, and found this soluble nucleotide to possess considerable
antiviral activity against DNA viruses. DHPG c-phosphate, 14, has the
broadest spectrum against DNA viruses of the acyclovir analogs. It is
active against HSV-1, HCMV, VZV, vaccinia, adenovirus type 2 in culture
and is active against HSV-1 and HSV-2 infections in mice(59). The
nucleotide, 14, is active against TK⁻HSV strains since DHPG c-phosphate is
taken up by the cell intact, converted intracellularly to the single
(S-)enantiomer of DHPG monophosphate(60). This is the same enantiomer obtained by phosphorylation of DHPG by HSV-1 thymidine kinase(61). Cellular
GMP kinase(60,61) phosphorylates the (S-) enantiomer much faster than the
(R-) to yield (S-)DHPG triphosphate which exhibited considerable potency
against HSV-1 DNA polymerase while the corresponding (R-)DHPG triphosphate was inactive. In addition, 14 would appear to have an additional
mode of antiviral action(60). The synthesis of 14 and related derivatives has recently been reviewed(62). Studies have shown that DHPG, 10,
is very effective in the treatment of herpetic corneal lesions in rabbits
(63). In a comparative response titration against acyclovir, 10 was 6.4
times as potent in this model. DHPG has been shown to be orally effec-

tive in preventing HSV-1 orofacial infection and HSV-2 genital infection of mice(64). Topical therapeutic application of DHPG prevented orofacial HSV-1 lesion development in mice and HSV-2 genital lesions in guinea pigs(64). Recent studies(65) on effects of acyclovir and DHPG on a lymphoblastoid cell line dually infected with Epstein-Barr virus and HSV-1 showed that HSV-1 rapidly developed resistance to acyclovir and rather more slowly to DHPG, 10. 100 µM of DHPG eliminated HSV-1 production curing the line of persistent infection(65). DHPG is more potent in vitro against Epstein-Barr virus than acyclovir (66), probably due to the higher level of phosphorylation over that of acyclovir. DHPG showed some effects against CMV infection in a guinea pig model(67). Recent treatment with DHPG of patients with AIDS and serious CMV infection(68) resulted in suppression of CMV replication and improvement in CMV retinitis in 7 of 8 patients. The major side effect was severe neutropenia which proved to be dose limiting(68). Although treatment of CMV retinitis appears beneficial(69) with DHPG, treatment of CMV viral pneumonia with DHPG has been disappointing(70,71).

(R)-9-(3,4-Dihydroxybutyl)guanine (DHBG, 11). Another acyclovir analog, DHBG, has been reported by Larsson et al.(72). DHBG is similar to acyclovir in its activity against HSV-1 keratitis in rabbits. With regard to HSV-2 infection in mice, 11 appears somewhat superior to acyclovir. DHBG, 11, however, is less effective than acyclovir against HSV-1 induced encephalitis in mice. Similarly, in mice infected intravaginally with HSV-2, DHBG systemically administered did not prevent the mortality and spread of virus to the brain(73). The metabolism and mode of action of DHBG has recently been studied in HSV-1 infected Vero cells. DHBG, 11, proved to be a good substrate for HSV-1 and HSV-2 thymidine kinases but not for cellular thymidine kinase(74). DHBG 5'-monophosphate was phosphorylated by cellular guanylate kinase in the infected cells. DHBG 5'-triphosphate was a selective and competitive inhibitor to dGTP as a substrate of the purified HSV-1 and HSV-2 DNA polymerases. Unlike acyclovir, data suggests DHBG 5'-triphosphate inhibition of these polymerases without incorporation into viral DNA(74). A series of analogs related to(R-)DHBG have recently been reported(75). In cell culture against HSV-1, 9-[4-hydroxy-3-(hydroxymethyl)butyl]guanine was superior to (R-)DHBG, however, against HSV-2 both drugs were essentially equivalent in potency(75). When employed topically against a cutaneous HSV-1 infection in guinea pigs the two drugs were essentially equivalent in efficacy(75).

Ara-carbocyclic Analog of 7-Deazaguanosine, (±)-2-Amino-3,4-dihydro-7-[(1α,2α,3β,4α)-2,3-dihydroxy-4-(hydroxymethyl)-1-cyclopentyl]pyrrolo-[2,3-d]pyrimidin-4-one, 15. The synthesis of 15 has recently been accomplished by Legraverend and coworkers(76). The nucleoside 15 studied against a HSV-1 encephalitis model at a total dose of 100 mg/kg given i.p. over a 48 hr period resulted in 4/15 survivors as compared to 0/30 for the controls (77). In this model in comparison, BVDU was only marginally active with 1/15 survivors. The nucleoside 15 at 25 µg/ml inhibited viral DNA synthesis four-fold over cellular DNA synthesis(77).

NUCLEOSIDE ANALOGS ACTIVE AGAINST RNA VIRUSES IN VIVO

I. Adenosine Analogs Active Against RNA Viruses. 3-Deazaadenosine (16).

The prototype nucleoside in this group is 3-deazaadenosine, 16, which was first prepared(78) in our laboratory and reported in 1966. 3-Deazaadenosine has been shown to inhibit replication of RNA type C virus, HL-23 and Rous sarcoma virus by Bodner, Cantoni and Chiang(79). Avian sarcoma virus B77, grown in chicken embryo fibroblasts, is 90 percent inhibited at 100 μM of **16.** 3-Deazaadenosine is also active in vitro against the vesicular stomatitis (VSV), parainfluenza 3 and reovirus at less than 10 μg/ml, which is 5-20-fold lower than the concentration cytoxic to the host cell. The mode of antiviral action of 3-deaza-adenosine has been attributed by Bader and coworkers(80) to the inhibition of S-adenosylhomocysteine (AdoHcy) hydrolase which results in prevention of methylation required for viral RNA transcription. Such inhibition of the methylation of the 5'-cap of Newcastle disease viral mRNA had been shown by Borchardt and coworkers(81) for analogs of S-adenosylhomocysteine. Since the accumulation of S-adenosylhomocysteine inhibits the methyl transferase reaction from S-adenosylmethionine (AdoMet) by a feedback mechanism, the viral mRNA is not transcribed due to inhibition of the methylation of the guanylate molecule at N-7 in the 5'-terminal position. 3-Deazaadenosine is not enzymatically deaminated or phosphorylated and is capable to act both as an inhibitor of S-adenosylhomocysteine(AdoHcy) hydrolase and a substrate of this enzyme to yield 3-deazaadenosylhomocysteine, which has been shown to inhibit methylations in vivo, itself acting as a feedback inhibitor of the viral methyl-transferases. Bodner, Cantoni and Chiang(79) have postulated that the accumulation of S-adenosylhomocysteine and 3-deazaadenosylhomocysteine leads to an inhibition of the methylation reactions essential to viral mRNA transcription.

3-Deazaadenosine **16**   (S)-DHPA **17**   (±)3-Deazaaristeromycin **18**   Neplanocin A **19**

(S)-9-(2,3-Dihydroxypropyl)adenine [(S)-DHPA, **17**]. The adenosine analog (S)-DHPA, **17**, was first synthesized by Holý(82). (S)-DHPA inhibits several DNA viruses in vitro including vaccinia, HSV-1 and HSV-2, as well as RNA viruses such as VSV, rabies, Ebola virus, rota viruses, parainfluenza 3 and measles. (S)-DHPA is inactive against polio virus, Coxsackie B-4, Newcastle disease and Sindbis virus in cell culture(83). In animal experiments, (S)-DHPA affords some protection against rabies and rotavirus infections. Against VSV viral infections in mice the final surviving number of mice treated with **17** i.p. at 135 mg/kg one hour pre-infection and daily following infection, increased from 37.5% in the control group to 67% in the treated animals. The in vivo activity is most pronounced against the rhabdoviruses such as VSV and the rabies virus(84). The antiviral mechanism of action of **17**, as with **16**, is due to the inhibition of S-adenosylhomocysteine hydrolase (84), and results in the cellular accumulation of S-adenosylhomocysteine, as in the case for 3-deazaadenosine. (S)-HPMPA, **7**, may be considered the 2-phosphonylmethyl derivative of (S)-DHPA. Unlike **17** however, **7** does not inhibit(37) S-adenosylhomocysteine(AdoHcy) hydrolase.

(±)3-Deazaaristeromycin (**18**). The carbocyclic analog of 3-deaza-adenosine [(±)3-deazaaristeromycin, **18**] was synthesized by Montgomery and coworkers (85) and reported to be active against HSV-1 and vaccinia virus and HL-23 in vitro. De Clercq and Montgomery(86) have shown that 18 is active against VSV in vitro and protected newborn mice against a lethal infection of VSV. When given one hour after infection mortality was reduced from 90% in the controls to 48% in the treated mice. De Clercq and coworkers(87) have, in addition, shown (±)3-deazaaristeromycin, 18, to be active in vitro against the RNA viruses Sindbis, measles, parainfluenza 3 and reo-1 viruses. The in vitro activity of 18 in general is superior to that of 3-deazaadenosine, 16; however like 16 and 17, (±)3-deazaaristeromycin, 18, is a potent inhibitor of AdoHcy hydrolase(85) and the antiviral activity appears to depend on this inhibition(87).

   Neplanocin A (19). The adenosine analog neplanocin A, was isolated in Japan and shown by Hayashi and coworkers(88) to have structure 19. Neplanocin A is an antitumor antibiotic which is readily deaminated by adenosine deaminase. A total synthesis of neplanocin has recently been published(89). Borchardt(90) has shown that neplanocin A is a potent inhibitor of vaccinia virus multiplication in mouse L-cells and is a potent inhibitor of AdoHcy hydrolase both in vitro and in vivo(90), resulting in a mechanism of antiviral action similar to that of 16, 17 and 18. Neplanocin A is active against parainfluenza, measles and VSV in vitro(91) and the mechanism of antiviral action confirmed by De Clercq (91). In young mice neplanocin A showed partial reduction in the mortality rate after inoculation intranasally with VSV(91). Neplanocin A did not protect newborn mice against a lethal Coxsackie virus type B-4 infection(91). The lack of antiviral efficacy of 19 could be due to deamination via adenosine deaminase or lack of selective inhibition of viral mRNA methylation since Glazer and Knode(92) have shown that neplanocin A inhibits cellular RNA methylation by formation of a nucleoside analog of S-adenosylmethionine. The fact that 20 μg per mouse of 19 was found lethal to newborn mice if injected intraperitoneally suggests a good therapeutic index could not be obtained due to toxicity(91).

   Sinefungin (20). Sinefungin, an antifungal antibiotic(93) is an analog of S-adenosylhomocysteine (AdoHcy) and an inhibitor of AdoHcy hydrolase. Sinefungin is a potent inhibitor of Newcastle disease virus (NDV) and vaccinia virus in mouse L-cells(94). Sinefungin, 20, is active against cutaneous HSV-1 and vaginal HSV-2 in guinea pigs(95), and systemic herpes viral infections in mice(95). Sinefungin appears to inhibit methylation of NDV and vaccinia viral mRNA(94). Sinefungin also was effective in preventing Rous sarcoma virus-induced transformation of chick embryo fibroblasts (CEF)(96). Unfortunately 20 is also an inhibitor of tRNA and protein methyltransferases in CEF cells which suggests it may be too toxic for human use.

Sinefungin   Ribavirin   Selenazofurin
20   21   22

AZOLE NUCLEOSIDES ACTIVE AGAINST RNA VIRUSES IN VIVO

1-β-D-Ribofuranosyl-1,2,4-triazole-3-carboxamide (Ribavirin, 21). As part of a research program to prepare broad-spectrum antiviral agents Witkowski et al.(97) synthesized and reported from our laboratory in 1972 the unique antiviral agent 1-β-D-ribofuranosyl-1,2,4-triazole-3-carbox-amide, 21 (ribavirin). This research was based on the concept that "initiation of virus specific protein synthesis and/or RNA synthesis may utilize unique viral enzymes which could be specifically inhibited"(97). Sidwell et al.(98) reported in 1972 on the broad antiviral spectrum of ribavirin against RNA and DNA viruses both in vitro and in vivo. Ribavirin was the first broad spectrum antiviral agent reported which possesses significant activity against both DNA and RNA viral infections in animal model systems(99). The 50% cytotoxic dose of 21 varies with the cell line and is approximately 200-1000 μg/ml as determined by microscopic examination, trypan blue uptake, plating efficiency of viable cells and total cellular protein(99). Drach and coworkers(100) have demonstrated that previous studies showing that ribavirin inhibited DNA synthesis based on thymidine uptake are in error since 21 inhibits thymidine phosphorylation. The cytotoxic effects of ribavirin on cellular metabolism observed in HeLa cells at high doses is readily reversible upon removal of the drug from the cells. No incorporation of ribavirin into DNA or RNA has been observed in mammalian or viral systems. For a discussion of the toxicology and pharmacology of ribavirin the reader is referred to recent reviews(101,102).

Ribavirin is rapidly phosphorylated by cellular adenosine kinase and then further phosphorylated to RTP presumably by cellular enzymes. Ribavirin and its 5'-phosphate derivatives are unique in that the amido group which should preferably be coplanar with the triazole ring, can be rotated so that ribavirin resembles adenosine for phosphorylation by adenosine kinase, or as in ribavirin 5'-triphosphate (RTP) the carboxamido group can be rotated to resemble either GTP or ATP. In fact in the inhibition of influenza RNA polymerase, RTP was shown to compete with both GTP and ATP but not CTP or UTP for the substrate site(103). The inhibition of cell proliferation by ribavirin in cell culture is believed to be due to the lowering of the pool sizes of GTP as noted for L-5178Y cells(104). This lowering of GTP pools has been confirmed by many investigators in numerous cellular systems and is due to the inhibition of IMP dehydrogenase by ribavirin 5'-phosphate, as first reported from our laboratory(105) in 1973. The structure of ribavirin as noted by single crystal X-ray studies of Sundaralingam(106), is strikingly similar to guanosine with the carbonyl oxygen and the amide nitrogen occupying stereochemically similar positions to the carbonyl oxygen and the amide ring in guanosine. The antiviral effect of ribavirin against measles in Vero cells is partially reversed by guanosine(105). Guanosine also reverses the cytostatic effect of ribavirin against L-5178Y cells and reverses the inhibition of hemagglutinin produced by influenza A virus in MDCK cells. Ribavirin 5'-monophosphate is 1000 times more potent than GMP acting as a feedback inhibitor of IMP dehydrogenase(105). (K_i = 2.5 x 10^{-7} compared to 2.2 x 10^{-4} for GMP). This potent inhibition prevents the formation of XMP and therefore GMP in the cell. In this sense there is a remarkable "self potentiation" effect of ribavirin since the natural GTP pool is considerably lowered which diminishes the reversing effect of GTP, since GTP competes with ribavirin 5'-triphosphate for the GTP sites on the more specific viral induced enzymes such as viral RNA polymerase. Linitskaya and coworkers(107) have shown that ribavirin at 50-100 μg/ml also inhibited virus induced synthesis of RNA-dependent RNA polymerase in chick fibroblast cultures infected with influenza A virus. John Oxford(108) first showed that in cells infected with influenza A in the presence of ribavirin, no new viral proteins were

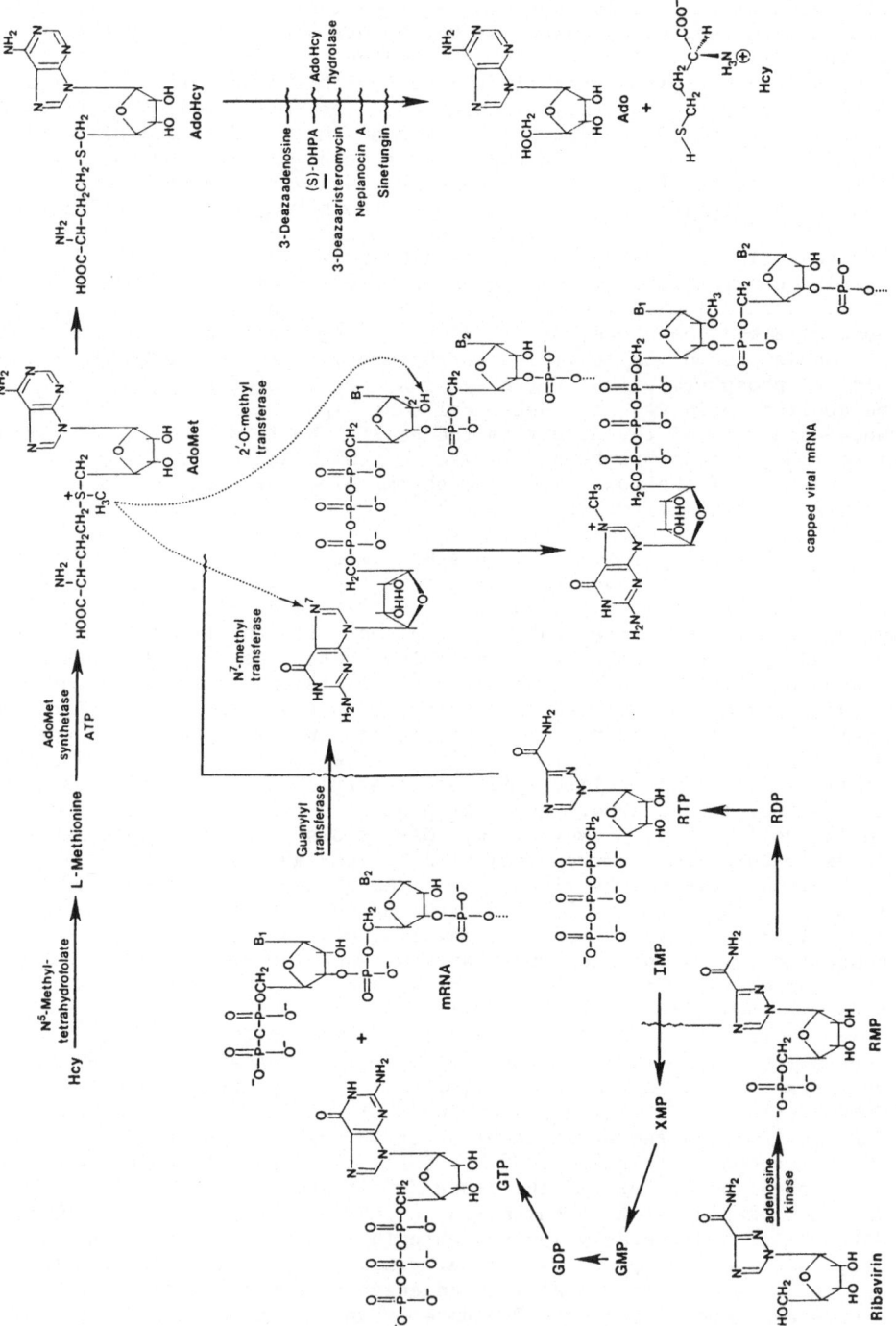

synthesized and suggested that the primary site of action of ribavirin was the inhibition of an early stage of viral RNA synthesis. In study of the inhibition of rabies virus by ribavirin in chick embryo cells, Bussereau and Ermine(109) have shown that the rate of cellular protein synthesis was not modified but the rate of viral protein synthesis decreased in a dose-dependent fashion which corresponded to a decrease in the rate of viral RNA synthesis. Scholtissek(110) showed that in the inhibition of influenza fowl plague virus by ribavirin in vitro, neither virion nor viral complementary RNA was produced. There was no interference with the normal incorporation of guanosine into chick embryo cells(110). Goswami, Smith et al.(111) showed that ribavirin 5'-triphosphate exhibits a potent inhibitory effect on the 5'-terminal guanylylation of in vitro synthesized uncapped viral mRNA of vaccinia virus. It has not been established whether ribavirin as the 5'-triphosphate is actually incorporated into the cap in place of GTP. Further studies (112) have shown that ribavirin 5'-triphosphate is an effective inhibitor of vaccinia viral mRNA guanine-N^7-methyltransferase. Ribavirin 5'-triphosphate at 25 µM gives a 70% inhibition of this viral enzyme. These results correspond to the observations of Katz and coworkers(113) who noted that vaccinia virus in the presence of ribavirin made viral polypeptides to some extent but new viral particles were not formed. It appears that many viruses have viral specific enzymes or viral encoded enzymes which are involved in capping viral mRNA required for initiation of viral protein synthesis. Canonico (101) has recently studied VEE infected BHK-21 cells and infected cells treated with ribavirin. Results show that with the VEE-infected cells treated with 300 µg/ml of ribavirin, the viral mRNA cap is reduced 10-fold and the viral mRNA is thereby defective. Recent studies on the mechanism of action of ribavirin against VSV infection in CHO cells indicates no effect on cellular protein synthesis and although VSV mRNA was synthesized, it was not functional and could not be translated. Such studies support the concept of interferences with viral 5'-cap formation (114). For a discussion of the biochemical differences in the eukaryote vs. viral enzymatic capping of mRNA the reader is referred to our recent paper(115).

In vivo antiviral activity of ribavirin against Toga-, Bunya- and Arena-viruses has been reviewed by Canonico and coworkers(116). Ribavirin has been shown in laboratory animal models to have significant inhibitory effects against Rift Valley fever virus, Punta Toro, Hantaan virus, Pichinde, Machupo and Lassa fever viral infections(116). Ribavirin gives greater than 90% protection in mice against a lethal dose of Rift Valley Fever virus. At 100 mg/kg/day ribavirin increases the survival rate from 10% to 90% of Punta Toro viral infected hamsters(116). Hantaan-infected suckling mice treated with ribavirin (50 mg/kg/day) on day 10 (following onset of clinical signs and demonstrable virus in serum and organs) saved 11 of 20 animals compared to 0 out of 70 controls(117). Surviving animals after 75 days showed no signs of disease. Pichinde virus-infected hamsters treated with ribavirin result in an increase of survival from 11% to 100%(116). In Machupo viral-infected rhesus monkeys, 20 mg/kg/day ribavirin significantly reduced viremia and protected rhesus monkeys from the hemorrhagic phase of the disease(116). Lassa fever virus-infected cynomolgus monkeys die with a mean time to death of 15 days. All monkeys treated with ribavirin beginning on days 0-4 survived. When treatment was delayed until day 7, four out of eight monkeys still survived(116). This remarkable protection with ribavirin, even late in the infection, prompted a major clinical trial of ribavirin in Sierra Leone, West Africa against Lassa fever. McCormick et al.(118) report that ribavirin is effective in the treatment of Lassa fever either intravenously or orally and may be used at any point in the illness. In high risk patients the fatality rate was reduced from 55% to 5% by ribavirin treatment(118). Clinical trials of ribavirin against Korean hemorrhagic fever are presently ongoing under the auspices of the U.S. Army Medical Research and Development Command. Ribavirin treatment of marmoset monkeys infected with Junin virus starting 6 days post infection (all animals viremic) increased the mean day of death from 18.3 days in controls to 35.6 days in the treated animals with 29% survivors compared to 0% in the controls(119). Ribavirin has been approved for use in the aerosol form in the U.S. and Canada for the treatment of respiratory syncytial virus (RSV) infections in infants(120,121). These clinical studies have recently been reviewed(122). Smee and Matthews(123) have recently shown that in RSV-infected cells there was a 2.6 fold greater amount of ribavirin 5'-triphosphate than in comparable uninfected cells which had been treated with similar concentrations of ribavirin. Of considerable interest is the recent report(124) of the increased survival time of mice infected with encephalomyocarditis virus when treated with ribavirin. Ribavirin effectively inhibited myocardial virus replication and reduced viral myocardial damage in the hearts of these test animals(124).

2-β-D-Ribofuranosylselenazole-4-carboxamide (Selenazofurin, 22). Selenazofurin, 22, synthesized in our laboratory(125) in 1983, is highly active in vitro against a wide spectrum of DNA and RNA viruses(126). Selenazofurin showed especially good in vitro antiviral activity against the RNA viruses, Yellow Fever virus, Junin virus, Rift Valley Fever virus, Sandfly Fever virus, Korean hemorrhagic Fever virus, and Venezuelan equine encephalitis viruses. Selenazofurin is highly active against influenza A and B viruses in vitro(127). Selenazofurin showed definite but marginal therapeutic effects against influenza A and B in mice(128). In combination with ribavirin against Pichinde viral infection in guinea pigs selenazofurin gave a survival rate of 4/5 animals when treatment was begun on day 3. The control animals died on day 14. Ribavirin alone gave 3/6 survivors with a prolongation of death to an average of 30 days. Selenazofurin alone was ineffective(129). In vitro data with combinations of ribavirin and selenazofurin have shown considerable synergism with a large number of viruses(130,131). A probable mechanism for this synergism has been discussed in a recent paper(115). Sidwell and coworkers(128) conclude that selenazofurin has a positive effect against murine hepatitis infections in mice.

NUCLEOSIDES ACTIVE AGAINST RETROVIRUSES AND PERSISTENT VIRAL INFECTIONS

Retroviruses are unique RNA viruses characterized by the transcription of their single-stranded RNA into double-stranded DNA of the host cell by the viral enzyme reverse transcriptase. Retroviruses as a class are often found to be responsible for persistent viral infections and because of integration of viral DNA into host cellular DNA present an unusual type of RNA viral infection. Hepatitis B, although a DNA virus, resembles retroviruses in that it has a reverse transcription step in the viral life cycle. Chronic hepatitis B infections afflict some 200 million people. Most recently the human immunodeficiency virus (HIV) has emerged as a major world wide health problem involving a persistent viral infection(132).

The potential of nucleosides and nucleotides and their analogs against retroviruses was reviewed by Robins(133) in 1984. 2',3'-Dideoxythymidine 5'-phosphate was first shown by Baltimore and coworkers(134) to inhibit reverse transcriptase from avian myeloblastosis virus. This inhibition is 100-fold greater than the inhibition of cellular α-DNA polymerase(135). Studies of 2',3'-dideoxyadenosine, 23, first prepared in our laboratory(136) in 1964, have shown significant inhibition of the retrovirus 334C MμLV in NIH Swiss 3T3 cells in culture(137). More re-

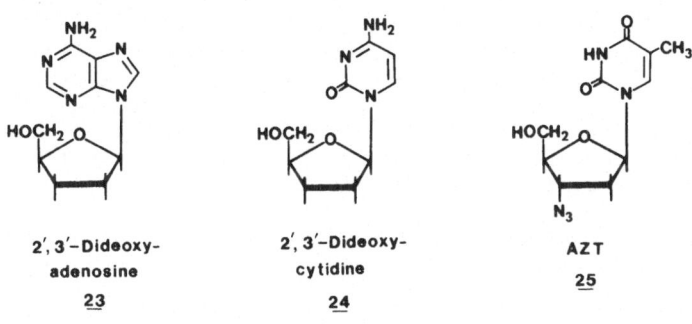

2′,3′-Dideoxy-
adenosine

23

2′,3′-Dideoxy-
cytidine

24

AZT

25

cently Mitsuya and Broder(138) have shown that 2′,3′-dideoxyadenosine at 10 μM completely protected ATH8 cells exposed to HIV and enabled the cells to survive and grow. Although 2′,3′-dideoxyadenosine is a potent DNA chain terminator of E. coli(139), resistance develops very rapidly. Similar inhibition is also shown by 2′,3′-dideoxyguanosine(137,138). 2′,3′-Dideoxycytidine, 24, first synthesized by Horwitz and coworkers(140) in 1967, has been reported (138) to inhibit HIV at 0.5 μM and is currently undergoing clinical trials at NIH in AIDS patients. The major problem with the dideoxynucleosides is that they may be converted within certain cells to the corresponding 2′,3′-dideoxynucleoside 5′-triphos-phate which could also terminate host DNA synthesis or be incorporated into host DNA(137,141). Recently Cooney et al.(141a) have shown that the conversion of 2′,3′-dideoxycytidine to the 5′-triphosphate takes place in many mammalian cells and is not solely a property of HIV infected T-lymphocytes. Starnes and Cheng(141b) have recently shown that 24 is readily incorporated into the DNA of a T-lymphoblastic cell line. No in vivo antiviral data has yet been published on 2′,3′-dideoxycytidine against retroviruses. The most studied nucleoside of the dideoxy type is 3′-azido-3′-deoxythymidine (AZT, 25), first prepared by Horwitz and co-workers(142) in 1964. The in vitro activity of AZT against HIV was first described by Mitsuya et al.(143). Furman et al.(144) report that AZT is nonselectively phosphorylated to AZT 5′-phosphate, which selectively binds to HIV reverse transcriptase. Chronic AZT treatment of mice in-fected with Rauscher murine leukemia virus prevents splenomegaly and suppresses viremia(145). Human trials with AZT in AIDS patients led to limited approval of AZT in the USA for treatment of advanced AIDS cases.

Ribavirin has been shown to significantly reduce splenomegaly and hepatomegaly induced by Friend leukemia virus in mice(146). Shannon(147) has shown ribavirin to inhibit Gross murine leukemia virus in vitro and to inhibit Rauscher murine leukemia induced splenomegaly in mice. Riba-virin is also active against bovine leukemia virus(148) and Rous sarcoma virus in vitro(149). This activity against various retroviruses prompted McCormick and coworkers(150) to study ribavirin against HIV in vitro. These investigators report viral suppression at 50 μg/ml and considerable lowering of reverse transcriptase activity in the presence of ribavirin. Phase I oral dose finding clinical studies of ribavirin in 23 patients with lymphadenopathy syndrome showed decreased viral replication and en-hanced immune function(151) at several dose levels. Phase II clinical trials of ribavirin in the treatment of patients with lymphadenopathy syndrome (LAS) have recently shown that at 800 mg/day of ribavirin in a randomized double blind placebo controlled trial, 52 patients did not develop AIDS during a 24 week study, whereas 10/56 patients developed AIDS in the placebo group during the same period(152). Open clinical studies in AIDS by Crumpacker and coworkers(153) in five AIDS patients treated with 1200 mg, BID for two weeks showed in 4/5 AIDS patients T4 cells doubled then returned to baseline on stopping the drug. Increased responses of PBL to PHA was seen in 4/5 ribavirin treated AIDS patients. Mitsuya and Broder(154) and Yarchoan and Broder(155) have recently summarized the concepts underlying nucleoside and nucleotide analogs as therapeutic agents against clinical HIV infections.

AIDS is an example of a slow or persistent viral infection with a long incubation period. Such viral infections may indeed be responsible for multiple sclerosis, subacute sclerosing panencephalitis, Alzheimer's disease and many other dementia of unknown etiology(156). The high incidence of AIDS dementia is believed to be due to direct brain infection by HIV(157). The concept of controlling such persistent viral infections by antiviral chemotherapy is indeed a most challenging one and would require penetration of the blood-brain barrier and prolonged treatment with a relatively nontoxic agent. The fact that ribavirin has recently been shown to penetrate the blood-brain barrier of AIDS patients(158) and has been well tolerated for over 24 weeks at 800 mg/day in the treatment of 52 LAS patients without significant side effects suggests this nucleoside may be a candidate for study in long term treatment of such viral diseases. The recent paper of Torre et al.(159) has shown that ribavirin eliminates foot and mouth disease virus (FMDV) from persistently infected cell cultures and is 10 times more active against the persistent line than the parent line of FMDV. It is known that ribavirin is highly active in vivo against the FMDV infection in mice(99). The fact that such persistence may make the virus more susceptible to inhibition is encouraging. Ribavirin is also active against acute and persistent mumps viral infections in BHK-21 cells(160).

EFFICACY, RESISTANCE AND TOXICITY IN NUCLEOSIDE ANTIVIRAL THERAPY

In vitro antiviral potency at relatively low multiplicity of viral infection can be a poor indicator of clinical efficacy. Indeed attempts to derive a therapeutic index from such cell culture data can be misleading and is very unlikely to be useful in extrapolation to animal models or to the clinical infection(1). Indeed Browne and coworkers(161) have pointed out that a better prediction of clinical efficacy might be in vitro data against a high multiplicity of viral infection which more likely approximates the clinical picture. Ribavirin stood up well against a high multiplicity of a series of influenza A strains and under these conditions proved to be much superior to amantadine. Significant data against model viral infections in animals would seem to be most important. A small increase in the number of surviving animals over controls may not be of sufficient import to justify clinical trials. Better criteria would be the use of a sufficient degree of viral infectivity which results in the death of all or nearly all of the test animals. Drug treatment should be delayed until 24 hours following viral infection and should result in close to 100% survivors. Such animal studies where death is an end point are the most reliable. The lack of efficacy in animals of a compound exhibiting a high degree of antiviral activity in cell culture may be due to rapid loss of potency against a high multiplicity of infection; rapid metabolism of the drug to an inactive metabolite; failure of the drug to transport to the target organ; too rapid drug excretion or failure of the drug to penetrate the specific cell type; or as may be the case with nucleosides, failure to be activated enzymatically within that particular infected cell (organ) to the 5'-triphosphate required for viral inhibition. Of course, inactivation of the nucleoside or nucleotide with a phosphorylase is much more likely to occur in vivo since the drug is subjected to contact with many cell types.

Resistance to nucleosides and nucleoside analogs can also develop for various reasons. For instance acyclovir resistance has been shown (162,163) in clinical isolates of HSV-1 to be due most often to loss of viral thymidine kinase activity. A recent clinical isolate of HSV-1 resistant to acyclovir proved also to be resistant to DHPG, BVDU and FIAC. These changes were found to correlate in either the amount or the affinity of viral thymidine kinase activity(163). Field(164) predicts that resistance to antiviral drugs will become a significant problem in the future. He suggests that drugs with multiple targets or sites of action in viral inhibition have the best chance to avoid viral resistance(164). These principles should be kept in mind in the design of new nucleoside analogs for potential use as antiviral agents. In this regard the general lack of the development of viral resistance to ribavirin is noteworthy(165). In particular it is difficult to extrapolate from in vitro data to clinical efficacy. The best example of this problem is the prophylaxis or therapy of rhino virus infections (the common cold). Numerous attempts to find a clinically useful agent based on considerable potency against a broad spectrum of rhino viruses evaluated in vitro have been uniformly unsuccessful to date. The problem of lack of a suitable animal model for the HIV virus greatly increases chances for clinical failure. A recent study of purine nucleoside analogs against various rhino viruses in vitro by De Clercq et al.(166) showed that several of the pyrrolopyrimidine nucleoside antibiotics such as tubercidin, toyocamycin and sangivamycin were the most potent inhibitors. Unfortunately these nucleosides proved cytotoxic for uninfected cells at similar concentrations. 3-Deazaguanine, 26, was identified to possess the best selectivity index (minimum toxic concentration, as determined by change in cell morphology divided by minimum concentration for 50% viral inhibition) of 50.

3-Deaza-guanine
26

3-Deaza-guanosine
27

Pyrazofurin
28

29

3-Deazaguanine, 26, prepared in our laboratory(167) was reported in 1977 to have significant in vitro antiviral activity against the RNA viruses, parainfluenza 1 and 3, five strains of rhino viruses, vesicular stomatitis virus (VSV) and a number of DNA viruses including adenovirus, HSV-1 and HSV-2, CMV, vaccinia virus and pseudorabies virus(168). 3-Deazaguanosine, 27(167,168), is similarly active in vitro. Both 26 and 27 show significant antiviral activity against parainfluenza and influenza A_2 infections in mice but animal toxicity noted prevented further consideration of these nucleosides for clinical development as antiviral agents (168). Recent studies indicate that 3-deazaguanine, 26, and 3-deazaguanosine, 27, are metabolized to the 5'-nucleotides(169,170) and are incorporated into DNA(171) which is believed to be responsible for the toxicity noted. 3-Deazaguanine,26,is currently undergoing Phase II clinical trials against breast cancer.

Pyrazofurin, 28, has been shown to exhibit excellent broad spectrum antiviral activity against both DNA and RNA viruses in vitro(172,173). Pyrazofurin shows a broad safety margin in vitro and is 10 to 100 times more potent in vitro than ribavirin. In mice inoculated with Rift Valley Fever virus, 0.25 mg/kg of pyrazofurin daily for 5 days with treatment initiated 1 day before infection resulted in a survival rate of 20%, only marginal efficacy(173). Similar results were obtained with 28, for treatment of Pichende virus infections in guinea pigs. Thus neither the potent in vitro efficacy nor the broad margins of safety as noted in vitro are reflected in appropriate animal models(173). Attempts to improve on the antiviral structure-activity relationship of pyrazofurin by synthesis of a number of pyrazofurin derivatives in our laboratory were unsuccessful(174).

Prusoff and coworkers(175) have discussed the problems of potential incorporation of nucleoside derivatives into viral and cellular nucleic acids. In general systemic treatment of a clinical viral infection which is itself systemic would suggest that incorporation into cellular DNA could very likely prevent clinical use due to toxicity problems. In those incidences where incorporation into viral DNA or RNA appears to be the major mode of antiviral action of the nucleoside, great care must be taken to ensure absence of incorporation into cellular DNA since even small amounts of the nucleoside analog could become cumulatively toxic, especially if the toxicity is not reversed rapidly after drug removal. A study of incorporation of Ara-C into DNA of HSV-1 suggests that this is probably the mechanism of the antiviral activity of Ara-C(176). Unfortunately Ara-C is also readily incorporated into cellular DNA with considerable resulting mutagenesis and toxicity(177). Contrary to a published review(175), there is no evidence for incorporation of ribavirin into viral or cellular RNA or DNA.

THE FUTURE OF NUCLEOSIDE ANALOGS AS ANTIVIRAL AGENTS

In addressing this subject at a meeting of the New York Academy of Sciences in 1974, R.K. Robins(178) stated, "Within the next ten to twenty years nucleosides will have become a major weapon against viral diseases". This rather brash statement at the time, today does not seem so unreasonable. It is clear we have made a good start but so much remains to be done. Oberg and Johansson(1) have discussed in some detail the merits and problems of nucleosides as antiviral agents. These investigators point out that it may be naive to believe that a nucleoside, which is incorporated into viral DNA as a mechanism of action, is absolutely specific since one must consider the effects of incorporation of nucleoside analogs into cellular DNA in the non-infected cell. This is true since there is always the possibility the nucleoside triphosphate can pass from an infected to an uninfected cell through gap junctions. Such incorporation into cellular DNA may not be a lethal or mutagenic event for the cell since DNA repair enzymes may rapidly excise the nucleotide and repair such DNA. However, as with all medicinal agents there is always a benefit to risk ratio to consider. One should not forget that viral infections are probably the greatest cause of mutations, cellular transformation, cellular destruction (brain cells) and may well be the major cause of problems due to aging or indeed the aging process itself.

It would appear from past studies on nucleosides that the further removed in chemical structure one gets from the naturally occurring nucleosides common to RNA and DNA, the less probability of incorporation into cellular RNA and DNA and the greater likelihood of antiviral specificity. One is struck by the antiviral specificity of acyclovir since a considerable number of modifications of the heterocyclic base (179,180) have resulted largely in inactive compounds. Similarly, modifications of ribavirin either in the base or carbohydrate moiety have in general resulted in inactive nucleosides(99,181). For example efforts to combine certain features of acyclovir and DHPG with the triazole carboxamide base of ribavirin have also provided nucleoside analogs with no antiviral effects(68,182).

Of considerable interest is the report(183) that a methoxyindole nucleoside antibiotic, 29, (SF-2140) is as active as amantadine in protecting mice against an APR-8 strain of influenza A, providing 60% survivors. This suggests that it may be possible to alter both the heterocyclic base and the carbohydrate moiety for greater specificity since no acute toxicity was observed by i.p. injection of SF-2140 at 2 g/kg. These studies must be confirmed since there were approximately 25% survivors in the control animals. The finding of Holý and De Clercq (37) is of great interest in that (S)-HPMPA, 7, a phosphonic acid analog of AMP, is active against DNA viral infections in animals. The fact that such a phosphate analog can penetrate cells sufficiently to circumvent the requirement for viral or cellular kinases is most remarkable indeed. Prisbe and coworkers(184) have synthesized the monophosphonate of DHPG, which is highly active in vivo against CMV infections(185). The concept of designing nucleotide analogs which would successfully penetrate cells and show activity without the stringent structural requirements for activation by viral or cellular kinase has previously been reviewed by R.K. Robins(133). It is quite possible that polar functions or groups that are less ionic than phosphate could be successfully employed to replace the 5'-phosphate functionality of nucleotides to readily enter the cell and specifically inhibit viral enzymes. Should one design analogs of the diphosphate or triphosphate moiety as well? Would such derivatives bind specifically to viral DNA and RNA polymerases?

Our recent advances in antiviral chemotherapy are due in large part to the rapid progress being made in molecular virology which can provide a more detailed mechanism of antiviral action. Once the active nucleoside has been identified, the medicinal chemist can continue to make substantial progress in preparation of the proper derivative which can aid in providing a better half life or distribution to a targeted organ under direct attack by the virus. An example is 2',3',5'-tri-O-acetylribavirin which was used by Koff et al.(186) to treat mice i.p. who had been inoculated intracranially with dengue virus type 2 to give a 44% increase in survival compared to 17% increase survival with ribavirin treatment and 12.5% for the control. Similarly Martin and coworkers(187) have recently shown that the bioavailability of DHPG can be increased 25 to 42% in mice by making the dipropionate ester. This diester is stable in monkey intestine homogenates and thus should pass intact from the intestine to the blood circulation. It is, however, hydrolyzed in human serum and monkey liver homogenates thus regenerating DHPG in the blood. This dipropionate ester may be a useful candidate for oral administration of DHPG(187).

The need for new antiviral agents world-wide continues to be documented (188) and widely discussed. It is quite predictable that new nucleoside and nucleotide antiviral agents, yet to be synthesized, will go a long way to fulfill this need.

REFERENCES

1. B. Öberg and N.G. Johansson, J. Antimicrob. Chemother., 14, Suppl. A, 5 (1984).
2. J. Hochsladt, Crit. Rev. Biochem. 2, 259 (1974).
3. R.D. Berlin and J.M. Oliver, Int. Rev. Cytol., 42, 287 (1975).
4. W.H. Prusoff, Biochim. Biophys. Acta, 32, 295 (1959).
5. E.C. Herrmann, Jr., Proc. Soc. Exp. Biol. Med., 107, 142 (1961).
6. H.E. Kaufman and E.D. Maloney, Arch. Ophthalmol., 68, 396 (1962).
7. H.E. Kaufman, Proc. Soc. Exp. Biol. Med., 109, 251 (1962).
8. A.S. Jones, G. Verhelst and R.T. Walker, Tetrahedron Lett., 4415 (1979).
9. E. De Clercq, J. Descamps, P. De Somer, P.J. Barr, A.S. Jones and R.T. Walker, Proc. Natl. Acad. Sci., USA, 76, 2947 (1979).
10. E. De Clercq and R.T. Walker, Pharmacol. Ther. 26, 1 (1984).
11. J. Wildiers and E. De Clercq, Eur. J. Cancer Clin. Oncol., 20, 471 (1984).
12. S. Olofsson, M. Lundström and R. Datema, Virology, 147, 201 (1985).
13. K.A. Watanabe, U. Reichman, K. Hirota, C. Lopez and J.J. Fox, J. Med. Chem., 22, 21 (1979).
14. R.F. Schnazi, J. Peters, M. K. Sokol and A.J. Nahmias, Antimicrob. Agents Chemother., 24, 95 (1983).
15. A. Feinberg, B. Leyland-Jones, M.P. Fanucchi, C. Hancock, J.J. Fox, K.A. Watanabe, P.M. Vidal, L. Williams, C.W. Young and F.S. Philips, Antimicrob. Agents Chemother., 27, 733 (1985).
16. B. Leyland-Jones, H. Donnelly, S. Groshen, P. Myskowski, A.L. Donner, M. Fanucchi, J.J. Fox and the Memorial Sloan-Kettering Antiviral Working Group, J. Infect. Dis., 154, 430 (1986).
17. C.W. Young, R. Schneider, B. Leyland-Jones, D. Armstrong, CTC Tan, C. Lopez, K.A. Watanabe, J.J. Fox, F.S. Phillips, Cancer Res., 43, 5006 (1983).
18. W. Bergman and R.J. Feeney, J. Am. Chem. Soc., 72, 2809 (1950). See also: J. Org. Chem., 16, 981 (1951).
19. J.J. Fox, N. Yung and A. Bendich, J. Am. Chem. Soc., 79, 2775 (1957).
20. G.E. Underwood, C.A. Wisner and S.D. Weed, Arch. Ophthalmol., 72, 505 (1964).
21. G.A. Gentry and J.F. Aswell, Virology, 65, 294 (1975).
22. O. Nakayama, H. Machida and M. Saneyoshi, J. Carbohydrates Nucleosides Nucleotides, 6, 295 (1979).
23. H. Machida, S. Sakata, S. Kuninaka, H. Yoshino, O. Nakayama and M. Saneyoshi, Antimicrob. Agents Chemother., 16, 158 (1979).
24. W.W. Lee, A. Benetz, L. Goodman and B.R. Baker, J. Am. Chem. Soc., 82, 2648 (1960).
25. F.A. Miller, G.J. Dixon, J. Ehrlich, B.J. Sloan and I.W. McLean, Jr., Antimicrob. Agents Chemother., 136 (1968).
26. M. P. de Garilhe and J. de Rudder, Compt. Rend. Acad. Sci., 259, 2725 (1964).
27. R.W. Sidwell, G.J. Dixon, F.M. Schabel Jr. and D.H. Kamp, Antimicrob. Agents Chemother., 148 (1968). See also: F.M. Schabel, Jr., Chemotherapy, 13, 321 (1968).
28. G.L. Vilde, F. Bricaire, A. Huchon and F. Brun-Vezinet, J. Med. Virol., 12, 149 (1983).
29. T.W. North and S.S. Cohen, Pharmacol. Ther., 4, 81 (1979).
30. C. Shipman, Jr., H.S. Smith, R.H. Carlson and J.C. Drach, Antimicrob. Agents Chemother., 9, 120 (1976).
31. W.E.G. Muller, R.K. Zahn, K. Bittlingmaier and D. Falke, Ann. N. Y. Acad. Sci., 284, 34 (1977).

32. W.E.G. Muller, R.K. Zahn, R. Beyer and D. Falke, *Virology*, *76*, 787 (1977).

33. J.C. Pelling, J.C. Drach and C. Shipman Jr., *Virology*, *104*, 323 (1981).

34. R. Vince and S. Daluge, *J. Med. Chem.*, *20*, 612 (1977).

35. R. Vince, S. Daluge, H. Lee, W.M. Shannon, G. Arnett, T. Schafer, T.L. Nagabhushan, P. Riechert and H. Tsai, *Science*, *221*, 1405 (1983).

36. A. Holý, *Chemica Scripta*, *26*, 83 (1986).

37. E. De Clercq, A. Holý, I. Rosenberg, T. Sakuma, J. Balzarini and P.C. Maudgal, *Nature*, *323*, 464 (1986).

38. M.J. Robins, Y. Fouron and W.H. Muhs, *Can. J. Chem.*, *55*, 1260 (1977).

39. E. De Clercq, J. Balzarini, D. Madej, F. Hansske and M.J. Robins, *J. Med. Chem.*, *30*, 481 (1987).

40. E. De Clercq and M.J. Robins, *Antimicrob. Agents Chemother.*, *30*, 719 (1986).

41. R.I. Glazer and A.L. Peale, *Biochem. Biophys. Res. Commun.*, *81*, 521 (1978). See also R.I. Glazer and A.L. Peale, *Cancer Lett.*, 6, 193 (1979).

42. G.B. Elion, P.A. Furman, J.A. Fyle, P. de Miranda, L. Beauchamp and H.J. Schaeffer, *Proc. Natl. Acad. Sci.*, USA, *74*, 5716 (1977).

43. H.J. Schaeffer, L. Beauchamp, P. de Miranda, G.B. Elion, D.J. Bauer and P. Collins, *Nature*, *272*, 583 (1978).

44. H.J. Schaeffer, *Am. J. Med.*, *73* (#1A), 4 (1982).

45. For reviews see: (a) A.P. Feddian, D. Brigden, J.M. Yeo and E.A. Hickmott, *Antiviral Res.*, *4*, 99 (1984); (b) R. Dolin, *Science*, *227*, 1296 (1985); (c) L. Corey and P. G. Spear, *New Eng. J. Med.*, *314*, 749 (1986); (d) B. Bean, *Postgraduate Med.*, *80*, 113 (1986).

46. D. Gold and L. Corey, *Antimicrob. Agents Chemother.*, *31*, 361 (1987).

47. R.J. Whitley, C.A. Alford, M.S. Hirsch, R.T. Schooley, J.P. Luby, F.Y. Aoki, D. Hanley, A.J. Nahmias and S-J. Soong, *N. Eng. J. Med.*, *314*, 144 (1986).

48. J. Andersson, S. Britton, I. Ernberg, U. Andersson, W. Henle, B. Sköldenberg and A. Tisell, *J. Infect. Dis.*, *153*, 283 (1986).

49. G.B. Elion, *Am. J. Med.*, *73*, 7 (1982).

50. J.A. Fyfe, P.M. Keller, P.A. Furman, R.L. Miller and G.B Elion, *J. Biol. Chem.*, *253*, 8721 (1978).

51. P.V. McGuirt, J.E. Shaw, G.B. Elion and P.A. Furman, *Antimicrob. Agents Chemother.*, *25*, 507 (1984).

52. A. Larsson, A. Sundquist and A.M. Parnrud, *Mol. Pharmacol.*, *29*, 614 (1986).

53. J.C. Martin, C.A. Dvorak, D.F. Smee, T.R. Matthews and J.P.H. Verheyden, *J. Med. Chem.*, *26*, 759 (1983).

54. D.F. Smee, J.C. Martin, J.P. Verheyden and T.R. Matthews, *Antimicrob. Agents Chemother.*, *23*, 676 (1983).

55. A.K. Field, M.E. Davies, C. DeWitt, H.C. Perry, R. Liou, J. Germerhausen, J.D. Karkas, W.T. Ashton, D.B.R. Johnston and R.L. Tolman, *Proc. Natl. Acad. Sci.*, USA, *80*, 4139 (1983).

56. K.K. Ogilvie, U.O. Cherujan, B.K. Radatus, K.O. Smith, K.S. Galloway and W.L. Kennel, *Can. J. Chem.*, *60*, 3005 (1982). See also, K.O. Smith, K.S. Galloway, K.K. Ogilvie and U.O. Cherujan, *Antimicrcb. Agents Chemother.*, *22*, 1026 (1982).

57. M.H. St. Clair, W.H. Miller, R.L. Miller, C.U. Lambe and P.A. furman, *Antimicrob. Agents Chemother.*, *25*, 191 (1984).

58. K.B. Frank, J.F. Chiou and Y.C. Cheng, *J. Biol. Chem.*, *259*, 1556 (1984).

59. R.L. Tolman, A.K. Field, J.D. Karkas, A.F. Wagner, J.

Germershausen, C. Crumpacker and E.M. Scolnick, Biochem. Biophys. Res. Commun., 128, 1329 (1985).

60. J. Germershausen, R. Bostedor, R. Liou, A.K. Field, A.F. Wagner, M. MacCoss, R.L. Tolman and J.D. Karkas, Antimicrob. Agents Chemother., 29, 1025 (1986).

61. J.D. Karkas, J. Germershausen, R.L. Tolman, M. MacCoss, A.F. Wagner, R. Liou and R. Bostedor, Biochem. Biophys. Acta, 911, 127 (1987).

62. M. MacCoss, R.L. Tolman, W.T. Ashton, A.F. Wagner, J. Hannah, A.K. Field, J.D. Karkas and J.T. Germershausen, Chemica Scripta, 26, 113 (1986).

63. M.E.M. Davies, J.V. Bondi, L. Grabowski, T.L. Schofield and A.K. Field, Antiviral Res., 7, 119 (1987).

64. A.K. Field, M.E.M. Davies, C.M. DeWitt, H.C. Perry, T.L. Schofield, J.D. Karkas, J. Germershausen, A.F. Wagner, C.L. Cantone, M. MacCoss and R.L. Tolman, Antiviral Res., 6, 329 (1986).

65. C.M. Vander Horst, J.C. Lin, N. Raab-Traub, M.C. Smith and J.S. Pagano, J. Virol., 61, 607 (1987).

66. J.C. Lin, D.J. Nelson, C.U. Lambe and E.I. Choi, J. Virol., 60, 569 (1986).

67. C.K.Y. Fong, S.D. Cohen, S. McCormick and G.D. Hsiung, Antiviral Res., 7, 11 (1987).

68. O.L. Laskin, C.M. Stahl-Bayliss, C.M. Kalman and L.R. Rosecan, J. Infect. Dis., 155, 323 (1987).

69. D. Felsenstein, D.J. D'Amico, M.S. Hirsch, D.A. Neumeyer, D.M. Cederberg, P. de Miranda and R.T. Schooley, Ann. Intern. Med., 103, 377 (1985).

70. D.H. Shepp, P.S. Dandliker, P. de Miranda, T.C. Burnette, D.M. Cederberg, L.E. Kirk and J.D. Meyers, Ann. Intern. Med., 103, 368 (1985).

71. E.C. Reed, P.S. Dandliker and J.D. Meyers, Ann. Intern. Med., 105, 214 (1986).

72. A. Larsson, B. Oberg, S. Alenius, C.E. Hagberg, N.G. Johansson, B. Lindborg and G. Stening, Antimicrob. Agents Chemother., 23, 664 (1983).

73. B. Lundgren, A.C. Ericson, M. Berg and R. Datema, Antimicrob. Agents Chemother., 29, 294 (1986).

74. K. Stenberg, A. Larsson and R. Datema, J. Biol. Chem., 261, 2134 (1986).

75. A. Larsson, K. Stenberg, A-C Ericson, U. Haglund, W-A. Yisak, N.G. Johansson, B. Öberg and R. Datema, Antimicrob. Agents Chemother., 30, 598 (1986).

76. M. Legraverend, R-M.N. Ngongo-Tekam, E. Bisagni and A. Zerial, J. Med. Chem., 28, 1477 (1985).

77. A. Zerial, M. Zerial, M. Legraverend and E. Bisagni, Ann. Inst. Pasteur/Virol., 137E, 317 (1986).

78. R.J. Rousseau, L.B. Townsend and R.K Robins, Biochemistry, 5, 756 (1966).

79. A.J. Bodner, G.L. Cantoni and P.K. Chiang, Biochem. Biophys. Res. Commun., 98, 476 (1981).

80. J.P. Bader, N.R. Brown, P.K. Chiang and G.L. Cantoni, Virology, 89, 494 (1978).

81. C.S.G. Pugh, R.T. Borchardt and H.O. Stone, Biochemistry, 16, 3928 (1977).

82. A. Holý, Coll. Czech. Chem. Commun., 40, 187 (1975).

83. E. De Clercq, J. Descamps, P. De Somer and A. Holý, Science, 200, 563 (1978).

84. I. Sodja and A. Holý, Acta Virol., 24, 317 (1980).

85. J.A. Montgomery, S.J. Clayton, H.J. Thomas, W.M. Shannon, G. Arnett, A.J. Bodner, I-K. Kion, G.L. Cantoni and P.K. Chiang, J. Med. Chem., 25, 626 (1982).
86. E. De Clercq and J.A. Montgomery, Antiviral Res., 3, 17 (1983).
87. E. De Clercq, D.E. Bergstrom, A. Holý and J.A. Montgomery, Antiviral Res., 4, 119 (1984).
88. M. Hayashi, S. Yaginuma, H. Yoshioka and K. Nakatsu, J. Antibiot., 34, 675 (1981).
89. M. Arita, K. Adachi, Y. Ito, H. Sawai and M. Ohno, J. Am. Chem. Soc., 105, 4049 (1983).
90. R.T. Borchardt, B.T. Keller and U. Patel-Thombre, J. Biol. Chem., 259, 4353 (1984).
91. E. De Clercq, Antimicrob. Agents Chemother., 28, 84 (1985).
92. R.I. Glazer and M.C. Knode, J. Biol. Chem., 259, 12964 (1984).
93. R.L. Hamill and M.M. Holhn, J. Antibiotics, 26, 463 (1973). Also see: R. Nagarajan, B.Chao, D.E. Dorman, S.M. Nash, J.L. Occolowitz and A. Schabel, Conf. on Antimicrob. Agents and Chemother., p.50 (1977).
94. C.S.G. Pugh, R.T. Borchardt and H.O. Stone, J. Biol. Chem., 253, 4075 (1978).
95. R. Nagarajan, U.S. Patent 4,158,056, 13 July 1978.
96. M. Vedel, F. Lawrence, M. Robert-Géro and E. Lederer, Biochem. Biophys. Res. Commun., 85, 371 (1978).
97. J.T. Witkowski, R.K. Robins, R.W. Sidwell and L.N. Simon, J. Med. Chem., 15, 1150 (1972).
98. R.W. Sidwell, J.H. Huffman, G.P. Khare, L.B. Allen, J.T. Witkowski and R.K. Robins, Science, 177, 705 (1972).
99. For a review of the antiviral activity of ribavirin see: R.W. Sidwell, G.R. Revankar and R.K. Robins in Viral Chemotherapy Volume 2, p. 49, International Encyclopedia of Pharmacology and Therapeutics, section 116, Ed. D.Shugar, Pergamon Press, 1985.
100. J.C. Drach, M.A. Thomas, J.W. Barnett, S.H. Smith and C. Shipman, Jr., Science, 212, 549 (1981).
101. P.G. Canonico in "Antibiotics VI, Modes and Mechanism of Microbial Growth Inhibitors", p. 161, Springer-Verlag, 1983.
102. I.W. Hillyard in "Ribavirin A Broad Spectrum Antiviral Agent", p. 59, Ed. R.A. Smith and W.A. Kirkpatrick, Academic Press, 1980.
103. B. Eriksson, E. Helgstrand, N.G. Johansson, A. Larsson, A. Misiorny, J.O. Noren, L. Philipson, K. Stenberg, G. Stening, S. Stridh and B. Öberg, Antimicrob. Agents Chemother., 11, 946 (1977).
104. T.P. Zimmerman and R.D. Deeprose, Biochem. Pharmacol., 27, 709 (1978).
105. D.G. Streeter, J.T. Witkowski, G.P. Khare, R.W. Sidwell, R.J. Bauer, R.K. Robins and L.N. Simon, Proc. Natl. Acad. Sci., USA, 70, 1174 (1973).
106. P. Prusiner and M. Sundaralingam, Nature New Biol., 244, 116 (1973).
107. G.L. Linitskaya, A.A. Yotsyna, N.L. Pushkarskaya and G.A. Galegov, Vopr. Med. Khim., 24, 699 (1978).
108. J.S. Oxford, J. Gen. Virol., 28, 409 (1975).
109. F. Bussereau and A. Ermine, Ann. Virol., 134E, 487 (1983).
110. C. Scholtissek, Arch. Virol., 50, 349 (1976).
111. B.B. Goswami, E. Borek, O.K. Sharma, J. Fujitaki and R.A. Smith, Biochem. Biophys. Res. Commun., 89, 830 (1979).
112. B.B. Goswami, O.K. Sharma, E. Borek and R.A. Smith, Current Chemotherapy and Immunotherapy (Proc. 12th Int. Cong. of Chemother., Florence, Italy 1981) Vol. 2, 1075.
113. E. Katz, E. Margalith and B. Winer, J. Gen. Virol., 32, 327 (1976).

114. P. Toltzis and A.S. Huang, Antimicrob. Agents Chemother., 29, 1010 (1986).
115. R.K. Robins, G.R. Revankar, P.A. McKernan, B.K. Murray, J.J. Kirsi and J.A. North, Adv. Enzyme Reg., 24, 29 (1986).
116. P.G. Canonico, M. Kende, B.J. Luscri and J.W. Huggins, J. Antimicrob. Chemother., 14, Suppl. A p. 27 (1984).
117. J.W. Higgins, G.R. Kim, O.M. Brand and K.T. McKee, Jr., J. Infect. Dis., 153, 489 (1986).
118. J.B. McCormick, I.J. King, P.A. Wegg, C.L. Scribner, R.B. Craven, K.M. Johnson, L.H. Elliott and R. Belmont-Williams, New. Eng. J. Med., 314, 1 (1986).
119. M.C. Weissenbacher, M.A. Calello, M.S. Merani, J.B. McCormick and M. Rodriguez, J. Med. Virol., 20, 261 (1980).
120. W.J. Rodriguez, H.W. Kim, C.D. Brandt, R.J. Fink, P.R. Getson, J. Arrobio, T.M. Murphy, V. McCarthy and R.H. Parrott, Pediatr. Infect. Dis., 6, 159 (1987).
121. D.A. Conrad, J.C. Christenson, J.L.Waner and M.I. Marks, Pediatr. Infect. Dis., 6, 152 (1987).
122. D.W. Barnes, Chest, 91, 246 (1987).
123. D.F. Smee and T.R. Matthews, Antimicrob. Agents Chemother., 30, 117 (1986).
124. A. Matsumori, H. Wang, W.H. Abelmann and C.W. Crumpacker, Circulation, 71, 834 (1985).
125. P.C. Srivastava and R.K. Robins, J. Med. Chem., 26, 445 (1983).
126. J.J. Kirsi, J.A. North, P.A. McKernan, B.K. Murray, P.J. Canonico, J.W. Huggins, P.C. Srivastava and R.K. Robins, Antimicrob. Agents Chemother., 24, 353 (1983).
127. R.W. Sidwell, J.H. Huffman, E.W. Call, H. Alaghamandan, P.D. Cook and R.K. Robins, Antimicrob. Agents Chemother. 28, 375 (1985).
128. R.W. Sidwell, J.H. Huffman, E.W. Call, H. Alaghamandan, P.D. Cook and R.K. Robins, Antiviral Res., 6, 343 (1986).
129. C.T. Liu, M.J. Griffin, P.B. Jahrling and C.J. Peters. Fed. Proc., 44, March 12, 1985, p. 1836, Abst. #8327.
130. J.W. Huggins, R.K. Robins and P.G. Canonico, Antimicrob. Agents Chemother., 26, 476 (1984).
131. J.J. Kirsi, P.A. McKernan, N.J. Burns III, J.A. North, B.K. Murray and R.K. Robins, Antimicrob. Agents Chemother., 26, 466 (1984).
132. For example see: "Confronting AIDS", a report of the Committee on a National Strategy for AIDS published Oct. 1986 by the National Academy of Sciences and Institute of Medicine, U.S.A.
133. R.K. Robins, Pharmaceutical Research, 1 (1984).
134. D. Smoler, I. Molleneux and D. Baltimore, J. Biol. Chem., 246, 7696 (1971).
135. P. Furmanski, G.J. Bourguignon, C.X. Bolles, J.D. Corombos, M.R. Das, Cancer Lett., 8, 307 (1980).
136. M.J. Robins and R.K. Robins, J. Am. Chem. Soc., 86, 3585 (1964).
137. M.A. Wagar, M.J. Evans, K.F. Manly, R.G. Hughes and J. A. Huberman, J. Cell. Physiol., 121, 402 (1984).
138. H. Mitsuya and S. Broder, Proc. Natl. Acad. Sci., USA, 83, 1911 (1986).
139. L. Toji and S.S. Cohen, Proc. Natl. Acad. Sci., USA, 63, 871 (1969); J. Bacteriol., 103, 323 (1970).
140. J.P. Horwitz, J.Chua, M. Noel and J.T. Donatti, J. Org. Chem., 32, 817 (1967).
141. (a) D.A. Cooney, M. Dalal, H. Mitsuya, J.B. McMahon, M. Nadkarni, J. Balzarini, S. Broder and D.G. Johns, Biochem. Pharmacol., 35, 2065 (1986). (b) M.C. Starnes and Y.C. Cheng, J. Biol. Chem., 262, 988 (1987).
142. J.P. Horwitz, J. Cha and M. Noel, J. Org. Chem., 29, 2076 (1964).
143. H. Mitsuya, K.J. Weinhold, P.A. Furman, M.H. St. Clair, S.N.

Lehrman, R.C. Gallo, D. Bolognesi, D.W. Barry and S. Broder, Proc. Natl. Acad. Sci., USA, 82, 7096 (1985).

144. P.A. Furman, J.A. Fyfe, M.H. St. Clair, K. Weinhold, J.L. Rideout, G.A. Freeman, S.N. Lehrman, D.P. Bolognesi, S. Broder, H. Mitsuya and D.W. Barry, Proc. Natl. Acad. Sci., USA, 83, 8333 (1986).

145. R.M. Ruprecht, L.G. O'Brien, L.D. Rossoni and S.N. Lehrman, Nature, 323, 467 (1986).

146. R.W. Sidwell, L.B. Allen, J.H. Huffman, J.T. Witkowski and L.N. Simon, Proc. Soc. Exp. Biol. Med., 148, 854 (1975).

147. W.M. Shannon, Ann. N.Y. Acad. Sci., 284, 472 (1977).

148. R.W. Sidwell and D.F. Smee, Antiviral Res., 1, 47 (1981).

149. F.J. Jenkins and Y.C. Chen, Antimicrob. Agents Chemother., 19, 364 (1981).

150. J.B. McCormick, S.W. Mitchell, J.P. Getchell and D.R. Hicks, Lancet, 1367 (Dec. 15, 1984).

151. R.B. Roberts, J. Laurence, D. Scavusso, Y. Kim and H.W. Murray, Clin. Res., 34, 678 (1986).

152. R.B. Roberts, P.N. Heseltine, P.W.A. Mansell, G.M. Dickinson, J.M. Leedom and K.M. Johnson, Clin. Res., 35, 616A (1987).

153. C. Crumpacker, G. Bubley, S. Hussey, L. Schnipper, W. Heagy, R. Finberg, M.F. McLane, J. Allen and M. Essex, Int. Conf. on AIDS, Paris, June 1986, Abst. #84, S14b.

154. H. Mitsuya and S. Broder, Nature, 325, 773 (1987).

155. R. Yarchoan and S. Broder, New Eng. J. Med., 316, 557 (1987).

156. A.T. Haase, J. Infect. Dis., 153, 441 (1986).

157. B.A. Navia, B.D. Jordan and R.W. Price, Ann. Neurol., 17, 517 (1986).

158. C. Crumpacker, G. Bubley, D. Lucey, S. Hussey and J. Connor, Lancet, 45 (July 5, 1986).

159. J.C. Torre, B. Alarcon, E. Martinez-Salas, L. Carrasco and E. Domingo, J. Virol., 61, 233 (1987).

160. J.R. McCammon, V.W. Riesser, Antimicrob. Agents. Chemother., 15, 356 (1979).

161. J.J. Browne, M.Y. Moss and M.R. Boyd, Antimicrob. Agents Chemother., 23, 503 (1983).

162. C.S. Crumpacker, L.E. Schnipper, S.I. Marlowe, P.N. Kowalsky, B.J. Hershey and M.J. Levin, New. Eng. J. Med., 306, 343 (1982).

163. C. McLaren, M.S. Chen, I. Ghazzouli, R.Saral and W.H. Burns, Antimicrob. Agents. Chemother., 28, 740 (1986).

164. H.J. Field, Brit. Med. Bull., 41, 345 (1985).

165. J.H. Huffman, L.B. Allen, R.W. Sidwell, Ann. N.Y. Acad. Sci., 284, 233 (1977).

166. E. De Clercq, R. Bernaerts, D.E. Bergstrom, M.J. Robins, J.A. Montgomery and A. Holy, Antimicrob. Agents Chemother., 29, 482 (1986).

167. P.D. Cook, R.J. Rousseau, A.M. Mian, P. Dea, R.B. Meyer, Jr., and R.K. Robins, J. Am. Chem. Soc., 98, 1492 (1976).

168. L.B. Allen, J.H. Huffman, P.D. Cook, R.B. Meyer, Jr., R.K.Robins and R.W. Sidwell, Antimicrob. Agents Chemother., 12, 114 (1977).

169. See: P.P. Saunders, M.T. Tan, C.D. Spindler, R.K. Robins and W. Plunkett, J. Biol. Chem., 261, 6416 (1986) and earlier references.

170. T.M. Page, S.J. Jacobsen, W.L. Nyan, J.H. Mangum and R.K. Robins, Int. J. Biochem., 18, 957 (1986).

171. R.O. Pieper, L.R. Barrows and H.G. Mandel, Cancer Res., 46, 4960 (1986).

172. E. De Clercq and P.F. Torrence, J. Carbohydr. Nucleosides Nucleotides, 5, 187 (1978).

173. P.G. Canonico, P.B. Jahrling and W.L. Pannier, Antiviral Res., 2, 331 (1983).

174. C.R. Petrie III, G.R. Revankar, N.K. Dalley, R.D. George, P.A. McKernan, R.L. Hamill and R.K. Robins, J. Med. Chem., 29, 268 (1986).
175. W.H. Prusoff, W.R. Mancini, T.S. Lin, J.J. Lee, S.A. Siegel and M.J. Otto, Antiviral Res., 4, 303 (1984).
176. D. Kufe, D. Herrick, C.S. Crumpacker and L. Schnipper, Cancer Res., 44, 69 (1984).
177. G.J. Bubley, C.S. Crumpacker and L. Schnipper, Antimicrob. Agents Chemother., 29, 716 (1986).
178. R.K. Robins, Ann. N.Y. Acad. Sci., 255, 597 (1975) (quote p. 598).
179. L.M. Beauchamp, B.L. Dolmach, H.J. Schaeffer, P. Collins, D.J. Bauer, P.M. Keller and J.A. Fyfe, J. Med. Chem., 28, 982 (1985).
180. C.K. Chu and S.J. Cutler, J. Heterocycl. Chem., 23, 289 (1986).
181. S. Harris and R.K. Robins in "Ribavirin-A Broad Spectrum Antiviral Agent" p. 1, Ed. R.A. Smith and W.A. Kirkpatrick, Academic Press, 1980.
182. T.L. Tsilevich, S.G. Zavgorodny, U. Marx, L.V. Ionova and V.L. Florentiev, Bioorg. Khim., 12, 819 (1986).
183. T. Ito, K. Ohba, M. Koyama, M. Sezaki, H. Tohyama, T. Shomura, H. Fukuyasu, Y. Kazuno, T. Niwa, M. Kojima, T. Niida, J. Antibiot., 37, 931 (1984).
184. E.J. Prisbe, J.C. Martin, D.P.C. McGee, M.F. Barker, D.F. Smee, A.E. Duke, T.R. Matthews, and J.P.H. Verheyden, J. Med. Chem., 29, 671 (1986).
185. A.E. Duke, D.F. Smee, M. Chernow, R. Boehme and T.R. Matthews, Antiviral Res., 6, 299 (1986).
186. W.C. Koff, R.D. Pratt, J.L. Elm, Jr., C.M. Venkateshan and S.B. Halstead, Antimicrob. Agents Chemother., 24, 134 (1983).
187. J.C. Martin, M.A. Tippie, D.P.C. McGee and J.P.H. Verheyden, J. Pharmaceutical Sci., 76, 180 (1987).
188. D.S. Freestone, Antiviral Res., 5, 307 (1985).

DESIGN OF ANTI-VIRAL AGENTS OTHER

THAN NUCLEOSIDE ANALOGUES

Stanley M. Roberts

Department of Chemistry
Exeter University
Exeter
Devon EX4 4QD, U.K.

INTRODUCTION

It is generally accepted that, by the very nature of the invading organisms and their interactions with the host, the search for anti-viral compounds is going to be much more difficult than the corresponding searches for anti-bacterial and anti-fungal substances. Having accepted this situation it is interesting to note that over the past ten years the only compounds to emerge as serious contenders for a place as commercially important anti-bacterials are more β-lactams (*e.g.* thienamycin, aztrenonam) and more nalidixic acid analogues (*e.g.* ofloxacin). The situation with regard to the establishment of new anti-fungal agents is even simpler; only imidazole derivatives (*e.g.* fluconazole) related to ketoconazole have emerged as compounds that may take up a noteworthy position in the market place. Not surprisingly, therefore, the establishment of promising anti-viral agents over the same time period has been painfully slow. Nevertheless acyclovir and various prodrugs, as well as some other nucleoside analogues with potent anti-viral properties, have been discovered and will be discussed in other Chapters. The opportunities afforded by inhibiting virally-coded enzymes with nucleoside analogues are well recognised and a number of interesting anti-viral agents have been 'designed' using the information (*viz.* mode of action, metabolism, etc.) gained from the leading compounds.

Unfortunately the same cannot be said for the area embracing the various anti-viral agents having non-nucleosoid structures. There have been a number of structural types that have displayed activity against different viruses. In almost all the cases the initial lead compound, generally found by a process of random screening, has not been displaced by

37

an analogue designed to potentiate the desired biological activity and/or to limit undesirable toxic effects. There are, of course, some exceptions to this generalisation and these exceptional cases will be described in some detail in the next pages.

This review will attempt to exemplify the different non-nucleoside structures that have shown anti-viral activity and I will comment on the modifications that have been made which led to the retention of some or all of the biological activity. Interesting anti-viral activity will be detailed particularly in the areas [herpes, influenza, common cold (rhino-virus), AIDS (HIV)] where the associated disease is sufficiently important to justify expenditure in research and development to attempt to bring a compound to the market place.

Phosphonoformic Acid, Phosphonoacetic Acid and Analogues

Phosphonoformic acid (PFA) (1) and phosphonoacetic acid (PAA) (2) inhibit herpes-specific DNA polymerases and both compounds inhibit the

$$(HO)_2P(O)CO_2H \qquad\qquad (HO)_2P(O)CH_2CO_2H$$

(1)   (2)

replication of herpes simplex virus in animals.[1] Both compounds show selective inhibition of DNA polymerase of various herpes viruses by inter-fering with pyrophosphate binding sites on the viral DNA polymerase. These analogues of pyrophosphate are also good inhibitors of the RNA transcriptase of influenza viruses.[2,3]

PAA is irritant when applied to skin,[4] a major disadvantage for a topical agent; PFA is relatively free from this defect but is accumulated into bones and teeth. PFA has recently undergone clinical evaluation in man for the treatment of recurrent genital herpes and AIDS and also cytomegalovirus (CMV) infection of bone marrow and renal transplant patients. The compound proved to be effective against genital herpes (when applied topically as a 3% cream) and against CMV (although some toxic effects were observed). Results have not been so encouraging when the compound was tried against AIDS and when the compound was administered as a 1% cream against genital herpes.

Both PFA and PAA are effective anti-herpes agents and both compounds have defects. In other circumstances this should provide a good basis for the preparation and testing of various analogues. However with molecules (1) and (2) there is scant possibility of major structural variation given that the carboxyl group and the phosphate group must remain unesterified.[3,5] Nevertheless efforts by several groups, including the elegant work by the

Astra group and by Hutchinson at Warwick, have uncovered some interesting results. For example, the thio-analogues (3) (4) of PFA and PAA are equally effective inhibitors of HSV-1 DNA polymerase than their oxygenated counterparts.[6] Phosphonoglycolic acid (5) is only slightly active against

$$(HO)_2\overset{\overset{\displaystyle S}{\|}}{P}CO_2H \qquad\qquad (HO)_2\overset{\overset{\displaystyle S}{\|}}{P}CH_2CO_2H \qquad\qquad (HO)_2\overset{\overset{\displaystyle O}{\|}}{P}CH(OH)CO_2H$$

(3)   (4)   (5)

HSV-1 *in vitro*, but distinct anti-viral effects occurred in cutaneous infection of guinea pigs after topical treatment. The effects were comparable with those seen with PFA.[7] Carbonyl bisphosphonic acid (6) has been shown to inhibit the reverse transcriptase of HIV.[8] Other pyrophosphate analogues such as the substituted methylene bisphosphonate (7) are good inhibitors of the RNA transcriptase of influenza viruses.[9]

Several keto phosphonates, phosphonoacetates and dialkyl phosphonates

$$[(HO)_2\overset{\overset{\displaystyle O}{\|}}{P}]_2CO \qquad\qquad\qquad [(HO)_2\overset{\overset{\displaystyle O}{\|}}{P}]_2CCl_2$$

(6)   (7)

containing (aryloxy)alkyl groups have been synthesized and evaluated for anti-herpetic activity. The ester (8) [related to arildone (*vide infra*)] shows activity *in vitro* against HSV-2 (MIC:6 µg/ml); the compound (9)

(8)   R = CO_2Et
(9)   R = COMe

is almost equi-active and furthermore the compound (10) with the ketone or ester group missing is still active. These results coupled with earlier studies led the Sterling-Winthrop workers to suggest that compounds with the structure (11) (where A = aromatic ring and C = β-ketoester or β-diketone

(10)   (11)

separated by a bridge of 5-8 methylene groups or an equivalent moiety) is a prerequisite for anti-viral activity.[10] The β-ketoester or β-diketone functionality can be replaced by certain heterocyclic ring systems. The water soluble compound (12) is active against the HSV-1 induced guinea pig

MeO—⟨benzene ring, Cl⟩—O(CH$_2$)$_6$—⟨pyrazole ring, Et, Et⟩ · CH$_3$SO$_3$H

(12)

skin infection and is effective against HSV-1 and HSV-2 when applied topically against the mouse genital infection. Compound (9) was evaluated in the mouse vaginal model against HSV-2 and led to an increased survival rate as well as an increased survival time.

Triterpenoid Compounds

Glycyrrhizin (13) is a selective inhibitor of VZV (selectivity index =

(13) R = ⟨sugar moiety with CO$_2$H, OH, OH, OH, OH, HO, CO$_2$H⟩

(14) R = -COCH$_2$CH$_2$CO$_2$H

(15) R = -CO-⟨ring, CO$_2$H⟩

30).[11] Carbenoxolone (14) and cicloxolone (15) are effective in the treatment of both oro-facial and genital herpes simplex infections in man. Available data suggest that the anti-viral activity derives wholly or largely from an apparent general effect upon cellular membranes such that normal properties are changed without killing the cell. Thus membrane-associated functions involved in HSV-protein synthesis, transport, and processing are all affected, directly or indirectly, and as a consequence, virus particle production is reduced and those virions which are made are of inferior quality.[12]

Thiosemicarbazones

2-Acetylpyridine thiosemicarbazone (16) has been shown to be a good inhibitor of HSV-1 and HSV-2.[13] It has been postulated that the mechanism of action involves chelation of a metal cofactor required by ribonucleotide reductase. The morpholinoethyl derivative of 2-acetylpyridine thiosemicarbazone potentiates the anti-viral activity of acyclovir.[14] However thiosemicarbazones have principally been used against pox viruses of the vaccinia family and one compound of this type, the isatin derivative (17) has been used prophylactically to prevent outbreaks of smallpox in persons who have had contact with the disease.[15] It is believed that this compound inhibits late protein synthesis in poxvirus-infected cells but the molecular mechanics relating to this property are not known,[16] though the binding of a metal ion may be a key factor.[17] Structure-activity studies tend to support this hypothesis; for example 3- and 4-acetylpyridine thiosemicarbazones have little anti-viral activity and replacement of the sulphur atom in the thiosemicarbazone by an oxygen atom also leads to a loss of activity. The anti-influenza activity of compound (16) has been recorded.[18]

(16)   (17)

Aliphatic Amines Including Amantadine and Rimantadine

Better known and more widely investigated anti-influenza agents are amantadine (18) and rimantadine (19).[19] Both compounds have been found to reduce the duration of influenza-A induced fever and malaise, and lessen viral shedding. Prophylactic treatment (e.g. 200 mg/day of amantadine) has been recommended for high risk patients.[20] It has been suggested that amantadine exerts its biological activity by interfering with a virally coded membrane protein hence disturbing the process of virus assembly.[21] Both amantadine and rimantadine have no effect against influenza B virus which may account, in part, for the fact that both compounds have had difficulty in being accepted as a valid treatment for influenza infections. In addition side effects (e.g. insomnia, drowsiness) affect $20 \pm 13\%$ of patients using amantadine. The incidence of side effects is less with rimantadine. Many derivatives of amantadine and rimantadine have been prepared and tested, some in clinical trials, but none has given overall results better than the original compounds.[22]

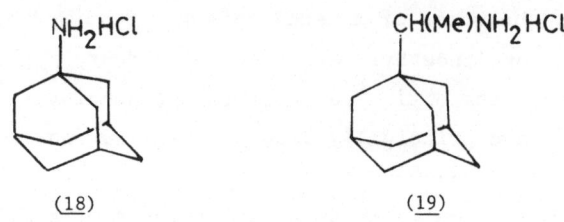

(18)   (19)

The compound (20) ICI 130685 represents a more distant relative of
rimantadine.   ICI 130685 is reported to show good prophylactic and therapeutic effects against influenza A2 in mice and ferrets at 50 mg/kg orally
twice daily, a distinctly better performance than shown by amantadine.
Clinical trials are in progress; in one study 91% protection was afforded
to volunteers dosed at 200 mg/day for seven days with ICI 130685 and
challenged with influenza A virus on day 3.

Interestingly, cyclooctylamine exhibits an *in vitro* activity similar
to amantadine whereas cyclopentylamine, like the aliphatic analogues (*e.g.*
octylamine) is inactive.[21]   The rigid diamine (21) also shows anti-
influenza properties.[23]

(20)

(21)   (22)

Modified Sugars

2-Deoxy-2,3-dehydro-N-trifluoroacetylneuraminic acid (FANA) (22) is a
competitive inhibitor of influenza - A neuraminidase.   The plaque formation
in cell culture of influenza virus was effectively inhibited by FANA but
animal experiments have been disappointing.[24]

Although the inhibitory effect of 2-deoxy-D-glucose and other glycosylation inhibitors on the replication of enveloped viruses has been known for several years,[25] efforts to use these compounds for virus chemotherapy have only recently been made. Therapeutic effects were obtained against herpes keratitis in rabbits after topical (20 µM) or subconjunctival (100 µM) administration of 2-deoxy-D-glucose. However, the compound failed to reduce lesion development in genital HSV-1 infection of guinea pigs after topical treatment and had little effect in models of influenza infection.[7]

ε-Aminocaproic acid has been reported to inhibit the proteolytic cleavage of influenza virus haemagglutinin and to exert significant protection to mice and chickens infected with a lethal dose of this virus.[26]

Ketones, Oximes and Surrogate Compounds

Arildone (23) and some other lipophilic β-diketones have *in vivo* activity against a number of RNA and DNA viruses. The most potent action is against poliovirus where it prevents the uncoating of the viral capsid.[27]

Arildone is marginally effective against the rhinoviruses. Other compounds from the arildone mould have been screened for this activity and compounds of the type (24) were discovered to have activity against the rhinovirus RV-2. Structure-activity studies with variation of the substituents on the phenyl ring and variation of the heterocyclic ring led to compounds (25) and (26) with optimized activity. Compound (25) was more effective than 4',6-dichloroflavan (see later) against RV-2 and poliovirus-2. When evaluated orally in mice infected intracerebrally with poliovirus-2 the compound showed some activity (53% survival rate at 31 mg/kg b.d.; *cf.* 22% survival rate with controls).[28]

(23)

(24)

(25)

(26)

(27)

One analogue of (26) is WIN 51711 (27) which shows potent anti-rhinovirus
and anti-poliovirus activity. WIN 51711 is effective in a prophylactic
regime against polio-induced death in mice at 100 mg/kg. (Thus WIN 51711
is a very promising compound for the inhibition of polio-virus but it will
probably remain purely of academic interest with respect to clinical treatment of this disease). Disappointingly, no significant effect was observed
when compound (27) was evaluated clinically using a human rhinovirus
challenge in man.[29]

Some fascinating results were obtained from X-ray diffraction studies
of compound WIN 51711 (27) and the analogue WIN 52084 (28) bound to crystals
of picornavirus HRV-14. The heteroaromatic group inserts into the hydrophobic interior of the VP1 β-barrel while the 4-oxazolinylphenoxy group

(28)

covers the entrance to an ion channel. Viral disassembly may be inhibited
by preventing the collapse of the VP1 hydrophobic pocket or by blocking the
flow of ions into the interior of the virus. The results can be used to
explain the different effects of compounds against human rhinoviruses
(e.g. HRV-2 and HRV-14) in terms of the amino-acid content of structural
viral proteins in the viral capsid. The binding of the compounds is
dependent on the substitution pattern in the heterocyclic rings and the
length of the inter-connecting chain. This study amounted to the first
description of the interaction of an anti-viral drug with a virion at
atomic resolution.[30]

Recent results suggest that further substitution in the phenyl ring can
also have a profound effect on anti-viral activity: for example the
compound (29) possesses only one-tenth ($\frac{1}{10}$) the activity of the homologue (30)

44

$$(29) \quad R = H$$
$$(30) \quad R = Me$$

against a range of rhinovirus serotypes.[31]

Enviroxime (31) is highly effective against many different strains of rhinovirus. When given intranasally to volunteers challenged with rhino-virus 9, fewer colds and cold symptons were noted.[32] An enviroxime surrogate called enviradene (32) has been shown to possess excellent anti-rhinovirus activity.

$$(31) \quad R = OH , X = N$$
$$(32) \quad R = Me , X = CH$$

Flavanoids and Chalcones

4',6-Dichloroflavan (33) and 4'-ethoxy-2'-hydroxy-4,6'-dimethoxy-chalcone (34) both inhibit the replication of rhinovirus in cell culture

(33)

$$(34) \quad R = H$$
$$(35) \quad R = PO(OH)_2$$

when present at the time of infection. Most probably they act by attaching to part of the virus particle required to initiate penetration into a susceptible cell thus blocking the process. Rhinovirus 9 is particularly sensitive to the flavan (33) but volunteers infected with this virus were not protected by oral administration of the compound. Similarly, the

chalcone (34) and the pro-drug (35) had no effect when given orally to
volunteers carrying rhinovirus infections.   It is quite possible that the
concentrations of flavan and chalcone obtained in the nasal mucosal cells
were not sufficiently high to afford protection.[33]

The chromone (36) was originally isolated from the Chinese medicinal

(36)

herb *Agastache rugoza* Kuntze.   The compound selectively inhibits the
replication of human picornaviruses (*e.g.* rhino- and coxsackie viruses) by
interfering with virus replication at a stage between uncoating and
initiation of viral RNA synthesis.   Oral administration of the diacetyl
derivative protected mice from lethal infections of coxsackie virus B1.[7]

Numerous other flavanoids have been reported to possess anti-viral
activity but little more than a preliminary report has been published
concerning these compounds.[34]

Miscellaneous Anti-picornavirus Compounds

The furan derivative (37) inactivates various strains of rhinovirus
(*e.g.* RV39) *in vitro* but was not active prophylactically in volunteers.[35]
Similarly, the guanidine derivative (38) was very active (0.2 - 1.6 μg/ml)
against a wide range of rhinoviruses *in vitro*.[36]   However, a clinical trial
in volunteers using frequent intranasal spray of a solution of the drug did
not prevent the development of colds.   Many structural analogues of the
guanidine derivative have been made; for example one phenyl ring can be
replaced by an heterocyclic or certain alicyclic ring systems.   High *in
vitro* activity was retained in a number of cases but none has shown
significant activity in humans.[22]

(37)

(38)

46

Many simple diphenyl and phenyl pyridyl ethers show considerable activity in cell culture against representative picornaviruses. Several of these ethers demonstrated broad spectrum activity against twenty human rhinovirus serotypes and enhanced the survival rate of mice given the compounds orally prior to an otherwise lethal challenge with coxsackie virus A21. Compound (39) was the most active in a series of seventy compounds.[37]

(39)   (40)

Diphenylthioureas [*e.g.* N-phenyl-N'-(3-hydroxyphenyl)thiourea (40)] which contain a nitrogen or oxygen atom in the 3-position in one aromatic ring exhibit marked anti-picornavirus activity *in vitro*.[38]

Interferon

An exciting development in recent years is the demonstration that a highly purified interferon (IFN), dosed intranasally, will prevent rhino-virus colds (RV-9) effectively (80–100% reduction in cold symptoms and reduced incidence in virus shedding).[39] The mild reversible inflammation of the nasal mucosa that occurs on administration of IFN-α militates against prophylactic treatment for prolonged periods but short term treat-ment may be tolerated. A different naturally occurring interferon (IFN-β or -γ) or a modified interferon may have a less irritant effect on the nasal passage;[33] such a compound would be a most interesting candidate for further development. Preliminary reports suggest that $IFN\alpha_2$ and IFN_γ show synergy with enviroxime; interferons may also act synergist-ically with the dichloroflavan (33). Other reports have suggested that IFN-α might be useful for short-term prophylaxis of influenza-A;[40] other studies have concentrated on the effect of interferons against different viral diseases (including VZV, CMV, hepatitis B, papilloma and adenovirus).[41] A full discussion on the clinical potential of interferons is featured in another Chapter.

Peptides

The use of peptides as anti-viral agents is an area of research that will probably attract a good deal of attention over the next few years. The work of Choppin and co-workers on the inhibition of myxoviruses and paramyxo-viruses using peptides exemplifies the possibilities.[42] Myxoviruses and paramyxoviruses require activation of a viral protein by a host cell protease before viral penetration into a cell can take place. The new N-terminus generated by the cleavage is highly conserved among different viral strains and is implicated in the membrane-fusing activity of the protein. Oligopeptides with similar hydrophobic sequences to the newly-formed N-terminus have been shown to be remarkably effective *in vitro* presumably by binding to the appropriate receptor on the surface of the cell. Using photon-affinity labelling techniques and compounds such as carbobenzoxy-D-Phe-L-Phe-L-azido-Phe it has been shown that the peptides bind to cells not to the virus. Thus the peptides appear to bind to the cellular receptor of the cleaved F-protein. Carbobenzoxy-D-Phe-L-Phe-Gly-D-Ala-D-Val-D-Ile-Gly is a potent inhibitor of the measles virus. Slight changes in the amino-acid sequence result in loss of activity. Carbobenzoxy-Gly-L-Leu-L-Phe-Gly and carbobenzoxy-Gly-L-Phe-L-Phe-L-Phe-Gly are active against the influenza virus. Radiolabelling studies suggest that the oligopeptides act on the target cell membrane.[7] Good *in vivo* efficacy has not been reported for peptides of this type.

The processing of certain viral proteins by proteolytic cleavage is a critical event for many viruses. Both cellular and viral encoded proteases can be involved in the process and some viral proteases have been singled out for inhibition studies. Carbobenzoxyleucine chloromethyl ketone (41) is an irreversible inhibitor of the poliovirus protease: both protein cleavage and virus production were inhibited.[43] Similarly, protinin, a polypeptide MW 6,000, has been found to inhibit the replication of myxovirus by inhibition of the protease required to activate this virus by cleavage of a viral glycoprotein.[44]

$$\underset{\text{PhCH}_2\text{OCNHCHCOCH}_2\text{Cl}}{\overset{\overset{\text{O}}{\underset{\|}{}} \quad \overset{\text{CHMe}_2}{\underset{|}{}}}{}$$

(41)

Several herpes viruses (*e.g.* HSV-1, HSV-2, and EBV) induce novel ribonucleotide reductase (RNR) activities. Evidence exists that HSV-1 RNR is virally coded and essential to the virus. HSV-1 RNR consists of two subunits which connect to obtain activity. The octapeptide H_2N-Tyr-Gly-Ala-Val-Val-Asn-Asp-Leu-CO_2H has seven amino-acid units common to the carboxy-terminus of the smaller subunit. The peptide binds to the large subunit preventing formation of active enzyme. The modified peptide H_2N-Tyr-Ala-Gly-Ala-Val-Val-Asn-Asp-Leu-CO_2H was five times more active than the octapeptide and the inhibitory activity is specified for virus-induced enzyme. However, exposure of infected cell monolayers to various concentrations of the nonapeptide did not reduce the yield of infectious virus probably because of lack of penetration of the peptide into the cell.[45]

The differentiation antigen T4 present on the helper/inducer subset of T-4 lymphocytes is thought to serve as the receptor for HIV. Part of the HIV glycoprotein gp120 which appears to be highly conserved was synthesised in the form of the octapeptide Ala-Ser-Thr-Thr-Thr-Asn-Tyr-Thr and was effective in preventing the binding of gp120 to the T4 antigen. D-Ala-Ser-Thr-Thr-Thr-Asn-Tyr-Thr was prepared as a more stable analogue and was found to be more potent. Both peptides partially blocked viral infection of human T-cells in the 0.1 nM range. It was suggested that these results point the way to a binding-assay assisted rational drug design programme.[46]

Miscellaneous Compounds Active against Human Immunodeficiency Virus (HIV)

Suramin (42) is a selective inhibitor of HIV but the safety margin is low. Evans Blue (43) behaves in a similar way to suramin while aurintricarboxylic acid (44) is another ionic compound with interesting activity against HIV. All three compounds probably exert their biological effect against HIV by reducing the affinity of the key reverse transcriptase enzyme for the template/primer molecule.[47]

(42)

(43)

(44)

Conclusions

A recent review[48] listing licensed anti-viral agents and experimental anti-viral agents showed one non-nucleoside (amantadine) in the former category and three non-nucleosides (rimantadine, phosphonoformate, and WIN 51711) in the latter category. Only WIN 51711 resulted from a conventional structure-activity based programme of research starting from arildone (Scheme). Further work in this area may be rewarded by the discovery of compounds more potent against rhinoviruses and hence candidates for use against the common cold.

Scheme

The ongoing work involving the detailed studies on the interaction of WIN 51711 and surrogate compounds with the rhinovirus capsid and the increasingly detailed knowledge of the constitution of virally coded enzymes and the make-up of virus receptors on the surface of cell membranes are signposts for the future. Some research groups are making significant headway in understanding at a molecular level some of the processes that are essential to the virus. These processes may be relevant to the attachment, import, replication, synthesis, assembly or maturation of the virus. Having understood the molecular mechanics of the process under investigation, proper drug design can begin. In anti-viral chemotherapy, as in most other areas of medicinal chemistry, the biologists, biochemists, analytical chemists, and often crystallographers, must make the headway before the medicinal chemists can be deployed in a sensible fashion.

There is potential for the use of modified peptides (peptoids?) as anti-viral agents.   The problem will not be finding oligopeptides active *in vitro* but will revolve around finding suitable mimics for the active peptides which will allow transport of sufficient quantities of the anti-viral agent to the site of action *in vivo*.   Organic chemists and specialists in computer graphics will have an important role in this drug-design phase and significant lessons can be learnt from recent endeavours in other areas of medicinal chemistry (*e.g.* the renin-angiotensin axis).

In conclusion, the search for non-nucleoside anti-viral agents is in an interesting and exciting phase and significant developments in the design of active compounds should be anticipated over the next five year period. At the same time random screening processes will probably uncover new leads in this vital area of research.

ACKNOWLEDGEMENTS

I thank Dr. J.C. Cameron (Glaxo Group Research, Greenford) for helpful advice and Glaxo Group Research for use of library facilities.

REFERENCES

1. N.L. Shipkowitz, R.R. Bower, R.N. Appell, C.W. Nordeen, L.R. Overby, W.R. Roderick, J.B. Schleicher, and A.M. von Esch, Suppression of herpes simplex virus infection by phosphonoacetic acid, Appl. Microbiol. 26:264 (1973); E. Helgstrand, B. Eriksson, N.G. Johansson, B. Lannerö, A. Larsson, A. Misiorny, J.O. Norén, B. Sjöberg, K. Stenberg, G. Stening, S. Stridh, B. Öberg, S. Alenius, and L. Philipson, Trisodium phosphonoformate, a new antiviral compound, Science 201:819 (1978).
2. P.A. Cload and D.W. Hutchinson, The inhibition of the RNA polymerase activity of influenza virus A by pyrophosphate analogues, Nucleic Acids Res. 11:5621 (1983).
3. E. Helgstrand and B. Öberg, Enzymatic targets in virus chemotherapy, Antibiotics Chemother. 27:22 (1980) and references therein.
4. B. Eriksson and B. Öberg, Phosphonoformate and phosphonoacetate, in: "Antiviral Drugs and Interferon: The Molecular Basis of Their Activity", Y. Becker, ed., Martinus Nijhoff, Boston, p. 127 (1984).
5. D.W. Hutchinson, M. Naylor, and G. Semple, Inhibition of viral nucleic acid synthesis by analogues of inorganic pyrophosphate, Chem. Scripta 26:91 (1986).
6. D.W. Hutchinson and S. Masson, The antiviral potential of compounds containing the thiophosphonyl group, I.R.C.S. Med. Sci. 14:176 (1986).
7. G. Streissle, A. Paessens, and H. Oediger, New antiviral compounds, in: "Advances in Virus Research", K. Maramorosch, F.A. Murphy and A.J. Shatlain, eds., Academic Press, Orlando (1985).
8. L. Vrang and B. Öberg, PPi analogs as inhibitors of human T-lymphotropic virus type III reverse transcriptase. Antimicrob. Agents Chemother. 29:867 (1986).
9. P.A. Cload and D.W. Hutchinson, The inhibition of RNA polymerase activity of influenza virus A by pyrophosphate analogues, Nucleic Acids Res. 11:5621 (1983); D.W. Hutchinson, M. Naylor, G. Semple,

and P.A. Cload, Pyrophosphate analogues as inhibitors of viral polymerases, <u>Biochem</u>. <u>Soc</u>. <u>Trans</u>. 13:752 (1985).

10. G.D. Diana, E.S. Zalay, U.J. Salvador, F. Pancic, and B. Steinberg, Synthesis of some phosphonates with antiherpetic activity, <u>J</u>. <u>Med</u>. <u>Chem</u>. 27:691 (1984).

11. M. Baba and S. Shigeta, Antiviral activity of glycyrrhizin against varicella-zoster virus in vitro, <u>Antiviral</u> <u>Res</u>. 7:99 (1987).

12. D.J. Dargan and J.H. Subak-Sharpe, The antiviral activity against herpes simplex virus of the triterpenoid compounds carbenoxolone sodium and cicloxolone sodium. <u>J</u>. <u>Antimicrob</u>. <u>Chemother</u>. 18 (Suppl. B):185 (1986); D.J. Dargan and J.H. Subak-Sharpe, The effect of triterpenoid compounds on uninfected and herpes simplex virus-infected cells in culture. I. Effect on cell growth, virus particles and virus replication, <u>J</u>. <u>Gen</u>. <u>Virol</u>. 66:1771 (1985).

13. C. Shipman Jr., S.H. Smith, J.C. Drach, and D.L. Klayman, Antiviral activity of 2-acetylpyridine thiosemicarbazones against herpes simplex virus, <u>Antimicrob</u>. <u>Agents</u> <u>Chemother</u>. 19:682 (1981).

14. T. Spector, D.R. Averett, D.J. Nelson, C.U. Lambe, R.W. Morrison Jr., M.H. St. Clair, and P.A. Furman, Potentiation of antiherpetic activity of acyclovir by ribonucleotide reductase inhibition, <u>Proc</u>. <u>Natl</u>. <u>Acad</u>. <u>Sci</u>. USA 82:4254 (1985).

15. C.J. Pfau, The thiosemicarbazones, <u>Handb</u>. <u>Exp</u>. <u>Pharmacol</u>. 61:147 (1982).

16. J.A. Cooper, B. Moss, and E. Katz, Inhibition of vaccinia virus late protein synthesis by isatin-β-thiosemicarbazone: characterization and <u>in</u> <u>vitro</u> translation of viral mRNA, <u>Virology</u> 96:381 (1979).

17. D.D. Perrin and A. Stünzi, Viral chemotherapy: antiviral actions of metal ions and metal-chelating agents, <u>Pharmac</u>. <u>Ther</u>. 12:255 (1981); D.W. Hutchinson, Metal chelators as potential antiviral agents, <u>Antiviral</u> <u>Res</u> 5:193 (1985).

18. J.S. Oxford and D.D. Perrin, Inhibition of the particle-associated RNA-dependent RNA polymerase activity of influenza viruses by chelating agents, <u>J</u>. <u>Gen</u>. <u>Virol</u>. 23:59 (1974).

19. A.W. Galbraith, Influenza - recent developments in prophylaxis and treatment, <u>Brit</u>. <u>Med</u>. <u>Bull</u>. 41:381 (1985).

20. S.L. Barriere, Antiviral therapy, <u>Pharm</u>. <u>Int</u>. 139 (1984).

21. A.J. Hay, A.J. Wolstenholme, J.J. Skehel, and M.H. Smith, The molecular basis of the specific anti-influenza action of amantadine, <u>EMBO</u> J. 4:3021 (1985).

22. D.L. Swallow and G.L. Kampfner, The laboratory selection of antiviral agents, <u>Brit</u>. <u>Med</u>. <u>Bull</u>. 41:322 (1985).

23. K.R. Bharucha, K.C. Tin, I. Ajdukovic, and D. Ajdukovic, Antiviral 1,2,3,4-tetrahydro-1,4-methanonaphthalene derivatives, US Patent 4,362,746 (1982).

24. P. Palese and J.L. Schulman, Inhibitor of viral neuraminidase as potential antiviral drugs in chemoprophylaxis and virus infections of the respiratory tract, Vol. 1, CRS Press, Cleveland, p. 189 (1977).

25. H.A. Blough, R. Kumarasamy, M. Massare, R.L. Giuntoli, S. Sprecher-Goldberger, and L. Thiry, Glycosylation inhibitors: mode of action and clinical efficacy, <u>in</u> "Herpes Viruses and Virus Chemotherapy", R. Kono and A. Nakajima, eds., Elsevier Science Publ., Amsterdam, p. 211 (1985); see also J.G. Mohanty and K.S. Rosenthal, 2-Deoxy-D-glucose inhibition of herpes simplex virus type-1 receptor expression, <u>Antiviral</u> <u>Res</u>. 6:137 (1986).

26. O.P. Zhirnov, A.V. Ovcharenko, and A.G. Bukrinskaya, Suppression of influenza virus replication in infected mice by protease inhibitors, <u>J</u>. <u>Gen</u>. <u>Virol</u>. 65:191 (1984).

27. J.J. McSharry, L.A. Caliguiri, and H.J. Eggers, Inhibition of uncoating of poliovirus by arildone, a new antiviral drug, <u>Virology</u> 97:307 (1979).

28. G.D. Diana, M.A. McKinlay, C.J. Brisson, E.S. Zalay, J.V. Miralles,

and U.J. Salvador, Isoxazoles with antipicornavirus activity, J. Med. Chem. 28:748 (1985).

29. G.D. Diana and M.J. Otto, Inhibitors of picornavirus uncoating as antiviral agents, Pharmac. Ther. 29:287 (1985).

30. T.J. Smith, M.J. Kremer, M. Luo, G. Vriend, E. Arnold, G. Kamer, M.G. Rossmann, M.A. McKinlay, G.D. Diana, and M.J. Otto, The site of attachment in human rhinovirus 14 for antiviral agents that inhibit uncoating, Science 233, 1286 (1986).

31. G.D. Diana, R.C. Oglesby, V. Akullian, P.M. Carabateas, D. Cutcliffe, J.P. Mallamo, M.J. Otto, M.A. McKinlay, E.G. Maliski, and S.J. Michalec, Structure-activity studies of 5-[[4-(4,5-dihydro-2-oxazolyl)phenoxy]alkyl]-3-methylisoxazoles: inhibitors of picornavirus uncoating, J. Med. Chem. 30:838 (1987).

32. R.J. Phillpotts, R.W. Jones, D.C. Delong, S.E. Reed, J. Wallace, and D.A.J. Tyrrell, The activity of enviroxime against rhinovirus infection in man, Lancet i:1342 (1981).

33. R.J. Phillpotts and D.A.J. Tyrrell, Rhinovirus colds, Brit. Med. Bull. 41:386 (1985).

34. I. Beladi, R. Pusztai, I. Mucsi, M. Bekay, and M. Gabor, Activity of some flavanoids against viruses, Ann. N.Y. Acad. Sci. 284:358 (1977); A. Wacker and H.G. Eilmes, Antiviral activity of plant components, Part 1, Flavanoids, Arzneim. Forsch. 28:347 (1978).

35. R.J. Ash, R.A. Parker, A.C. Hagan, G.D. Mayer, RMI 15,731 (1-[5-tetradecyloxy-2-furanyl]-ethanone), a new antirhinovirus compound, Antimicrob. Agents Chemother. 16:301 (1979).

36. D.L. Swallow, R.A. Bucknall, W.E. Stanier, A. Hutchinson, and H. Gaskin, A New antirhinovirus compound, ICI 73602: structure-properties and spectrum of activity, Ann. N.Y. Acad. Sci. 284:305 (1977).

37. L.D. Markley, Y.C. Tong, J.K. Dulworth, D.L. Steward, C.T. Goralski, H. Johnston, S.G. Wood, A.P. Vinogradoff, and T.M. Bargar, Antipicornavirus activity of substituted phenoxybenzenes and phenoxypyridines, J. Med. Chem. 29:427 (1986).

38. A.S. Galabov, B.S. Galabov, and N.A. Neykova, Structure-activity relationships of diphenylthiourea antivirals, J. Med. Chem. 23:1048 (1980).

39. G.M. Scott, R.J. Phillpotts, J. Wallace, D.S. Secher, K. Cantell, and D.A.J. Tyrrell, Purified interferon as protection against rhinovirus infection, Brit. Med. J. 284:1822 (1982); G.M. Scott, R.J. Phillpotts, J. Wallace, C.L. Gauci, J. Greiner, and D.A.J. Tyrrell, Prevention of rhinovirus colds by human interferon alpha-2 from Escherichia coli, Lancet ii:186 (1982).

40. R.J. Phillpotts, P.G. Higgins, J.S. Willman, D.A.J. Tyrrell, D.S. Freestone, and W.M. Shepherd, Intranasal lymphoblastoid interferon ("Wellferon") prophylaxis against rhinovirus and influenza virus in volunteers, J. Interferon Res. 4:535 (1984).

41. D.O. White, Antiviral chemotheapy, Med. J. Australia, 715 (1984).

42. P.W. Choppin, C.D. Richardson, and A. Scheid, Oligopeptides as specific antiviral agents, in: "Targets for the Design of Antiviral Agents", E. De Clercq and R.T. Walker, eds., Plenum Press, New York and London, p. 287 (1984); C.D. Richardson and P.W. Choppin, Oligopeptides that specifically inhibit membrane fusion by paramyxoviruses: studies on the site of action, Virology 131:518 (1983); see also P.W. Choppin, Basic virology, in: "Virology in Medicine", H. Rothschild and J.C. Cohen, eds., Oxford University Press, Oxford, p. 24 (1986).

43. W.H. Prusoff, T.-S. Lin, and M. Zucker, Potential targets for antiviral chemotherapy, Antiviral Res. 6:311 (1986).

44. O.P. Zhirnov, A.V. Ovcharenko, and A.G. Bukrinskaya, Myxovirus replication in chicken embryos can be suppressed by aprotinin due to the blockage of viral glycoprotein cleavage, J. Gen. Virol. 66:1633 (1985); see also O.P.Zhirnov, High protection of animals lethally infected with influenza virus by aprotinin-rimantadine combination, J. Med. Virol. 21:161 (1987).

45. B.M. Dutia, M.C. Frame, J.H. Subak-Sharpe, W.N. Clark, and H.S. Marsden, Specific inhibition of herpesvirus ribonucleotide reductase by synthetic peptides, Nature 321:439 (1986).; E.A. Cohen, P. Gaudreau, P. Brazeau, and Y. Langelier, Specific inhibition of herpesvirus ribonucleotide reductase by a nonapeptide derived from the carboxy terminus of subunit 2, Nature 321:441 (1986).

46. C.B. Pert, J.M. Hill, M.R. Ruff, R.M. Berman, W.G. Robey, L.O. Arthur, F.W. Ruscetti, and W.L. Farrar, Octapeptides deduced from the neuropeptide receptor-like pattern of antigen T4 in brain potently inhibit human immunodeficiency virus receptor binding and T-cell infectivity, Proc. Natl. Acad. Sci. USA 83:9254 (1986).

47. E. De Clercq, Chemotherapeutic approaches to the treatment of the acquired immune deficiency syndrome (AIDS), J. Med. Chem. 29:1561 (1986).

48. M.E. Gore and P. Selby, Antiviral chemotherapy, Brit. J. Hosp. Med. 24: in press (1987).

ANTIVIRALS FOR HIGH HAZARD VIRUSES

Peter G. Canonico

Department of Antiviral Studies
U.S. Army Medical Research Institute of Infectious Diseases
Fort Detrick, Frederick, Maryland 21701-5011

INTRODUCTION

In large areas of the world there exist extremely high hazard viruses for which there are no vaccines for prophylaxis and no effective drugs for therapy (Canonico et al., 1984). Examples of such viruses are Ebola, Argentine, Bolivian, Crimean-Congo, and Korean hemorrhagic fevers, Rift Valley fever, Venezuelan equine encephalitis, Japanese B encephalitis and Lassa fever. With the notable exception of the yellow fever vaccine, immunoprophylactic approaches have not been developed for most of these "exotic" viral diseases. Even when effective vaccines are available, vaccination does not provide absolute protection. To compensate for this inadequacy, we recognized the need to develop antiviral drugs which are effective against a broad-range of viruses and which can be used either prophylactically or therapeutically, as required. However, the development of chemotherapy for viral diseases has been slow. The search for substances specifically targeted to viral replication processes has been difficult and shares some parallels with the development of antibiotics.

Although Fleming discovered penicillin in 1929, the prevailing medical opinion of the time opposed the development of chemotherapy for microbial diseases. It had been a general rule of medicine that any substance toxic to a microorganism would also be toxic to the host. Proponents of the immunoprophylactic approach to the treatment and prevention of bacterial diseases prevailed, and penicillin lay idle for over a decade. It was only the interest of military leaders concerned with the treatment of acute combat traumas that spurred the development and ultimate acceptance of penicillin as a antimicrobial chemotherapeutic agent.

History appears to be repeating itself in the development of the chemotherapeutic agents for pathogenic viruses. The antiviral drug development program at USAMRIID has discovered a number of substances with various levels of toxicity that selectively inhibit the replication of exotic viruses. In this paper, I plan to present antiviral data on selected compounds evaluated in our antiviral and immunomodulator program, as well as results from a clinical study of ribavirin prophylaxis of Sandfly fever infection in man. I will begin with an overview of several high hazard exotic viruses having great medical and economic impact.

OVERVIEW OF HIGHLY PATHOGENIC VIRUSES

Lassa Fever

Lassa fever is a West African disease, found particularly in Liberia, Sierra Leone and Nigeria, although related viruses have been isolated in Mozambique, Zimbabwe, and the Central African Republic. A member of the arenaviridae family, Lassa was first isolated in 1969 from the blood of a missionary nurse during a hospital outbreak of disease which originated from an obstetrical patient residing in Lassa, Nigeria (McCormick and Johnson, 1978). In certain hospitals Lassa appears to be the most important cause of adult medical deaths. From 1977 to 1979, two hospitals in Sierra Leone registered 461 confirmed cases of Lassa fever. Deaths from the Lassa fever cases represented 30% of all medical deaths at these two facilities. Lassa occurs throughout the year, with a peak in the number of reported cases occurring during the dry season. With antibody prevalence rates of 10-50% in many villages, the actual number of cases in Western Africa may well number in the tens of thousands per year.

As with all other currently known arenaviruses pathogenic for man, Lassa fever is a rodent virus and Mastomys natalensis appears to be its natural reservoir. Mastomys sustain chronic viremia and virus shedding without apparent histological lesions or any other deleterious effect. Thus, the epidemiology of Lassa fever is controlled by factors that affect the spread of the virus in rodent populations and that bring man in proximity to infected rodents. The chronically infected rodents shed virus in oropharyngeal secretions, urine, and feces. This may represent the major route of infection since these viruses are stable and infectious as aerosols.

The typical incubation period for Lassa fever ranges from 10 to 14 days, though symptoms may occur as early as five to seven days and possibly as late as three weeks after exposure (Jahrling and Peters, 1985). The onset of the disease is usually insidious and patients typically experience progressive fever, malaise and myalgia. Other symptoms commonly include weakness, nausea, vomiting, diarrhea, chest pain, abdominal pain, headache, sore throat, cough and dizziness. The illness may last one to four weeks but typically survivors begin to recover by two to three weeks. Disorders of consciousness, coma and convulsions, are often found in severely ill patients and indicate a poor prognosis. Frank bleeding tendencies may develop during the evolution of the illness, particularly in severely ill patients. Marked elevations of serum enzymes also are associated with a poor prognosis. Patients with SGOT activities of 150 IU/ml or greater have a 50% mortality compared with 22% in patients with lower values. In the later stages of illness, frequently as recovery begins, patients may experience pericarditis or deafness. Finally, abrupt onset of relentless shock leads to death in the second or third week of illness in approximately 15% of hospitalized cases. Survivors require several weeks before they are fully restored to health and no longer complain of weakness and dizziness (Peters, 1985).

Rift Valley Fever

Rift Valley fever virus, a member of the Bunyaviridae family, was originally isolated in the Rift Valley in Kenya during a 1931 epizootic in sheep (Daubney et al., 1931). Since that time, there have been repeated epidemics infecting herds of sheep, cattle and goats, which typically result in mortality rates of 10% to 30%. Until 1977, the disease was restricted to man and domestic animals in sub-Saharan Africa, with epizootics in Kenya, South Africa, Southern Rhodesia, Sudan, Uganda, United Republic of Tanzania, and Zambia (Shope et al., 1982). Other parts of sub-Saharan Africa appear to harbor RVF virus, as demonstrated by virus isolation from, or the presence of antibody in man, however, epizootics have not been reported. The extension of Rift Valley fever into Egypt in 1977 radically changed established concepts of Rift Valley fever virus. The introduction of RVF in Egypt resulted in an epidemic of 18,000 human cases with nearly 600 deaths officially reported. If the results of a 1978 serosurvey are extrapolated to the official population figures, then there may have been more than a million infections of man (Lupton et al., 1982). The epidemic in Egypt was also unique in that it resulted in many patients with severe and fatal complications including retinal vascular disease, encephalitis, and hemorrhagic fever. By 1979, there were only occasional anecdotes of outbreaks and a few RVF virus isolates, and by 1980 there was little evidence of disease in Egypt. The factors responsible for the introduction of RVF in Egypt remain unclear. Whether RVF will cease to be a problem in Egypt or whether it is temporarily dormant and will erupt to cause major epidemics in the future is impossible to predict. Nevertheless, the potential for RVF infection in Egypt continues to exist. Control of disease in Africa or in areas of the Americas or Europe with large livestock populations and high vector density could be difficult or impossible (Lupton et al., 1982). It should be noted that uncomplicated Rift Valley fever and Rift Valley fever retinitis have both been identified in travelers returning from Africa to North America or Europe.

Rift Valley fever virus has been isolated from several genera of mosquitoes and it is usually assumed that the virus is mosquito borne. The recent findings in Kenya by Linthicum (Linthicum et al., 1985) that Aedes mosquitoes emerging from flooded depressions called "damboes" are already infected with the virus provide evidence for transovarial transmission of Rift Valley fever virus in nature and have led to the following sequence of events to explain the epizoology of the disease. Periods of sustained high rainfall flood depressions where dormant eggs of certain floodwater Aedes species are triggered into development. The adult mosquitos, already infected, emerge soon afterwards and transmit virus to domestic livestock or other amplifying hosts. Susceptible sheep and cattle experience extremely high viremias, and these virus concentrations can serve to infect even inefficient mosquito vectors which would sustain the epizootic. Rift Valley fever virus is also highly infectious by aerosol, and human infections have been well-documented in the research laboratory.

Uncomplicated clinical illness follows a two- to six- day incubation period, ushered by the abrupt onset of fever, chills, and a sandfly fever type of illness (Peters and LeDuc, 1985). Incapacitating prostration, myalgia, and fever typically resolve after two to three days, but may last a week. Physical findings are confined to conjunctival and pharyngeal injections. Viremia occurs through the acute phase and may exceed titers of 10^6 to 10^8 PFU/ml. Although Rift Valley fever is usually a temporarily prostrating, nonfatal illness, about 1% of cases will develop hemorrhagic fever, encephalitis, or eye lesions.

Hemorrhagic disease begins as classical Rift Valley fever, but in the first few days of illness patients develop petechial, mucous membrane, and gastrointestinal bleeding. Patients become jaundiced and die in shock. Disseminated vascular damage and hepatic failure probably contribute to death.

Encephalitis has been added recently to the clinical spectrum of human disease. Typically, encephalitic patients first recover from the acute febrile illness, then three days to two weeks later, present with headache, meningismus, confusion and often a recrudescent fever. Decreased visual acuity can also be a late manifestation of Rift Valley fever. Vision usually returns days to weeks later as the edema and exudates resolve, but half the patients experience some degree of permanent loss of sight.

Hemmorrhagic Fever with Renal Syndrome (HFRS)

Hantaan is the name given to the prototype member of a group of viruses recently identified, which cause febrile diseases accompanied by hematologic and renal abnormalities. These disorders were first brought to the attention of Eastern physicians in the early 1950's during the Korean conflict (McKee et al., 1985a). Widely recognized across northern Asia, the disease is characterized by a hemorrhagic phenomenon, severe renal failure, and appreciable mortality. There are suggestions that HFRS was observed in China as early as 960 AD. In recent history, the disease was first noted by Soviet workers in 1913, with sporadic outbreaks recorded in Far Eastern USSR during the 1920's and 1930's. Between 1932 and 1935, annual outbreaks of this "new" disease were noted in the Amur River Valley region by Soviet scientists, while simultaneous observations of a similar illness among troops stationed in nearby Manchuria were being made by the Japanese (Gajdusek, 1956). Presumptive evidence for both an infectious and viral etiology of HFRS was obtained independently by investigators from both the USSR and Japan in the early 1940's. It was not until 1976, however, that efforts to identify the causative agent were rewarded when convalescent human sera were used to demonstrate antigen in the lungs of the striped field mouse, Apodemus agrarius (Lee and Lee, 1976). Subsequent isolations of the viral agent and adaption of the virus to cell culture provided the tools to define the virus/disease relationships. Ultimately, Hantaan was revealed to be an enveloped RNA virus with properties similar to that of members of the family Bunyaviridae (White et al., 1982; Schmaljohn and Dalrymple, 1983).

HFRS probably results from infection with any of 3 distinct viruses: Hantaan virus, Puumala virus or Seoul virus. All 3 viruses are antigenically closely related, and along with Prospect Hill virus, form a proposed new genus of viruses within the family Bunyaviridae. Hantavirus has been suggested as the genus name in recognition of the prototype Hantaan virus. In addition to Korea, China, Japan, and the USSR, HFRS has been recognized in Sweden, Finland, Norway, Denmark, France, Belgium, Hungary, Czechoslovakia, Rumania, Bulgaria, Yugoslavia, and Greece (McKee et al., 1985b). The distribution of Hantaan virus, associated with the Far Eastern form of HFRS, coincides with the distribution of its principal rodent host, Apodemus species. Seasonal patterns of the occurrence of human disease correspond to peaks of rodent reproductive activity, extent of natural rodent infection, and rodent feeding and sheltering behavior. The presumed route of natural infection is through the accidental or casual contact with contaminated excreta from infected rodents. Human-to-human transmission of HFRS has never been documented (McKee et al., 1985a). An urban HFRS, associated with the Seoul virus, is becoming increasingly appreciated as a significant disease. The urban reservoir appears to be Rattus rattus and R. norvegicus. The infections may be clinically milder than those contracted in rural areas, although fatalities have been reported.

The observations that Rattus rodents in Asia harbor Hantaan-like viruses associated with HFRS led to the investigation of the global distribution and prevalence of antibody to Hantaan-related viruses (LeDuc et al., 1986). Over 1700 rodent sera were collected and assayed from more than 20 different sites around the world between 1981 and 1983. Hantaan-like viruses have been detected by immunofluorescent antibody tests in nearly all locations sampled except Antartica. Overall, greater than 20% of all sera tested gave positive results, with Baltimore, MD, USA, and Belem, Brazil having the highest rates of 64 and 56%, respectively. The results of this study clearly show that the Hantaan-related viruses have a worldwide distribution. This widespread distribution raises the possibility that HFRS may exist, albeit undiagnosed, in regions beyond its traditionally recognized enzootic boundaries. A search for HFRS in man in such areas is in progress (LeDuc et al., 1986).

Clinically severe HFRS disease progresses through a series of five stages arbitrarily designated as the febrile, hypotensive, oliguric, diuretic, and convalescent phase (McKee et al., 1985a).

Disease is usually initiated with the abrupt onset of chills, high fever ($>40^{o}C$), malaise, myalgia, myasthenia, headache, eye pain, dizziness, and anorexia. Within one to three days, large extravasation of plasma into the peritoneum and retroperitoneal space results in severe back and abdominal pain. This phase lasts for three to seven days, and toward the end petechiae appear on the face, neck, axillae, chest, and soft palate.

In the hypotensive phase, hypotension develops abruptly, lasting from several hours to several days coincident with defervescence. Systemic symptoms increase in severity, particularly nausea, vomiting and wretching. Persistence of fever at this point is an ominous prognostic sign. The classic picture of shock quickly develops, with tachycardia and hypotension. A third of all deaths occur at this stage, apparently as a result of relative hypovolemia.

In the oliguric phase, the vascular derangements characterizing the initial stages of disease begin to normalize during the second week of illness. Blood pressure returns to its baseline level, and many patients experience a transient hypertension, presumably as a result of relative hypervolemia. Almost all patients experience a reduction in urinary output at this stage. BUN and creatinine rise to peak levels, and metabolic expressions of renal failure appear. A third of the patients will show hemorrhagic manifestations of varying degree. Nearly 50% of all deaths occur in this phase. Pulmonary infections and/or edema, electrolyte disturbances, late shock, and central nervous system hemorrhage are precipitating events in a fatal outcome. The oliguric phase generally lasts three to seven days. Towards the end, correction of the bleeding diathesis and return of renal function are seen.

With the onset of the diuretic phase, clinical recovery is usually initiated. The volume of urine excreted is dependent upon the severity of the disease, but urinary outputs of three to six liters daily are the rule. Patients with mild to moderate disease begin subjective improvement. On the other hand, severely ill patients, particularly those dehydrated as a result of antecedent clinical difficulties, are at extreme risk at this point. Negative fluid balances may result in shock, while persistent metabolic derangements, together with secondary infection, hazard survival.

The convalescent phase often extends for three or more months. Anemia and hyposthenuria may persist for months to years. Aside from those patients who sustained central nervous system hemorrhages, complete recovery is the usual course.

Mortality in Far Eastern HFRS has decreased from 10 to 15% to around 5%, due primarily to a more physiologic approach to patient management and supportive care and to the introduction of renal dialysis. Estimates of the incidence of Far Eastern HFRS encompass 500 to 1000 cases per year in the Soviet Union. In China, incidence rates of up to 283 per 100,000 have been recorded in some communes. The total number of cases in China may exceed 400,000 cases per year with 20,000 to 40,000 deaths. In the Republic of Korea, 500 to 1000 cases have been recognized each year for the last 25 years.

Congo-Crimean Hemorrhagic Fever Virus

The first clinical description of Crimean hemorrhagic fever dates back to the World War II (1944-1945) when a severe outbreak of hemorrhagic fever occured in Crimea (Chumakov, 1974). During this outbreak, 200 soviet soldiers and a number of peasants were affected by the disease with a case-fatality rate of approximately 10%. The causative agent of Crimean hemorrhagic fever was first isolated in 1967, although the viral etiology of this disease was suspected earlier. Another virus isolated in 1956 from a patient native of Belgian Congo (now Zaire) (Simpson et al., 1967) was found to be indistinguishable from Crimean virus, thus the current name: Crimean-Congo hemorrhagic fever virus (CCHF) (Chumakov et al., 1969).

CCHF is one of the most widely distributed among arthropod borne viruses and the most widely distributed virus of medical importance with the exception of Dengue virus. It has been isolated throughout sub-Saharan Africa, Eastern Europe, the Middle East, the Indian subcontinent and as far east as the Xinjiang Province of China (Hoogstraal, 1979). In addition, virologically documented, naturally acquired, human disease is known to occur in the Soviet Union, Bulgaria, Yugoslavia, Pakistan, Iraq, Dubai, Zaire, Uganda and South Africa. However, CCHF occurs sporadically, both spacially and from year to year. It has been shown that the disease continues to occur with 8 outbreaks in 1985 and 5 outbreaks in 1986 (Watts et al., 1987).

Ticks appear to be the major reservoir and vector for the virus but CCHF is unique among tick-borne arboviruses in that a very large number of tick species have been implicated. CCHF has been isolated from ticks belonging to two families and 25 different species or sub-species although Hyalomma ticks are thought to be the most common vector.

Although CCHF does not appear to cause disease in mammals other than man, it has been isolated from a number of wild (hares, hedgehogs) and domestic animals (cattle, sheep, goats, camels) (Causey et al., 1970; Kemp, 1976; Woodall et al., 1965) and a variety of wild animals develop a viremia upon experimental inoculation (Chumakov et al., 1974; Zarubinski et al., 1976). It is likely that small mammals serve as the principal virus-amplifying hosts.

Man acquires the disease via two different routes: through tick bite and via contaminated blood products. The disease appears to be more severe following aerosol exposure than after tick bite, with case mortality rates approaching 50% (Burney et al., 1980; Bulynin and Poshekhonov, 1959; Blyakher et al., 1971; Van Eeden et al., 1985a). Laboratory infections due to exposure to infected materials by inhalation or mucosal adsorption have also been reported (Baladov et al., 1974). The incubation period following a tick bite or exposure to contaminated blood varies between 2 and 12 days (Van Eeden et al., 1985b).

The disease CCHF is characterized by a sudden onset with fever exceeding 39°C. The initial symptoms include malaise, anorexia, chills, headache, back and abdominal pain, vomiting and loose stools, photophobia and stiffness (Bilibin, 1950; Van Eeden et al., 1985a,b). The disease progresses to include hyperemia of the face, bradycardia, relative hypotension, leukopenia and low platelet counts. Three to five days after the onset of disease the hemorrhagic phase begins. CCHF usually has more flagrant hemorrhagic manifestations than other hemorrhagic fevers. Overt bleeding from the nose, gastrointestinal tract, and genitourinary system are not uncommon with hemoglobin and erythrocyte counts reaching less than half of normal in severe cases. Patients usually die of multiple organ failures. Those patients destined to recover begin to show signs of improvement 9 to 10 days after the onset of the disease but convalescence may be extended. The major pathophysiologic event is the leakage of erythrocytes and plasma through the vascular congestion. There are no pathognomonic lesions for CCHF; however, edema, focal necrosis and hemorrhage are found in the brain, liver and heart (Joubert et al., 1985).

Viral antigen, as monitored by fluorescent antibody techniques, can be found in the liver and in scattered macrophage-like cells throughout many organs. Based on this observation, it has been hypothesized that the vascular lesions seen with CCHF may be triggered indirectly through viral effects on macrophages. The viremia in humans can be quite high and persists for a week or slightly longer. The infection of pregnant women usually results in a spontaneous abortion. Antibodies to CCHF are detectable by 15 days after the onset of illness but only 33-42% of convalescent patients are seropositive after one year. It appears that antibodies persist longer in severe cases than in mild cases.

The impact of these and other viral diseases throughout the world is significant. Unfortunately, the principal endemic areas for these diseases do not represent viable commercial markets, hence, the development of therapeutic agents has been ignored to a large extent by the pharmaceutical industry.

ANTIVIRALS

The perception of the pharmaceutical industry that exotic viruses do not represent profitable ventures for commercial companies leaves potential development of antivirals for these diseases to the category of orphan drugs or products too costly to develop. Responding to this void in antiviral drug research, we established a collaborative program between our institute, other government agencies, universities, pharmaceutical, biotechnical and chemical companies. Under these collaborative agreements, participants submit compounds for in vitro evaluation and testing against a battery of viruses (Table 1).

TABLE 1: ANTIVIRAL DRUG SCREENING MODELS AT USAMRIID

IN VITRO MODELS	ANIMAL MODELS
RNA	
Bunyavirus	
Rift Valley Fever (RVF)	Mouse, Hamster, Monkey
Sandfly Fever* (SF)	**
Punta Toro* (PT)	Mouse, Hamster
Hantaan* (KHF)	Mouse**
Alphavirus	
Venezuelan Equine Encephalomyelitis* (VEE)	Mouse, Hamster, Monkey
Flavivirus	
Yellow Fever* (YF)	Monkey**
Japanese Encephalitis* (JE)	Mouse, Monkey**
Banzi	Mouse
Dengue 1 and 3 (2 and 4)	Monkey**
Arenavirus	
Pichinde* (PIC)	Guinea Pig, Hamster
Lassa	Guinea Pig, Hamster, Monkey**
Junin	Guinea Pig, Hamster, Monkey
Rhabodovirus	
Vesicular Stomatitis Virus* (VSV)	
Filovirus	
Ebola	Guinea Pig, Monkey
Retrovirus	
Human Immunodeficiency Virus (HIV)	
DNA	
Poxvirus	
Vaccinia*	
Adenovirus	
Human Adenovirus Type 2*	

*Denotes viruses currently used in the initial battery of tests
 for assessing the potential activity of candidate compounds.
**Potential for clinical studies

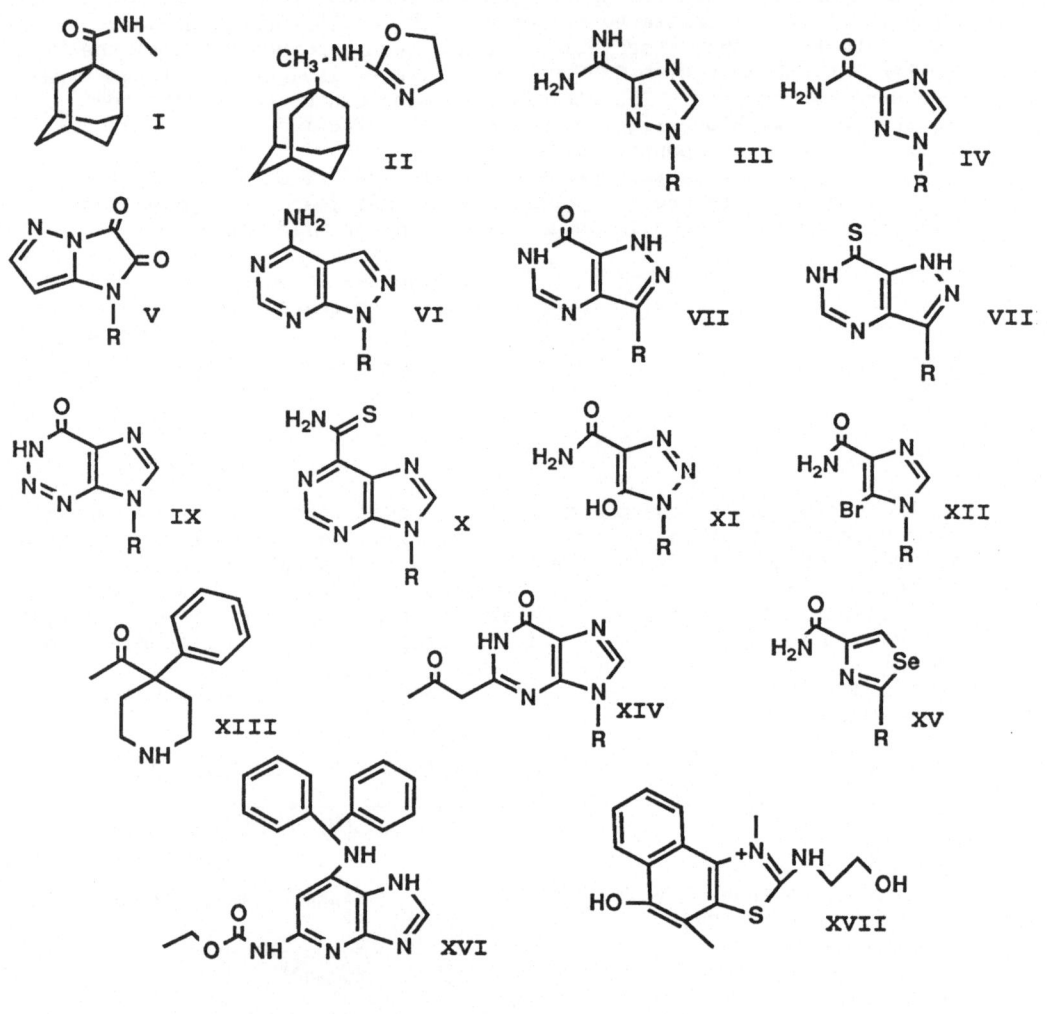

I   N-methyl-1-adamantanecarboxamide
II   N-[1-(adamantyl)-ethyl]-4,5-dihydro-2-oxazolamine
III   1-Beta-D-ribofuranosyl-1,2,4-triazole-3-carboxamidine
IV   1-Beta-D-ribofuranosyl-1,2,4-triazole-3-carboxamide
V   1-Beta-D-ribofuranosyl[1,2-b]pyrazole-2,3-dione
VI   4-amino-1-(Beta-D-ribofuranosyl)pyrazole[3,4-D]pyrimidine
VII   3-(Beta-D-ribofuranosyl)pyrazole[4,3-D]pyrimidine-7-one
VIII   3-(Beta-D-ribofuranosyl)pyrazole[4,3-D]pyrimidine-7-thione
IX   2-azainosine
X   9-(Beta-D-ribofuranosyl)purine-6-thiocarboximide
XI   5-hydroxy-1-(Beta-D-ribofuranosyl)1,2,3-triazole-4-carboxamide
XII   5-bromo-1-(Beta-D-ribofuranosyl)imidazole-4-carboxamide
XIII   4-acetyl-4-phenylpiperidine
XIV   2-acetonylinosine
XV   2-(Beta-D-ribofuranosyl)selenazole-4-carboxamide
XVI   6-diphenylmethylamino-2-aminoethyl carbamate-1-deazapurine
XVII   2-(2'-hydroxy)ethylamino-3,7-dimethyl-6-hydroxybenzo[e]benzothiazole

Figure 1. Antiviral compounds with in vitro activity against high hazard virus

Promising compounds are then tested in any of a number of animal models. Through this program a great many compounds have been found active against one or more of these viruses in vitro. These include compounds with a broad range of activity (Table 2). Compound IV, ribavirin, has been found to be active in the treatment of exotic viruses in man including Lassa fever and hemorrhagic fever with renal syndrome. The amidine analog II of ribavirin has been shown to be effective in various animal systems with little, if any, toxicity. Compounds such as XV, while highly effective in vitro, has been found to have an unacceptable level of toxicity in animal studies. Other compounds such as XI failed to exhibit acceptable efficacy in animals. Numerous compounds with highly specific antiviral activity have also been identified (Table 2).

TABLE 2: INHIBITORY DOSE $(ID_{50})^a$ OF SELECTED COMPOUNDS AGAINST EXOTIC RNA VIRUSES

VIRUS

Compound[b]	RVF[c]	PT	SF	KHF	VEE	JE	YF	PIC	VSV	HIV
I	56	-	49	-	38	‡	49	33	67	-
II	2	8	9	97	737	445	111	391	‡	0.1
III	1	-	‡	‡	‡	‡	‡	‡	‡	‡
IV	10	-	28	-	9	11	-	8	18	-
V	40	2.1	5	15	190	42	46	61	‡	0.3
VI	9	1	‡	0.1	98	0.1	59	50	-	-
VII	‡	182	564	240	‡	4	‡	‡	‡	-
VIII	504	15	384	179	‡	58	‡	10	‡	-
IX	‡	7	‡	3	‡	27	24	‡	‡	-
X	2	270	‡	344	‡	78	347	143	‡	-
XI	754	262	17	‡	‡	‡	54	28	551	-
XII	‡	‡	304	13	‡	450	‡	‡	-	-
XIII	‡	‡	‡	‡	‡	‡	‡	‡	‡	0.1
XIV	-	‡	0.3	-	-	-	‡	-	-	-
XV	3	3	1	-	7	3	0.2	4	‡	-
XVI	15	-	-	-	‡	6	-	0.3	-	-
XVII	-	4	1	-	-	-	‡	-	‡	-

a = ug/ml
b = refers to compounds in Figure 1
c = refers to viruses in Table 1

‡ = Not active at non-toxic concentrations
- = Not tested

IMMUNOMODULATORS

An important aspect of our program is the evaluation of immunoenhancing drugs and biologicals as antivirals. The relative antiviral activity of a series of known immunoenhancers is given in Tables 3 and 4. A number of these compounds have been found to be effective when given either prophylactically or therapeutically. Combinations of immunomodulators with antivirals such as ribavirin provide an even greater degree of efficacy. For example, to increase the rate of survival of Rift Valley fever infected mice when treatment is initiated late in the course of the disease combinations of high doses of Poly(ICLC) and ribavirin are required. For maximum efficacy, 6 treatments of 20 μg of Poly(ICLC) and 100 mg/kg of ribavirin given between days 2 and 11 of infection result in 80% survivals, compared to 20% survivals in placebo-treated controls. Complete protection is obtained when treatment is initiated within 24 hrs of infection. The use of such high doses, however, produces severe toxicity that argues against their use in man. Treatment with lower doses of Poly (ICLC) and ribavirin combinations can be equally effective. Multiple treatment beginning 24 hrs after challenge with either 1 μg of Poly (ICLC) or 50, 25 or 12.5 mg/kg of ribavirin alone results in 17% survival for Poly (ICLC) and 33,8,0% longterm survivors for ribavirin, respectively. However, a greater efficacy is obtained when 1 μg of Poly(ICLC) is combined with 50, 25 or 12.5 mg/kg of ribavirin yielding 75, 92, and 58% rates of survival, respectively. These data make it clear that enhancement of therapeutic efficacy by combination therapy with compounds given at low dosages can be achieved. The importance of these findings is that adverse side affects are not expected during 7-14 day therapy with such small doses of Poly(ICLC) and ribavirin.

RIBAVIRIN PROPHYLAXIS OF SANDFLY FEVER

Ribavirin currently stands alone among known substances as a broad-spectrum antiviral chemotherapeutic agent demonstrating activity against a wide range of toga-, bunya-, and arenaviruses (Huggins et al., 1984; Jahrling, Peters and Stephen 1984; Smith and Canonico, 1984). The drug has been evaluated in a number of clinical trials and has been found to be well tolerated in adults as well as in infants. We evaluated the prophylactic efficacy of ribavirin against sandfly fever (SF) virus (Sicilian strain) infection in a prospective, randomized, double-blind, placebo-controlled, phase II clinical trial.

Sandfly fever infection is a self-limiting, febrile illness transmitted by biting insects of the genus Phlebotomus. As early as 1887, it was recognized that, during the summer, army recruits in western Yugoslavia had a high incidence of a mild febrile illness now known to be sandfly fever. The clinical manifestations of sandfly fever were studied in 1944 by Dr. Albert Sabin in more than 100 cases of experimentally induced disease. Sandfly fever is sudden in onset and characterized by frontal and retroorbital headache, photophobia, generalized malaise, arthralgia and lower back pain. Approximately 65% of Dr. Sabin's patients developed temperatures of 102°F or more with a duration of 1 to 3 days in 85% of individuals.

Sixteen subjects were selected from among consenting Medical Research Volunteer Subject personnel. Test subjects were randomized into three experimental groups. Six subjects each were assigned to the placebo and drug group (Groups I and II, respectively), while the remaining 4 subjects were selected to participate in group III, the pharmacokinetic phase of the study. Subjects in groups I and II were inoculated intravenously with 0.5ml of a 10^{-1} dilution of plasma (Brownell) previously shown to contain SF virus. Group III subjects were inoculated with 0.5ml of sterile saline.

TABLE 3: PROPHYLACTIC ANTIVIRAL ACTIVITY OF IMMUNOMODULATORS

Model	Virus		p(ICLC)	Ampligen	CL-246,738	MTP-PE	hrIFNa	mrIFNg
Mouse	Bunya-	Rift Valley Fever	****	**	****	***		
		Punta Toro	****	****		****		
	Flavi-	Yellow Fever			0		0	0
		Banzi	****	****	****		****	**
	Alpha-	Semliki	****	****	****		****	**
		VEE	****	****	****			****
	Herpes	Herpes Pneumonitis	***		***	****		
		Herpes Encephalitis	****	****	****		****	
Guinea Pig	Arena-	Pichinde	0					
Monkey	Bunya-	Rift Valley Fever					••••	
	Flavi-	Yellow Fever	••••					

Sources: p(ICLC) = stabilized, double-stranded polyriboinosiniç-polyribocytodylic acid [NIAID,NIH]; Ampligen = double stranded polyriboinosinic-polyribocytidilic acid, where every 12th cytidine is replaced by uridine [Johns Hopkins U]; Cl-246,738 [Lederle]; AVS-1300 [Riker]
Percent survivors: 0, none; *, 0-25%; **, 26-50%; ***, 51-75%; ****, 76-100% [Empty block = not tested]
Viremia: •••• = none

TABLE 4: THERAPEUTIC ANTIVIRAL ACTIVITY OF IMMUNOMODULATORS

Model	Virus		p(ICLC)	Ampligen	CL-246,738	MTP-PE	hrIFNa	mrIFNg
Mouse	Bunya-	Rift Valley Fever	****	**	****	***		
		Punta Toro				*		
	Flavi-	Yellow Fever						
		Banzi		**	**		****	
	Alpha-	Semliki				'		
		VEE	**	**	**		****	*
	Herpes	Herpes Pneumonitis	****		**	****		
		Herpes Encephalitis	*****		****	****		
Guinea Pig	Arena-	Pichinde	*	0	0	0		
Monkey	Bunya-	Rift Valley Fever						
	Flavi-	Yellow Fever	••••				••••	

Sources: p(ICLC) = stabilized, double-stranded polyriboinosinic-polyribocytodylic acid [NIAID,NIH]; Ampligen = double stranded polyriboinosinic-polyribocytidilic acid, where every 12th cytidine is replaced by uridine [Johns Hopkins U]; Cl-246,738 [Lederle]; AVS-1300 [Riker]
Percent survivors: 0, none; *, 0-25%; **, 26-50%; ***, 51-75%; ****, 76-100% [Empty block = not tested]
Viremia: •••• = none

Ribavirin was administered orally in 100 mg capsules. Group I was given placebo capsules, while group II subjects received four (4) ribavirin capsules at 08:00, 16:00, and 24:00 hr beginning 1 day prior to virus inoculation and continuing for 8 days. Group III received ribavirin capsules as above through one additional dose following the eighth day.

Subjects were monitored for vital signs throughout the test period. Blood and urine samples were systematically collected for an expanded standard set of clinical chemistry analyses, hematology, antigen quantitation, lymphocytes subset studies, and interferon determinations. Additional blood and urine samples for pharmacokinetic analyses were obtained from group III.

Scores assigned to various clinical signs and symptoms were minimal and principally associated with the expected course of Sandfly fever in the placebo-treated group. In this group scores were greatest on days 3 to 5 reaching cumulative values that were more than double the scores recorded for the ribavirin treatment group.

Placebo-treated subjects (group I) became febrile at approximately 3.5 days after virus inoculation. Fever persisted for about 60 hr reaching a maximum of $102^{o}F$ in 4 of the 5 placebo-treated subjects that sero-converted. In contrast, all ribavirin-treated subjects exhibited normal body temperatures and a normal diurnal temperature cycle throughout the hospitalization period.

Other viral signs, consisting of diastolic and systolic pressure, pulse and respiratory rate, and body weight were unremarkable, and no differences were observed between the placebo- and ribavirin-treated subjects.

Urinalysis findings were unremarkable. There were no effects observed which could be attributed to ribavirin treatment. There were no statistically significant differences observed for 2 or more consecutive days between placebo- and ribavirin-treated groups for any of the 30 clinical chemistry parameters measured.

Total bilirubin was the only chemical parameter which was significantly elevated for 2 consecutive days. Values of 1.14 mg/dl and 1.19 mg/dl were obtained for ribavirin-treated subjects on days 5 and 6 compared to 0.46 mg/dl and 0.41mg/dl for the placebo group ($P < .009$ and $.006$, respectively).

Hematological changes seen in the placebo group were consistent with a typical response to viral infections. These changes included a drop in absolute granulocyte counts from 3.62×10^3 on day 3 to 1.12×10^3 on day 8, with a rise to 2.60×10^3 by day 12 (Fig. 2). Statistically significant differences between the placebo and ribavirin groups were found on days 7-11. The relative percentage of granulocytes also dropped in the placebo group, from a value of 70% on day 4 to 35% on day 8. While there were no statistically significantly differences between the placebo- and ribavirin-treated groups with regard to absolute lymphocyte counts, the percentage of lymphocytes increased significantly in the placebo group during days 6-9. Total WBC counts decreased significantly compared to the ribavirin-treated group on days 7-11, with a nadir of $3.3 \times 10^3 \pm 0.64$ on day 8 compared to $7.60 \times 10^3 \pm 0.52$ for the ribavirin-treated group (Fig. 3).

None of the red cell indices, including red cell numbers (Fig. 4), hematocrit (Fig. 5), and hemoglobin differed significantly between placebo- and ribavirin-treated groups. All 3 of these indices, however, showed a trend to lower values from the time of admission through day 10 of the study. A reticulocytosis did develop in the ribavirin-treated group, that was statistically significant from the placebo group on day 5-9 and 11-12. Peak reticulocyte counts reached a value of $4.62/1000 \pm 0.76$ on day 11.

GRANULOCYTE COUNTS IN SANDFLY FEVER PATIENTS

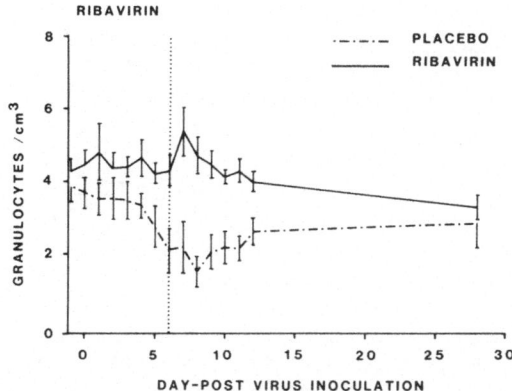

Figure 2. Mean neurophil counts (10^3/cmm) for days -2 to 28 for ribavirin (dashed line) and placebo-treated (solid line) SF virus-inoculated volunteers.

WBC COUNTS IN SANDFLY FEVER PATIENTS

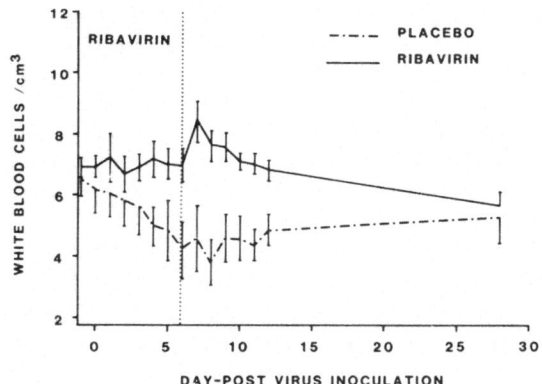

Figure 3. Mean white blood cell counts (10^3/cmm) over days -2 to 28 for infected, placebo- and ribavirin-treated volunteers (symbols as in Figure 2).

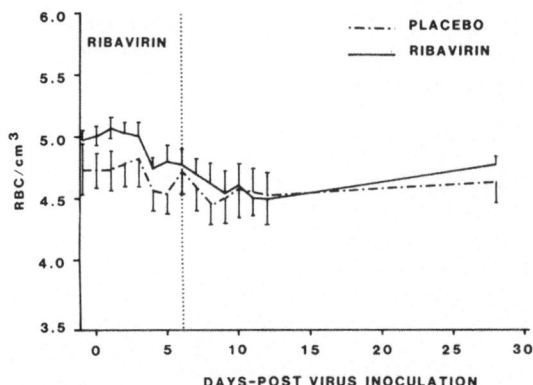

Figure 4. Mean red blood cell counts
(10^6/cmm) over days -2 to 28 for placebo-
and ribavirin-infected volunteers
(symbols as in Figure 2.).

HEMATOCRIT VALUES IN SANDFLY FEVER PATIENTS

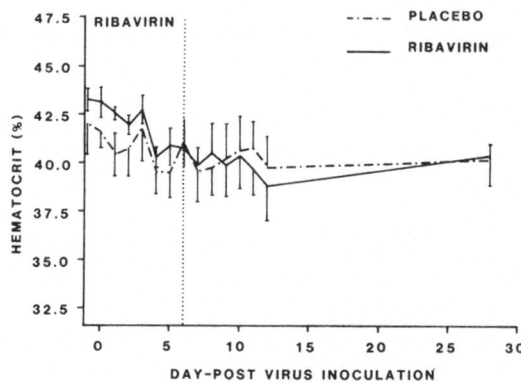

Figure 5. Mean hematocrit values (%) over
days -2 to 28 for placebo- and ribavirin-
treated volunteers (symbols as in Figure 2).

A consistent change in platelet values was observed between drug- and placebo-treated groups. Platelet counts in the placebo group decreased during the febrile period from values above 200,000 to approximately 160,000 on days 5-8. On the other hand, platelet values in the ribavirin group increased, reaching a peak of 377,000 on day 11. The differences in platelet counts between groups I and II attained statistical significance on days 5-12.

IgG antibody titers were elevated in the placebo group beginning on day 7 after virus inoculation, reached a peak by day 14, then declined to a titer of 1/3000 by day 126. In contrast, IgM titers rose sharply on day 21 in the ribavirin-treatment group, and declined on day 64 through day 126 to titers which were comparable to the placebo group.

IgM titers were detectable in both groups on day 28. Antibody levels were substantially higher in the placebo group and persisted through day 124. IgG titers in the ribavirin-treatment group were only 1/10 the magnitude that was measured in the placebo group. These titers were maintained through day 124.

Alpha and gamma interferon titers were detected in significant levels in the placebo group on days 3 to 8. No interferon of either type was observed in the ribavirin-treated group. Antigenemia was assessed with an ELISA-based assay for SF antigens. Circulating SF antigens were detected in 4 of the 5 responders in the placebo group on days 3 to 5.

Plasma ribavirin concentrations were measured in the 4 pharmacokinetics subjects for the first 4 days of ribavirin administration and in 3 subjects thereafter. A plateau of about 5-6 μM was achieved by 96 hr, with a maximum peak of 9.94 μM 2 hr after dosing on day 4. Clearance of ribavirin was relatively slow, with serum levels of about 3.3 μmoles remaining 72 hr after administration of the final dose.

Ribavirin appears to concentrate in red cells, attaining levels 50-80 times greater than in the plasma. A plateau of about 300-500 μM was reached in red cells after about 4 days of dosing. Clearance of ribavirin from red cells appeared relatively slow where values in excess of 300 μM were measured 9 days after drug treatment. On day 28, ribavirin levels in red cells of the 3 subjects ranged from 139 to 265 μM.

Infection by SF virus in the placebo group was confirmed by the significant decrease in total WBC, the occurrence of neutrophilia followed by neutropenia, fever, antigenemia on days 3-5 after inoculation, and the presence of both alpha and gamma interferon. In contrast, the prophylactic use of ribavirin averted all overt clinical symptoms of the disease. Remarkably, ribavirin did not prevent the development of an immune response to SF. Ribavirin was well tolerated, causing only mild, transient alterations in a limited number of hematological indices.

These data indicate that ribavirin is a relatively safe drug, that can prevent clinical disease while not suppressing the immune response to a live virus exposure. These are important qualities for a prophylactic antiviral drug.

Sandfly fever virus is classified as a Phlebovirus in the Bunyaviridae family. Other important members of this family are Rift Valley fever, Crimean-Congo, and Hazara. The demonstration of ribavirin's antiviral activity against human sandfly fever has broader implications than simply the treatment of this high morbidity, but self-limiting, disease. The possiblity that this broad-spectrum antiviral drug will be active against several related viruses is increased by extending the distinctly encouraging rodent and non-human primate data against Rift Valley fever and other hemorrhagic fever viruses (Stephen et al., 1980).

REFERENCES

Baladov, M. E., V. N. Lazarev, E. K. Koimchidi and G. A. Karinskaya. 1974. In a translation of "Crimean Hemorrhagic Fever", Papers at the third regional workshop at Rostov-on-don, 1970. Miscellaneous publications, Entomol. Soc. of Am. 9:160-161.

Bilibin, A. F. 1950. Omsk and Crimean hemorrhagic fevers. In: Symptoms and diagnosis of infections. Medzig; Moskva, pp. 200-208.

Blyakher, I. A., T. P. Pak and A. V. Yasinsky. 1971. Hospital outbreak of Crimean hemorrhagic fever in Tadzhikistan. In: Viral hemorrhagic fevers. Crimean Hemorrhagic fever, Omsk Hemorrhagic fever, Hemorrhagic fever with renal syndrome, edited by Chumakov, M. P. Trudy Inst. Polio. Virus. Entsef Akad. Med. Nauk. SSSR. 19:130-133.

Bulynin, V. I. and S. A. Poshekhonov. 1959. The contagiousness of hemorrhagic fever in the Stravropol territory. J. Med. Exp. Immunol. 10:182-183.

Burney, M. I., A. Ghafoor, M. Saleen, P. Webb and J. Casals. 1980. Nosocomial outbreak of viral hemorrhagic fever caused by Crimean hemorrhagic fever-Congo virus in Pakistan, January 1976. Am. J. Trop. Med. Hyg. 29:941-947.

Canonico, P. G., M. Kende, B. J. Luscri and J. W. Huggins. 1984. In vivo activity of antiviral against exotic RNA viral infections. J. Antimicrob. Chemother. Suppl. A, pp.27-41.

Causey, O. R., G. E. Kemp, M. H. Madbouly, and T. S. David-West. 1970. Congo virus from domestic livestock, African hedgehogs, and arthropods in Nigeria. Am. J. Trop. Med. Hyg. 19:846.

Chumakov, M. P. 1974. On thirty years of investigation of Crimean hemorrhagic fever. Tr. Inst. Polio. Virusn. Entsefalitov. Akad. Med. Nauk SSSR. 22:5 (in Russian). (In English, NAMRU3-T950).

Chumakov, M. P., S. E. Smirnova and E. A. Tkachenko. 1969. Antigenic relationships between the soviet strains of Crimean hemorrhagic fever virus and the Afro-Asian Congo virus strains, Mater. 16 Nauchn. Sess. Inst. Polio. Virus Entsefalitov. (Moscow, October 1969), Z, 152 (In Russian) (In English, NAMRUS - T614)

Chumakov, M. P., S. K. Andreeva, T. I. Zavodova, G. N. Zgurskaya, N. V. Kostetsky, L. I. Martyanova, A. M. Nikitin, K. M. Sinyak, S. E. Smirnova, L. I. Turta, E. D. Ustinova and S. P. Chunikin. 1974. In "Medical Virology", M. P. Chumakov, ed., Works Inst. of Polio. and Virus Encephalitis Moscow. pp. 24.

Daubney, R., J. R. Hudson and P. C. Garnham. 1931. Enzootic hepatitis or Rift Valley fever: An undescribed disease of sheep, cattle and man from East Africa. J. Path. Bact. 34:545-579.

Gajdusek, D. C. 1956. Hemorrhagic fevers in Asia: A problem in medical ecology. Geogr. Rev 41(1):20-42.

Hoogstrall, H. 1979. The epidemiology of tick-borne Crimean-Congo hemorrhagic fever in Asia, Europe, and Africa. J. Med. Ent. 15:307.

Huggins, J. W., P. Jahrling, M. Kende, P. G. Canonico. 1984. Efficacy of Ribavirin Against Virulent RNA Virus Infections. In: Clinical Applications of Ribavirin. Smith, R.A., Knight, V., Smith, J.A.D. (eds) Academic Press, New York, pp.49-63.

Jahrling, P. B. and C. J. Peters. 1985. Arenaviruses. In: Laboratory diagnosis of viral infections (E. H. Lennette, ed.) Marcel Dekker, Inc., N.Y., pp. 171-189.

Jahrling, P. B., C. J. Peters, and E. L. Stephen. 1984. Enhanced Treatment of Lassa Fever by Immune Plasma Combines with Ribavirin of Lassa Fever by Immune Plasma Combines with Ribavirin in Cynomolgus Monkeys. J. Infect. Dis. 149(3):420-7.

Joubert, J. R., J. B. King, D. J. Rossouw, R. Cooper. 1985. A nosocomial outbreak of Crimean-Congo hemorrhagic fever at Tygerberg hospital. Part III. Clinical patholgoy and pathogenesis. S. Afr. Med. J. 68:722-728.

Kemp, G. E., O. R. Causey, H. W. Setzer and D. L. Moore. 1976. Isolation of viruses from wild mammals in West Africa, 1966-1970. J. Wildl. Dis. 10:279.

Leduc, J. W., G. A. Smith, J. E. Childs, F. P. Pinheiro, J. I. Maiztegui, B. Niklasson, A. Antoniades, D. M. Robinson, M. Khin, K. F. Shortridge, M. T. Wooster, M. R. Elwell, P. L. T. Ilbery, D. Koech, E. S. T. Rosa and L. Rosen. 1986. Global survey of antibody to Hantaan-related viruses among peridomestic rodents. Bull. WHO. 64(1):139-144.

Lee, H. W. and P. W. Lee. 1976. Korean hemorrhagic fever. I. Demonstration of causative antigen and antibodies. Korean J. Intern. Med. 19:371-383.

Linthicum, K. J., F. G. Davies, A. Kairo and C. L. Bailey. 1985 Rift Valley fever virus (family Bunyaviridae, genus Phlebovirus). Isolations from Diptera collected during an inter-epizootic period in Kenya. Journal of Hygiene, Cambridge. 95:197-209.

Lupton, H. W., C. J. Peters and G. A. Eddy. 1982. Rift Valley fever: Global spread or global control? Proc. 86th Ann. Meeting U.S. Animal Health Assoc., Nashville, TN pp. 261-275.

McCormick, J. B. and K. M. Johnson. 1978. Lassa Fever: Historical review and contemporary investigation. In: Ebola virus hemorrhagic fever (S. R. Pattyn, ed.). North Holland Biomedical Press, N. Y. pp. 279-283.

McKee, K. T., C. MacDonald, J. W. LeDuc and C. J. Peters. 1985a. Hemorrhagic fever with renal syndrome-A clinical perspective. Military Medicine. 150:640-647.

McKee, K. T., Jr., C. J. Peters, R. B. Craven and D. B. Francy. 1985b. Other viral hemorrhagic fevers and Colorado Tick Fever. In: Textbook of Human Virology (R. B. Belshe, ed.) PSG Publishing Co., Inc., MA pp. 547-597.

Peters, C. J. 1985. Arenaviruses. In: Textbook of Human Virology (R. B. Belshe, ed.) PSG Publishing Co., Inc., MA pp. 547-597.

Peters, C. J. and J. W. LeDuc. 1985. Bunyaviruses, Phleboviruses and related viruses. In: Textbook of Human Virology. (R. B. Belshe, ed.) PSG Publishing Co., Inc., MA pp. 547-597.

Schmaljohn, C. S. and J. M. Dalrymple. 1983. Analysis of Hantaan virus RNA: Evidence for a new genus of Bunyaviridae. Virol. 131:482-491.

Shope, R. E., C. J. Peters and F. G. Davies. 1982. The spread of Rift Valley fever and approaches to its control. Bull. WHO. 60(3):299-304.

Simpson, D. I. H., E. M. Knight, G. H. Courtois, M. C. Williams, M. P. Weinbren and J. W. Kibukamusoke. 1967. Congo virus: A hitherto undescribed virus occuring in Africa. Part I. Human isolations. Clinical Notes. E. Afr. Med. J. 44:87.

Smith, R.A., P.G. Canonico. Ribavirin. 1984. In: Antiviral Drugs and Interferon: The Molecular Basis of their Activity. Y. Becker (ed) Martinus Nijhoff Publishing, Boston.

Stephen, E.L., E.E. Jones, C.J. Peters, G.A. Eddy, P.S. Loizeaux, and P.B. Jahrling. 1980. Ribavirin treatment of toga-, arena- and bunyavirus infections in subhuman primates and other laboratory animal species. In: Ribavirin A broad spectrum antiviral agent. R.A. Smith, W. Kirkpatrick (Eds). Academic Press, New York. pp.169-183.

Van Eeden, P. J., J. R. Joubert, B. W. Van de Wal, J. B. King, A. De Kock and J. H. Groenewald. 1985a. A nosocomial outbreak of Crimean Congo hemorrhagic fever at Tygerberg hospital. Part I. Clinical Features. S. Afr. Med. J. 68:711-717.

Van Eeden, P. J., S. F. Van Eeden, J. R. Joubert, J. B. King,
    B. W. Van de Wal and W. L. Michell. 1985b. A nosocomial outbreak
    of Crimean-Congo hemorrhagic fever at Tygerberg hospital. Part
    II. Management of patients. S. Afr. Med. J. 68:718-721.
Watts, D. M., T. G. Ksiazek, K. J. Linthicum and H. Hoogstraal. 1987.
    Arboviruses: Epidemiology and ecology, Vol III (T. P. Monath,
    ed.). CRC Press, Inc., Boca Raton, Fl. U.S.A. (in press)
White, J. D., F. C. Shirey, G. R. French, et al: 1982. Hantaan Virus,
    etiological agent of Korean hemorrhagic fever, has Bunyaviridae-
    like morphology. Lancet. 1:768-771.
Woodall, J. P., M. C., Williams, D. I. H. Simpson, P. Ardoin, M. Lule and R.
    West. 1965. The Congo group of agents. Rep. E. Afr. Virus. Res.
    Inst. 14:34.
Zarubinski, V. Ya., V. F. Kondratenko, N. M. Blagoveshchenskaya,
    L. V. Zarubina and V. V. Kuchin. 1976. Tezisy Dolk. 9. Vses.
    Konf. Prirod. Ochag. Bolez. Chelov. Zhivot. Omsk. pp. 130-131.

HUMAN RETROVIRUSES AND ANTI-VIRAL THERAPIES

R.C. Gallo and M.S. Reitz, Jr.

Laboratory of Tumor Cell Biology
National Cancer Institute
National Institutes of Health
Bethesda, Maryland 20892, U.S.A.

INTRODUCTION

At least four types of human retroviruses are currently identified. They are causally involved in several different kinds of diseases in humans, including leukemia and lymphoma immunodeficiency diseases and neurologic and central nervous system diseases. Their identification has led to the ability to easily detect them and to a basic understanding of their life cycle. This knowledge makes possible various strategies for prevention and treatment of diseases due to infection by these viruses.

HUMAN RETROVIRUSES

The first human retrovirus was identified from a cutaneous T-cell lymphoma and was called human T-cell leukemia virus (later, type I) (HTLV-I)[1]. This virus is endemic to certain parts of the world, including parts of Africa, southwestern Japan, the southeastern United States and the Caribbean, and has been shown to be the etiologic agent for adult T-cell leukemia (ATL) and tropical spastic paraparesis, a central nervous system disease somewhat similar to multiple sclerosis. Infection by HTLV-I also leads to immunosuppression, occasionally causing a disease clinically indistinguishable from the acquired immunodeficiency syndrome (AIDS), and appears to be able to indirectly cause B cell lymphomas, probably by chronic immune stimulation. The second human retrovirus was isolated from a patient with a T-cell variant of hairy cell leukemia after being identified by a serologic cross-reactivity with HTLV-I and was called HTLV-II[2].

HTLV-II does not appear to be nearly as prevalent as HTLV-I and is not currently associated with any disease, although its true distribution may be obscured somewhat by the difficulty in easily distinguishing it from HTLV-I. The third human retrovirus originally called HTLV-III[3] or LAV-1[4] but currently called HIV-1 (for human immunodeficiency virus type I) is the cause of the current AIDS pandemic. A fourth type of human retrovirus, which is distantly related to HIV-1 but is closely related to a retrovirus in African monkeys[5], and which has been called LAV-2[6] and HTLV-4[7], has been identified and is widely distributed in Western Africa. Viruses of this group, which has recently been designated HIV-2, seem to cause AIDS, but much less efficiently than HIV-1. Further study is needed to clarify this point, however. For a more complete review of these viruses, see reference 8.

PREVENTION

The first line of prevention comes from knowledge about the mode of transmission of these viruses. All seem to be transmitted in the same manner, namely by exchange of body fluids. Thus, transmission can be sexual, by transfusion of blood products and by intravenous drug abuse. Avoidance of sexual promiscuity and the use of condoms can limit sexual transmission; avoiding the sharing of unsterilized needles can eliminate transmission by the route of drug abuse. Assays which detect the presence of viral antibodies in blood products have greatly decreased the possibility of transmission by transfusion, although a small percentage of infected people (those recently infected, for example) may be antibody-negative. Use of all the above practices, however, can greatly limit the further spread of all of these viruses.

A future way to prevent spread of these viruses is the development of vaccines. This would seem to be particularly urgent in the case of HIV-1. There are some problems, however. For one thing, the envelope protein of HIV-1 shows an unusually great degree of strain variability and this will probably have to be taken into account in the design of a vaccine. Secondly, even though infected individuals have low levels of neutralizing antibody to HIV-1, this does not appear to be protective against disease, and it may be that humans cannot elaborate an effective immune response to the virus.

The retroviral life cycle is shown schematically in Figure 1. First the viral outer envelope protein binds to a specific cellular receptor. The virus then penetrates the cell membrane and the viral core is uncoated. The genomic RNA is then reverse-transcribed into a double-stranded DNA copy by the viral DNA polymerase. This DNA moves to the nucleus, is circularized, and is integrated into the host cell genome by the integrase specified by part of the viral messenger RNA (mRNA). The proviral DNA is then transcribed to genomic RNA by cellular enzymes and translated into viral proteins. The gag and pol precursor polypeptides are cleaved, probably by the viral protease, also specified within the pol gene. The envelope precursor polypeptide is also cleaved, perhaps by a cellular enzyme, and the viral components are assembled at the cell membrane, where they bud off into the exterior of the infected cell to begin another round of infection.

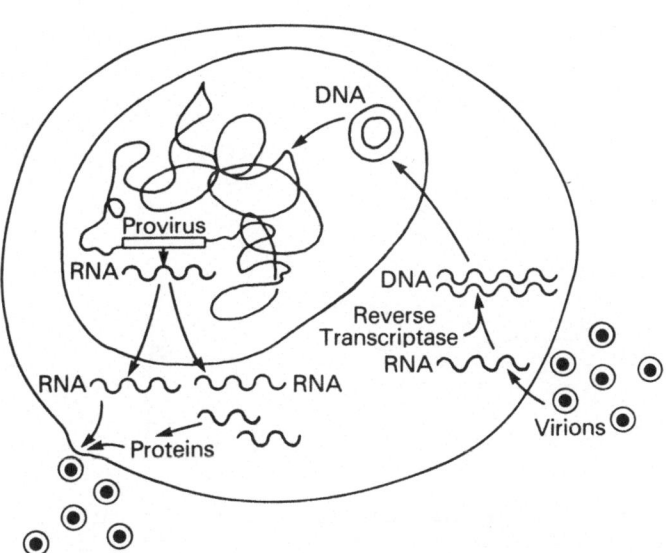

Fig. 1. Life cycle of a retrovirus.

In addition to the genes present in other retroviruses, the genomes of the human retroviruses contain extra genes, as shown in Figure 2. All four appear to contain a gene which has been called tat, for transactivator of translation/transcription. This gene, which is contained in different regions in HTLV-I and HIV-1, produces a protein which increases the efficiency of translation of viral mRNA and seems also to increase the levels of mRNA. The tat mRNA is double spliced and contains a second gene, called trs or art, which is necessary for the expression of unspliced and singly spliced viral mRNA. Trs perhaps acts by blocking the complete processing of all mRNA to the doubly spliced form. Trs is thus necessary for synthesis of the virion proteins. HIV-1 also has genes called sor, 3'-orf, and R which have been identified on the basis of the DNA sequence of viral DNA and the presence of antibodies in infected people. Their functions are not known, but sor is known to be necessary for the infection of cells by cell-free virions, while the 3'-orf is not necessary and may function in vivo as a negative regulatory element. HIV-2 has yet another open reading frame, called X, whose function, if any, is unknown. All of these proteins may present targets for intervention in the viral life cycle.

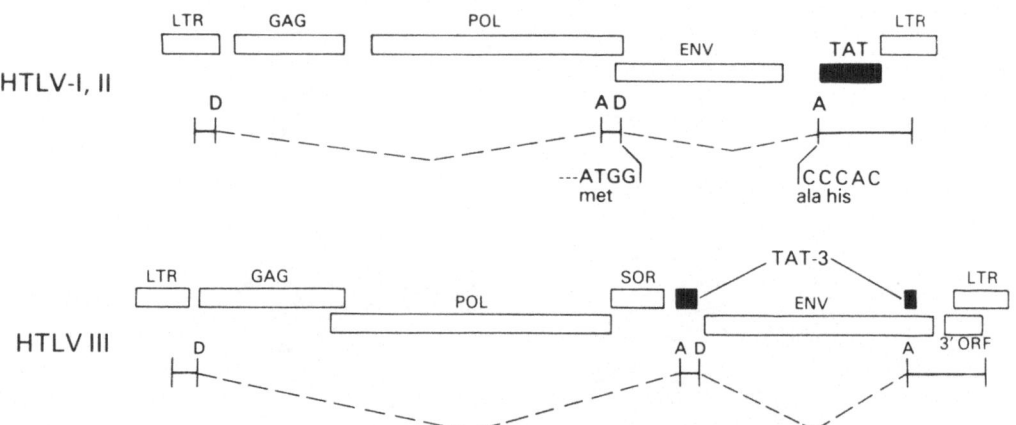

Fig. 2. Genome structure of the human retrovirus. The three exons of the tat mRNA are shown with the coding regions in solid black. The initiating methionine codon is shown for tat-1.

The necessity for the binding of the viral envelope to the cellular receptors provides a possible target for anti-viral strategies. Elicitation of neutralizing antibodies by vaccines are, of course, one of the ways of preventing this early step, and could prevent infection of the exposed individual. Furthermore, if virus spread during the course of infection is necessary for disease progression, passive antibodies to the viral envelope would be effective in people who are already infected. Furthermore, antibodies or synthetic peptides which bind to the virus receptor on the surface of the target cell (the CD4 antigen in the case of HIV-1 and -2) could block infection. The virion envelope structure is composed in part of a lipid bilayer into which the envelope protein is inserted. Disruption of this structure by some sort of surface active agent could conceivably block infection.

Another step in the virus life cycle which is available as a potential anti-viral target is the reverse transcription of the genomic RNA. This could be affected by inhibitors which are specific for reverse transcriptase. A number of these compounds have already been tried, and the most promising appear to be derivatives of 2'-deoxyribonucleoside triphosphates which lack a 3'-hydroxyl group and hence prematurely terminate DNA transcription. The viral DNA polymerase, unlike the analogous enzyme in the host cell, is unable to remove these bases because it lacks a proof-reading function, providing a basis for viral specificity. Some of these compounds have been shown to protect cells from infection at doses below those toxic to the cells. One such compound, 3'-azido-2',3'-dideoxythymidine (AZT) has been used with some success in preventing and to some extent reversing the progression of patients with AIDS. Another similar compound, 2',3'-dideoxycytidine, is currently ready for clinical trials. These drugs have the advantage of good oral availability and are able to cross the blood-brain barrier. This is of major importance as virus is present in the central nervous system of many AIDS patients or other HIV-1-infected people. For a fuller discussion of this class of compounds, see reference 9.

Another possible target in the initial infection process is the step of integration. No current inhibitors of the viral integrase have been identified, however. Furthermore, in HIV-1-infected cells the majority of the viral DNA is unintegrated, and it is not clear whether integration is a requirement in the life cycle of this virus.

After the establishment of the provirus in the infected cell, viral mRNA is formed by the transcriptional factors of the infected cell, including the cellular RNA polymerase II. Full length viral RNA is processed by splicing to form the mature viral mRNA. In the case of env mRNA, a single splice removes the gag and pol sequences to form the mature mRNA. For most of the other genes, multiple splicing events are involved in mRNA maturation. The viral mRNA is then translated using the normal translation processes and factors of the host cell. The viral proteins are proteolytically processed to their mature forms, using either the viral protease (in the case of the gag and pol gene products) or cellular proteases (likely the case with the env proteins). These viral proteins are then assembled at the cell surface around the viral RNA, which contains a packaging recognition site just before the initiation codon of the gag gene. The virion then buds from the cell at the cell surface. Several of the viral proteins are also modified post-translationally. These modifications include the addition of glycosyl groups to the envelope proteins and the myristylation of the gag polyprotein.

Drugs which selectively interfere with any of these processes could theoretically inhibit virus expression from infected cells. Many of them, however, are so intricately associated with normal cellular processes that finding the basis for a selective drug will be difficult. Potential attractive targets would be the tat and trs gene products. Perhaps drugs could be found which would bind to either of these proteins selectively, since they do not appear to have cellular counterparts.

Another possible way in which the expression of virus may be selectively inhibited is through the use of antisense RNA; that is, RNA (or DNA) which is complementary to the viral mRNA. This type of mechanism has been shown in other systems, both eukaryotic and prokaryotic, to be capable of inhibiting gene expression. The theoretical mechanism of action of antisense RNA is that it can hybridize to mRNA, thus blocking the elongation of the nascent peptide chain during translation. In practice, however, it has been observed that inhibition occurs at other levels, such as that of expression of RNA, in addition to that of translational arrest. Furthermore, sometimes no inhibition is achieved. There are two ways of delivering antisense nucleic acids. One is to add or administer oligodeoxynucleotides extracellularly and allow the target cell to incorporate these compounds. Although the normal oligonucleotides are soon degraded, modified derivatives (thiophosphates, for example) can be made which are stable but are still internalized by the target cells (M. Matsukura et al., submitted for publication). These compounds seem to inhibit the reverse transcription of genomic RNA early in infection with HIV-1, perhaps by binding to the genomic RNA and blocking elongation of viral DNA.

A second way to introduce antisense RNA is to use retroviral cloning vectors to infect cells which are already infected with one of the human retroviruses or are possible target cells. The retroviral vector then integrates the antisense information in a form which is capable of being transcribed at high levels. The constitutive expression of antisense mRNA then would theoretically be able to block viral gene expression, preventing further spread of the virus. Possible levels of inhibition by this procedure would include translation arrest, inhibition of the splicing of viral mRNA, and inhibition of packaging of viral genomic RNA into virus particles.

Although there are many possible ways in which to inhibit virus infection, expression, and spread, it is likely that no single method will be sufficient, and that many different methods will need to be explored. Although this is in some respects a pessimistic assessment, several approaches have shown either promise or a degree of actual success, and it is, therefore, highly possible that a combination of approaches outlined above and others not mentioned will give us the ability to ultimately contain and control human retroviral diseases.

REFERENCES

1. B.J. Poiesz, F.W. Ruscetti, A.F. Gazdar, P.A. Bunn, J.D. Minna, and R.C. Gallo, Detection and isolation of type C retrovirus particles from fresh and cultured lymphocytes of a patient with cutaneous T-cell lymphoma, Proc. Natl. Acad. Sci USA 77:7415 (1980).
2. V.S. Kalyanaraman, M.G. Sarngadharan, M. Robert-Guroff, I. Miyoshi, D. Blayney, D. Golde, and R.C. Gallo, A new subtype of human T-cell leukemia virus (HTLV-II) associated with a T-cell variant of hairy cell leukemia, Science 218:571 (1982).
3. R.C. Gallo, S.Z. Salahuddin, M. Popovic, G.M. Shearer, M. Kaplan, B.F. Haynes, T.J. Palker, R. Redfield, J. Oleste, B. Safai, G. White, P. Foster, and P.D. Markham, Frequent detection and isolation of cytopathic retroviruses (HTLV-III) from patients with AIDS and at risk for AIDS, Science 224:500 (1984).

4. F. Barré-Sinoussi, J.C. Chermann, F. Rey, M.T. Nugeybe, S. Chamaret, J. Gruest, C. Dauguet, C. Axler-Blin, F. Brun-Vezinet, C., Rouzioux, W. Rozenbaum, and L. Montagnier, Isolation of a T-lymphotropic retrovirus from a patient as risk of acquired immune deficiency syndrome (AIDS), Science 220:868 (1983).

5. P.J. Kanki, J. Alroy, and M. Essex, Isolation of T-lymphotropic retrovirus related to HTLV-III/LAV from wild-caught African green monkeys, Science 230:951 (1985).

6. F. Clavel, D. Guetard, F. Brun-Vezinet, S. Chamaret, M.A. Rey, M.O. Santos-Ferreira, A.G. Laurent, C. Dauguet, C. Katlama, C. Rouzioux, D. Klatzmann, S.L. Champalimand, and L. Montagnier, Isolation of a new human retrovirus from West African patients with AIDS, Science 233:343 (1986).

7. P.J. Kanki, F. Barin, S. M'Boup, J.S. Allan, J.L. Romet-Lemonne, R. Markink, M.F. McLane, T.H. Lee, B. Arbeille, and M. Essex, New human T-lymphotropic retrovirus related to simian T-lymphotropic virus type III (STLV-III$_{AGM}$), Science 232:238 (1986).

8. F. Wong-Staal and R.C. Gallo, Human T-lymphotropic retroviruses, Nature 317:395 (1985).

9. H. Mitsuya and S. Broder, Strategies for antiviral therapy in AIDS, Nature 325:773 (1987).

EPSTEIN-BARR VIRUS

Joseph S. Pagano

Lineberger Cancer Research Center 237H
University of North Carolina at Chapel Hill
Chapel Hill, NC   27514

INTRODUCTION

    Infection with human herpesviruses poses a problem for
treatment with antiviral drugs, stemming from the latent
infection that viruses produce. There are now drugs, in use
or experimental, that are effective against replication of
herpesviruses in vitro and in vivo. None of these drugs
affects latent herpetic infection. Until recently we have
lacked even a theoretical approach to treatment of latency.
The Epstein-Barr virus is ideal for consideration of the
different issues presented by herpetic infection and how to
study the approaches to treatment in vitro. Well
characterized in vitro systems for EBV replication provide
excellent models for testing and study of drugs active
against productive infection. In addition the only cellular
model for herpesviral latency is provided by EBV. In cells
latently infected with EBV we have known or at least
inferred that different DNA polymerases are utilized in
productive versus latent infection, with only the virus-
specified polymerase being susceptible to the action of any
of the known antiviral drugs. Now however we are beginning
to understand in molecular detail in EBV cellular models how
latent EBV genomes are maintained in latently infected
cells. This new knowledge opens the first conceptual
approaches to antiviral treatment of latent infection. In
this chapter I review biological and molecular features of
EBV infection needed for an understanding of the rational
use of antiviral drugs for this viral infection. I also
review briefly drugs active against EBV replication and
conclude with some predictions about what results antiviral
therapy might be expected to yield.

    The Epstein-Barr virus (1) causes infection
universally. Excreted in the saliva the virus is
transmitted by close contact. Primary infection with EBV is
usually silent; infectious mononucleosis is a disease
largely of adolescents and young adults. Virus shedding
reflects the primary sites of infection: the epithelial
cells of the oropharynx, posterior nasopharynx and parotid
gland; these cells are permissive for EBV replication. The

secondary target cell for the virus is the B-lymphocyte. The C2R receptor serves as receptor for the virus in both cell types. B-lymphocytes become latently infected; they are not destroyed by infection, but are stimulated to grow. The hallmark of EBV infection is polyclonal B-lymphocytic proliferation. This lymphoproliferative phase is checked by normal host immune responses, but a subset of latently infected lymphocytes or their progeny persist.

After primary infection excretion of virus may continue for years, but the notable feature of EBV infection is latency: latent infection persists for life and is subject to reactivation. Latent infection of B-lymphocytes and epithelial cells in the posterior nasopharynx may have the delayed consequence of development of malignancies arising in these cells.

Morphologically the Epstein-Barr virus is indistinguishable from other herpesviruses. The large linear double-stranded DNA, which is contained within an enveloped hexameric nucleocapsid, is replicated by the virus-encoded DNA polymerase (Figure 1). The genome consists of approximately 170×10^3 nucleotide base-pairs and has multiple reiterations including direct tandem repeats at both termini (TR) and several other sets of

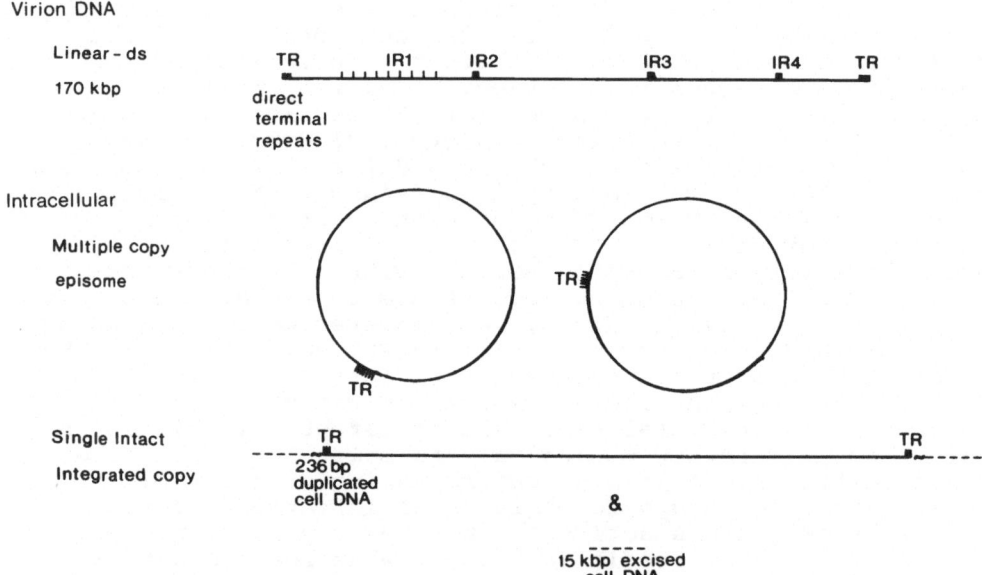

Figure 1. Forms of the Epstein-Barr virus genome.

The linear double-stranded form of EBV DNA is associated with virus production, is replicated by EBV encoded DNA polymerase, and is encapsidated in virions. There are two intracellular forms of the EBV genome: the EBV episome, which is a unit length circularized form of the genome that is replicated by host DNA polymerase, and single unit length genomes that have been detected integrated into the cellular DNA of a cell line derived from Burkitt's lymphoma tissue.

direct tandem reiterations positioned internally in the genome (IR 1,2,3,4). The complementary terminal repetitions can join to cause circularization of the genome after infection resulting in an episomal intracellular form (2,3). EBV episomes are replicated by host cell polymerase rather than by the viral polymerase; antiviral drugs effective against the viral DNA polymerase do not affect the copy number of viral episomes (4). Once latent infection is established the number of viral episomes is constant, and viral expression and replication are strictly regulated; the few genes expressed do not include those necessary to make virus.

Integrated copies of the EBV genome have been detected in some cell lines (5). The sequence of viral-cellular junction fragments has been analyzed in the Namalwa cell line, which contains a single intact copy of the EBV genome integrated into chromosome 1 with both duplication and deletion of cellular sequences at the site of integration. However there does not appear to be a unique site for integration.

EBV can be propagated in vitro by cultivation of continuously infected B-lymphoblastoid cells obtained from infected tissues. Viral infection confers on the cells the ability to grow indefinitely in culture (6). In some infected cell lines a fraction of cells produce virus and viral antigens: early antigen (EA), viral capsid antigen (VCA) and membrane antigen (MA). Other cell lines such as Raji are latently infected, do not produce virus and disclose another set of intranuclear antigens (EBNA). After treatment of Raji cells with certain DNA base analogs, sodium butyrate or phorbol esters, induction of EA but not viral DNA synthesis occurs in an abortive infection (7,8). Superinfection of Raji cells approximates lytic permissive infection. The polypeptides produced can be classified into immediate-early, early, or late classes. The synthesis of seven or more of the late polypeptides is impaired by inhibitors of viral DNA synthesis such as acyclovir and results in overproduction of at least six early polypeptides (9). Sequential synthesis of viral polypeptides and repression of synthesis of early proteins by late proteins resemble the cascade regulation of HSV with its α, β and γ classes of polypeptides. EBV early antigen may correspond to β HSV polypeptides with enzymatic functions involved in viral DNA replication.

In viral transcription in permissive infection at least 75 additional RNAs (10). Early functions defined as those RNAs whose synthesis occurs despite presence of an inhibitor of viral DNA synthesis and therefore of late mRNA synthesis have been identified. Immunoprecipitation of in vitro translation products shows that most of these early functions are precipitable with sera containing EA antibodies. Sequences encoding components of VCA and the major structural glycoproteins, gp350/220, or membrane antigen, have also been identified.

In latently infected lymphocytes at least seven polyadenylated mRNAs and two small nonpolyadenylated RNAs are transcribed in vitro (11,12). The viral genes expressed in these cells are of great interest as they may be determinants of growth transformation and maintenance of viral episomes. Four of the mRNAs encode components of the EBV nuclear antigens, EBNA. Most human antisera with antibodies to EBNA react with the protein encoded by the BamHI K fragment of EBV, EBNA 1, which contains the rare repeated sequence of glycine and alanine residues (13). The 3.7 kb mRNA which encodes EBNA 1 is transcribed from left to right on the genome, is intricately spliced and may originate 70 kbp upstream within the large internal repeat sequence IR1. However the full-sized protein is encoded by an open reading frame within the BamHI K fragment. Pinpointed within the BamHI C fragment at the left end of the genome is the origin of replication (ori-P) for the plasmid form of the genome (14,15,16). EBNA 1 protein binds to specific sequences within ori-P allowing for maintenance of the EBV episomal forms (17). EBNA 1, which acts in trans to maintain the replication of the EBV episome, may be the key virally encoded determinant of latent infection.

A second nuclear protein, EBNA 2, is encoded by a 3.0 kb mRNA transcribed from the BamHI W, Y, and H fragments of EBV (18). This protein is implicated in the initiation of transformation because the only nontransforming EBV isolates are produced by HR1 cells which contain virus deleted for these sequences. The parental Jijoye line retains the sequences, expresses EBNA 2, and the virus produced is transformation-competent (19). Recombinants between the Raji and HR1 genomes that have regained these sequences are transformation-competent. Two allelic forms of EBNA-2 have been identified; the DNA sequences and the predicted amino acid sequences are highly divergent.

Recently this same region has been linked to an additional protein which may be encoded by multiple copies of two exons within the internal repeat sequence, IR1, and additional sequences within BamHI Y (19). This protein is heterogeneous varying in size by increments that may reflect differing numbers of copies of the IR1 exons. The protein has been referred to as leader protein, LP, or EBNA 4.

A third nuclear protein, EBNA 3, of approximately 145,000 daltons is encoded by a 4.5 kb mRNA (20). This mRNA is also spliced; however the protein is largely encoded by an open reading frame within BamHI E. There are three similarly sized open reading frames within BamHI E which may encode three distinct EBNAs.

A fourth region transcribed in latent infection at the right end of the genome is transcribed from right to left from the opposite strand of the genome (21). The sequence of the 2.6 kb mRNA predicts a hydrophobic protein with six domains envisaged as winding through cell membrane. The 60 kd protein has been identified in the membrane fraction and on the surface of latently infected lymphocytes and is referred to as latent membrane protein or LMP.

A fifth region of the genome encodes the two small nonpolyadenylated RNAs, the EBERs, and two highly spliced mRNAs of 2.3 and 2.0 kb. The protein products of the latter RNAs have not been identified, nor have their functions in lymphocytes been assigned. Expression of LMP in Rat-1 cells causes focus formation and confers on the cells the ability

to grow in nude mice (22). Expression of EBNA 2 in Rat-1
cells reduces serum requirements but does not result in
morphologic changes.

Burkitt's Lymphoma and Nasopharyngeal Carcinoma.
Lymphomas of B-lymphocytic origin are relatively uncommon,
and except for Burkitt's Lymphoma mostly arise in
immunocompromised hosts. Burkitt's lymphoma is a
distinctive B-lymphocytic malignancy, rare in Western
civilization but endemic in the equatorial belt of Africa.
In Africa BL is a childhood disease, presenting as a tumor
of the jaw. The tumor is usually fatal but responds to
chemotherapy and radiotherapy. It is a monoclonal
malignancy as established by immunoglobulin surface markers.

BL has characteristic chromosomal translocations (8;14,
8;22, 2;8). These translocations involve the immunoglobulin
heavy chain locus on chromosome 8, the k light chain locus
on chromosome 22, the j light chain locus on chromosome 2
and the c-myc locus on chromosome 14 (23,24). The most
frequent translocation involves the Ig heavy chain locus
with breakpoints within the constant regions for l , a and c
heavy chains and within variable genes. The breakpoints on
chromosome 14 may occur 5' to c-myc within the first non-
structural exon, within the first intron, or within
sequences flanking the 5' end. The translocation of c-myc
appears to turn on its expression by the new proximity to
immunoglobulin regulatory sequences such as the Ig enhancers
or by structural changes within the c-myc gene. Altered
expression of c-myc and the chromosomal translocations are
common to both endemic and nonendemic BL. However only the
African variety of BL regularly contains EBV genomes
(22a,25). EBV is thought to contribute to the pathogenesis
of BL by stimulating B-cell growth, increasing the
likelihood of the critical translocations.

Patients with BL have elevated antibody titers to VCA
and EA(r) and a range of titers of EBNA antibodies.
Elevated EA(r) titers also occur in normal persons, perhaps
reflecting sporadic viral reactivation. Children in Africa
BL will later develop, when followed prospectively beginning
early in life tend to have higher than usual titers of EBV
antibodies (26).

Carcinoma of the posterior nasopharynx is a malignancy
of epithelial origin arising in Waldeyer's ring. This
disease is endemic in Southern China and occurs with high
frequency in Chinese who originate from this area. The
disease is also relatively common in North Africa and in
Alaskan eskimos, and it occurs sporadically at low incidence
in Caucasians in the United States and in Europe (26).

The association of NPC with EBV infection is striking,
marked by high EBV antibody titers and by presence of EBV
DNA in specimens from every part of the world (27,28,29).
Co-factors that might explain the high incidence of NPC in
Southern China have yet to be identified; genetic
predisposition may play a role. All forms of NPC ranging
from anaplastic to well differentiated contain EBV genomes.
The infiltration with lymphocytes characteristic of the
anaplastic form of NPC may suggest that the tumor is
expressing an EBV surface antigen that provokes a cytotoxic

T-cell response or a lymphokine such as interleukin-1 which
would attract T-cells. The viral DNA is in the epithelial
elements of the tumor as episomes; whether integrated
genomes co-exist with the characterized episomal forms is
unknown. Recent analyses of viral DNA structure within the
tumor indicate that NPC, like BL, is a monoclonal malignancy
and that viral infection precedes the cellular proliferation
(30).

Distinctive in the development of NPC is the appearance
of IgA antibodies to VCA and EA(d) that precede and may
herald the appearance of the tumor (26,27). This specific
antibody response is a harbinger of malignant disease,
although not all persons in whom EBV IgA antibodies develop
get NPC. Ideas about the genesis of these unusual responses
are based on the fact that infection with EBV in patients
with NPC has taken place many years earlier in childhood
with the infection being dormant for 40 to 60 years and then
reactivated 1 to 2 years before the appearance of NPC.
Recent evidence suggests that the site of viral reactivation
might be in the tumor tissue itself as indicated by the
detection of viral transcripts and viral DNA-restriction
patterns consistent with viral replication, abortive or
complete (30,31).

Other tumors of epithelial origin are now linked to EBV
infection. EBV genomes have been found in supraglottal and
laryngeal carcinomas, in a thymoma of epithelial origin and
in parotid carcinomas (32). All of these tissues including
the nasopharynx have a common origin in the primitive
oropharynx. EBV particles have been visualized by electron
microscopy in hairy leukoplakia of the tongue, a condition
found in AIDS (32); usually leukoplakia is a pre-malignant
lesion. The parotid gland is also a site of EBV
replication. Thus EBV is becoming implicated in other
carcinomas of the upper airway, and virus has been detected
in pulmonary tissue in rare infants with immune deficiency.

EBV infection in acquired immunodeficiency. The
development of lymphoproliferative syndromes is related to
intensity of immunosuppression. Manifestations of EBV-
associated lymphoproliferative disease in graft recipients
range from progressive IM to infiltrative polyclonal B-cell
hyperplasias to fatal monoclonal lymphoma. Although
cytogenetic abnormalities occur, the translocations
characteristic of BL have not been described in graft
recipients. Polyclonal proliferations may regress if the
immunosuppression is reduced. EBV-containing B-
proliferations have been documented in renal, cardiac, bone
marrow and thymic transplant recipients.

In acquired immunodeficiency syndrome (AIDS) EBV-
associated lymphoproliferation including authentic Burkitt's
lymphoma with the chromosomal translocations are relatively
common. The rarity of BL in allograft recipients contrasts
with its relative frequency in AIDS. This may reflect more
severe T-cell abnormalities in AIDS, or it may point to a
facilitative interaction at the molecular level between EBV
and HIV in coinfected B-cells. In fact in vitro an EBV
immediate-early gene product transactivates the HIV master
transcriptional promoter contained in the LTR portion of the
HIV genome (33).

86

Drugs that inhibit EBV replication. The first drugs effective in vitro were phosphonoacetic acid and phosphonoformic acid (34,35). Their action depends on their ability to interact at the pyrophosphate binding site, directly inhibiting viral DNA polymerase, without requiring prior phosphorylation. Most inhibitors of EBV replication are base analogs. Adenosine arabinoside inhibits EBV replication in vitro (36). Acyclovir is an effective inhibitor with an ED50 of 0.3 micromole (μm) (37,38). E-5-(2-bromovinyl)-2'-deoxyuridine (BVDU) is 6-10 times more effective in vitro than acyclovir but has not been used clinically (39,40). 9-(1,3-dihydroxy-2-propoxymethyl)guanine (DHPG) is about six times more active in vitro than acyclovir with an ED50 of 0.05 μm. The halogenated nucleoside analogs, 1-(2-deoxy-2-fluoro-β-D-arabinofuranosyl) 5-iodocytosine (FIAC) and 1-(2-deoxy-2-fluoro-β-D-arabinofuranosyl)-5-methyluracil (FMAU), are approximately ten times more active than DHPG and 50 times more active than ACV in vitro (41). 6-deoxyacyclovir is prodrug of acyclovir that is administered orally and metabolized to acyclovir by the action of xanthine oxidase in the gut wall and liver (42). The blood levels of acyclovir achieved with the prodrug are comparable to those produced by intravenous administration of acyclovir. A second new drug, 2-acetylpyridine-thiosemicarbazone (BW A723U), is an effective inhibitor of herpes simplex virus replication in vitro (43). The drug acts through inhibition of herpesvirus-specified ribonucleotide reductase. This inhibitor acts synergistically with acyclovir against HSV replication so that the dosage of BW A723U can be reduced. Possibly the drug combination may prove useful for treatment of ACV-resistant HSV mutants.

Recently an antiviral drug that inhibits replication of retroviruses has emerged as the leading drug for treatment of AIDS due to HIV infection. This drug, Azidothymidine (AZT), is a thymidine analog with high affinity for HIV reverse transcriptase (44). It functions as a DNA chain terminator. We have discovered that, unexpectedly, AZT inhibits EBV replication, although it has no effect on the other human herpesviruses. HSV 1, HSV 2, CMV and VZV (45). The mechanism of action against EBV is unknown. Presumably EBV DNA polymerase shares features in common with HIV reverse transcriptase that favor binding of AZT-triphosphate to these viral enzymes. Since active EBV infection is common in AIDS the dual action of AZT may have clinical utility.

Mechanism of Acyclovir. In herpes simplex virus infection, acyclovir is first monophosphorylated by virus-encoded thymidine kinase and then further phosphorylated by cellular enzymes (46). The triphosphate is the active form of the drug which interacts with high specificity with HSV DNA polymerase; ACV-triphosphate has at least 100-fold greater affinity for the viral than the α cellular polymerase. In EBV-infected cells, monophosphorylation of ACV seems to be accomplished by cellular kinases, as yet unidentified, rather than by EBV-encoded TK, although the gene for an EBV encoded TK has recently been cloned and sequenced (47,48,49). Phosphorylation of ACV in EBV-infected cells is much less efficient than it is in HSV-infected cells. ACV's specificity of action and nontoxicity therefore depend on either selective phosphorylation or preferential affinity for viral versus cellular polymerases or on both (50a, 50b). The importance of the various steps in the action of the drug differs with the different herpesviruses (50).

The favorable therapeutic ratio of Acyclovir is also due to the concentration of the active form of the drug in infected tissue inasmuch as ACV triphosphate cannot permeate normal cell membranes; it reaches its greatest concentration in infected cells since here phosphorylation is more efficient and viral polymerase which has high affinity for the drug triphosphate is present. However, all these effects are relative, and small amounts of the drug are phosphorylated in normal tissue.

ACV triphosphate interacts with herpesvirus polymerase as a competitive inhibitor of dGTP. With HSV polymerase the interaction appears to be nonreversible; ACV triphosphate is incorporated into the DNA causing chain termination and binding of the viral polymerase (51). Incorporation into

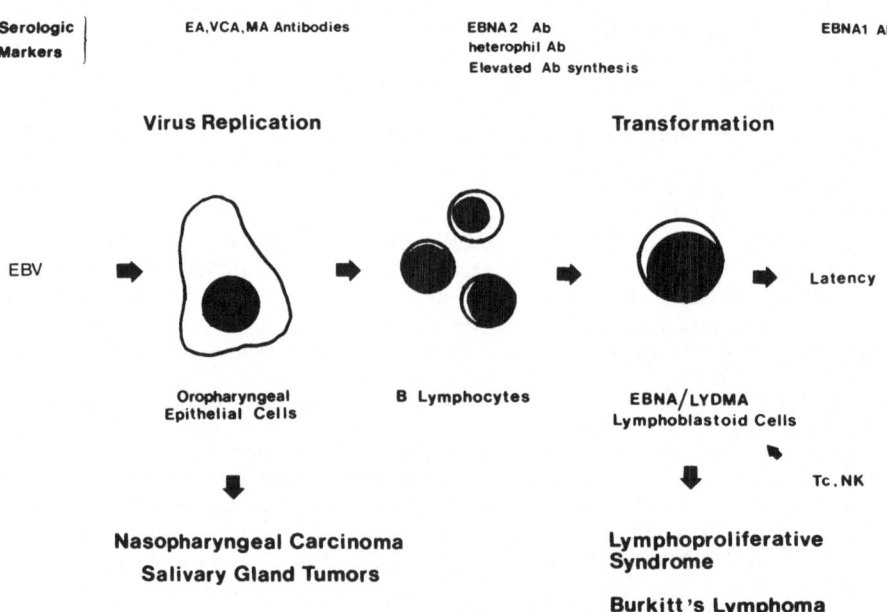

Figure 2. Pathogenesis of EBV infection.

EBV infects both epithelial cells and B lymphocytes, causing productive infection and cytolysis in the epithelial cells and polyclonal lymphoproliferation and viral latency in the B lymphocytes. Antibodies to early antigens (EA), viral capsid antigen (VCA), membrane antigen (MA), Epstein-Barr virus nuclear antigens types 1 and 2 (EBNA 1 and 2), and heterophile antibodies appear in the sera of infected persons. There are also prominent cell-mediated immune responses including appearance of cytolytic T cells (Tc) and natural killer (NK) cells . In addition to infectious mononucleosis, which is the natural outcome of the process depicted, EBV is associated with several neoplastic conditions.

EBV DNA probably also occurs. Incorporation into viral DNA is much more likely than into cellular DNA because of the greater affinity of the drug for the viral polymerase and the presence of triphosphorylated drug in greater concentration in virus-infected tissue, at least in HSV-infected cells.

Recently new adenosine analogs have been shown to inhibit EBV replication (52).

Reversal of Drug Action. Because EBV infection states are persistent, the events that occur after drug is removed from infected cells are relevant to treatment. When ACV is applied to a chronically EBV-producing cell line, free linear viral DNA replication is abolished and virus production stops. However, some viral DNA in the form of episomes remains, detectable by hybridization, in the treated cell cultures (37). When the drug is removed, viral replication resumes rapidly and is restored to pre-treatment levels (38). Virus replication is suppressed as long as drug is present. This reversibility of inhibition suggests either that the drug triphosphate is reversibly bound to EBV DNA polymerase or that new polymerase molecules are generated and become functional upon removal of the drug.

Bearing on this important issue are observations that the duration of drug effect differs from drug to drug. DHPG is more inhibitory than ACV, and FMAU even more so. In addition, after removal of DHPG from the virus-producing cell line, the kinetics of recovery of virus production follow a slower course before restoration to pre-drug exposure levels. Even more remarkable is the persistent drug effect with FMAU: inhibition of viral replication continues for more than 58 days after removal of drug (41). These different kinetics of recovery of virus replication point to differences in mode of action of the drugs, as yet unknown. The different persistent effects seem unlikely to be due merely to differences in drug metabolism (53) may be related to consequences of incorporation of drug into DNA.

Latent infection. None of the drugs so far tested has any effect on replication of the episomal form of the EBV genome in Raji cells, regardless of differences in potency and mode of action. This lack of effect is quite independent of drug phosphorylation. Neither drugs that are nonphosphorylated in EBV-infected cells or efficiently phosphorylated affect episomal replication or cure the virus-producing cells. Clearly, a new approach for treatment of latent EBV infection is needed. As brought out earlier in this chapter binding of the EBNA 1 protein appears to be an essential function for maintenance of EBV episomal forms in infected cell cultures. Small amounts of this protein bind directly to specific sites in the region of the EBV genome containing the origin for plasmid replication (ori-P). Therefore is seems reasonable to believe that drugs that might interfere with the binding of EBNA 1 to viral DNA might disrupt EBV latency by interfering specifically with replication of the EBV episomes.

Predictions about Therapy. All of the drugs discussed would be expected to interfere with virus synthesized in epithelial cells and shed in the oropharynx, with replication being suppressed as long as drug is administered. The drugs would have no effect on the latently infected B-lymphocytes which are infected early. Preliminary results of a trial of acyclovir in patients with acute infectious mononucleosis have confirmed these predictions (54). In these patients administration of acyclovir suppressed virus excretion but did not abolish it. There was no effect on establishment of latency as assessed by the ability to establish EBV-infected B-lymphocyte lines from the blood of patients being given acyclovir. Probably, however, suppression of virus production would reduce the number of additional B-lymphocytes that become infected with EBV during the course of infection, perhaps with favorable effects. A respite in virus replication might tip the balance in favor of host mechanisms and aid recovery even though virus replication resumes. In fact, silent oropharyngeal shedding of virus continues for years even in untreated patients. Finally, some of the clinical manifestations of acute EBV infection such as Guillain-Barre syndrome, hepatitis, and suppression of hematopoiesis may be caused by secondary immune responses to latently infected lymphocytes. These immunologically based manifestations should be indifferent to antiviral drugs except insofar as amplification of number of EBV-infected lymphocytes by spread of infection would be reduced. The ultimate antiviral drug that could abolish latent as well as productive infection awaits the development of a wholly new class of agents that would interfere specifically with binding of the trans-acting protein thought to be needed for maintenance of latent EBV infection.

References

Pagano, J.S. A perspective on treatment of Epstein-Barr virus infection states. Epstein-Barr virus and associated diseases. In Proceedings of the First International Symposium on Epstein-Barr Virus and Associated Malignant Diseases. Loutraki, Greece. (P.H. Levine, D.V. Ablashi, G.R. Pearson, and S.D. Kottaridis, eds.). Martinus Nijhoff Publishers, 619-630, 1985.

Pagano, J.S. Epithelial cell interactions of the Epstein-Barr virus. In Concepts in Viral Pathogenesis, chap. 42, 307-314. (A Notkins, M. Oldstone, eds.) Springer-Verlag, 1984.

Pagano, J.S. The Epstein-Barr virus plasmid. In Extrachromosomal DNA, [ISBN (0-12-198780-9)], Academic Press, Inc., New York, 1979, 235-248.

Pagano, J.S. and J.-C. Lin. Cellular tranformation by the herpesviruses and antiviral drugs. (S. Goff, ed.). For Pharmacology and Therapeutics. XXVIII, 135-161, Pergamon Press, Ltd., 1985.

Pagano, J.S. and S.M. Lemon. The Herpesviruses. In Infectious Diseases and Medical Microbiology, 2nd Edition. (A. Braude, ed.) W.B. Saunders Co. 470-477, 1986.

Raab-Traub, N. and J.S. Pagano. Hybridization of viral nucleic acids: Newer methods on solid media and in solution. In Methods in Virology, VII. (K. Maramorosch, H. Koprowski, eds.). Academic Press, New York. 1-38, 1984.

Sixbey, J.W. and J.S. Pagano. New perspectives on the Epstein-Barr virus in the pathogenesis of lymphoproliferative disorders. In Current Clinical Topics in Infectious Diseases, Vol. 5. J. Remington and M. Schwartz, ed., 146-176, McGraw Hill, New York, 1984.

The Epstein-Barr Virus (M.A. Epstein and B.G. Achong, eds.), Springer-Verlag, Inc., New York, 1979.

Literature Cited

1. Epstein, M.A., B.G. Achong, and Y.M. Barr. Virus particles in cultured lymphoblasts from Burkitt's lymphoma. Lancet 1:702-703, 1964.

2. Nonoyama, M. and J.S. Pagano. Separation of Epstein-Barr virus DNA from large chromosomal DNA in non-virus producing cells. Nature 238:169-171, 1972.

3. Adams, A. and T. Lindahl. EBV genomes with properties of circular DNA molecules in carrier cells. Proc. Natl. Acad. Sci. USA 72:1477-1481, 1975.

4. Sixbey, J.W. and J.S. Pagano. Epstein-Barr virus transformation of human B-lymphocytes despite inhibition of viral polymerase. J. Virol., 52:299-301, 1985.

5. Matsuo, T., M. Heller, L. Petti, E. O'Shiro, and E. Kieff. Persistence of the entire Epstein-Barr virus genome integrated into human lymphocyte DNA. Science 226:1322-1325, 1984.

6. Pope, J., M. Horne, and W. Scott. Transformation of fetal human leukocytes in vitro by filtrates of a human leukemic cell line containing herpes like virus. Int. J. Cancer 3:857-866, 1968.

7. Klein, G., J. Zenthen, P. Terasaki, R. Billing, R. Hoing, M. Jondal, A. Westman, and G. Clements. Inducibility of the Epstein-Barr virus (EBV) cycle and surface marker properties of EBV-coinfected sublines. Int. J. Cancer 18:639-652, 1976.

8. Weigle, R. and G. Miller. Major EB virus-specific cytoplasmic transcripts in a cellular clone of the HR-1 Burkitt lymphoma line during latency and after induction of viral replicative cycle by phorbol esters. Virology 125:287-298, 1983.

9. Feighny, R.J., B.E. Henry II, and J.S. Pagano. Epstein-Barr virus polypeptides: Effect of inhibition of viral DNA replication on their synthesis. J. Virol. 37:61-71, 1981.

10. Hummel, M. and E. Kieff. Epstein-Barr virus RNA. VIII. Viral RNA in permissively infected B95-8 cells. J. Virol. 43:262-272, 1982.

11. King, W., A. Thomas-Powell, N. Raab-Traub, M. Hawke, and E. Kieff. Epstein-Barr virus RNA. VI. Viral RNA in restringently and abortively infected Raji cells. J. Virol. 38:649-660, 1981.

12. Rosa, M.D., E. Gottlieb, M.R. Lerner, and J.A. Steitz. Striking similarities are exhibited by two small Epstein-Barr virus-encoded ribonucleic acids and the adenovirus-associated ribonucleic acids VAI and VAII. Mol. Cell. Biol. 1:785-796, 1981.

13. Hennessy, K., M. Heller, V. Van Santen, and E. Kieff. Simple repeat array in Epstein-Barr virus DNA encodes part of the Epstein-Barr nuclear antigen. Science 220:1396-1398, 1983.

14. Yates, J.L., N. Warren, and B. Sugden. Stable replication of plasmids derived from Epstein-Barr virus in various mammalian cells. Nature 313:812-815, 1985.

15. Sugden, B., K. Marsh, and J. Yates. A vector that replicates as a plasmid and can be efficiently selected in B-lymphoblasts transformed by Epstein-Barr virus. Mol. Cell. Biol. 5:410-413, 1985.

16. Pesano, R.L. and J.S. Pagano. Herpesvirus papio contains a plasmid origin of replication that acts in cis interspecies with an Epstein-Barr virus trans-acting function. J. Virol., 60:1159-1162, 1986.

17. Milman, G. and E.S. Hwang. Epstein-Barr virus nuclear antigen forms a complex that binds with high concentration dependence to a single DNA-binding site. J. Virol. 61:465-471, 1987

18. Hennessy, K. and E. Kieff. A second nuclear protein is encoded by the Epstein-Barr virus in latent infection. Science 227:1238-1239, 1985.

19. Rabson, M., L. Gradoville, L. Heston, and G. Miller. Non-immortalizing P3J-HR-1 Epstein-Barr virus: a deletion mutant of its transforming parent, Jijoye. J. Virol. 44:834-844, 1982.

20. Hennessy, K., S. Fennewald, M. Hummel, T. Cole, and E. Kieff. A third viral nuclear protein in lymphoblasts immortalized by Epstein-Barr virus. Proc. Natl. Acad. Sci USA, 82:5944-5948, 1985.

21. Hennessy, K., S. Fennewald, M. Hummel, T. Cole, and E. Kieff. A membrane protein encoded by Epstein-Barr virus in latent growth transforming infection. Proc. Natl. Acad. Sci. USA 81:7202-7211, 1984.

22. Wang, D., D. Liebowitz, and E. Kieff. An EBV membrane protein expressed in immortalized lymphocytes transforms established rodent cells. Cell 43:831-840, 1985.

22a. Nonoyama, M. and J.S. Pagano. Homology between Epstein-Barr virus DNA and viral DNA from Burkitt's lymphoma and nasopharyngeal carcinoma determined by DNA-DNA reassociation kinetics. Nature 242:44-47, 1973.

23. Croce, C.M., M. Shandler, J. Martinis, L. Cicurel, G.G. D'Ancona, T.W. Dolby, and H. Koprowski. Chromosomal location of the human immunoglobulin heavy chain genes. Proc. Natl. Acad. Sci. USA 76:3416-3419, 1979.

24. Dalla-Favera, R., M. Bregni, J. Erikson, D. Patterson, R.C. Gallo, and C. Croce. Human c-myc oncogene is located on the region of chromosome 8 that is translocated in Burkitt lymphoma cells. Proc. Natl. Acad. Sci. USA 79:7824-7827, 1982.

25. Pagano, J.S., C.S. Huang, and P. Klein. Absence of Epstein-Barr viral DNA in Burkitt's lymphoma. New Engl. J. Med. 289:1395-1399, 1973.

26. de The, G. Epidemiology of Epstein-Barr virus and associated diseases in man. The Herpesviruses, B. Roizman (ed.), 1982.

27. Henle, G. and W. Henle. Epstein-Barr virus-specific IgA serum antibodies as an outstanding feature of nasopharyngeal carcinoma. Int. J. Cancer 17:1-7, 1976.

28. Desgranges, C., H. Wolf, G. de The', K. Shanmugaratnam, R. Ellouz, N. Cammoun, G. Klein, and H. zur Hausen. Nasopharyngeal carcinoma X. Presence of Epstein-Barr virus genomes in epithelial cells of tumors from high and medium risk areas. Int. J. Cancer 16:7-15, 1975.

29. Pagano, J.S., C.-H. Huang, G. Klein, G. de The, K. Shanmugaratnam, and C.-S. Yang. Homology of Epstein-Barr virus DNA in nasopharyngeal carcinomas from Kenya, Taiwan, Singapore, and Tunis. In Proceedings of the Second International Symposium on Oncogenesis and Herpesviruses, Nuremburg, Germany. IARC, Lyon, France, 1975. Publication No. 11, Part II, 179-190.

29a. Nonoyama, M., C.-H. Huang, J.S. Pagano, G. Klein, and S. Singh. DNA of Epstein-Barr virus detected in tissue of Burkitt's lymphoma and nasopharyngeal carcinoma. Proc. Natl. Acad. Sci. USA 70:3265-3268, 1973.

30. Raab-Traub, N. and K. Flynn. The structure of the termini of the Epstein-Barr virus as a marker of clonal cellular proliferation. Cell 47:883-889, 1986.

31. Raab-Traub, N., R. Hood, C.S. Yang, B. Henry, and J.S. Pagano. Epstein-Barr virus transcription in nasopharyngeal carcinoma. J. Virol. 48:580-590, 1983.

32. Greenspan, J.S., D. Greenspan, E.T. Lennette, D.I. Abrams, M.A. Conant, V. Petersen, and U.K. Freese. Replication of Epstein-Barr virus within the epithelial cells of oral "hairy" leukoplakia, an AIDS-associated lesion. N. Engl. J. Med., 313:1564-1571, 1985.

33. Kenney, S., J. Kamine, D. Markovitz, R. Fenrick, and J.S. Pagano. An EBV Immediate-Early gene product transactivates gene expression from the Human Immunodeficiency Virus (HIV) Long Terminal Repeat (LTR). Submitted, 1987.

34. Summers, W.C. and G. Klein. Inhibition of Epstein-Barr Virus DNA synthesis and late gene expression by phosphonoacetic acid. J. Virol. 18:151-155, 1976.

35. Datta, A.K. and R.E. Hood. Mechanisms of inhibition of Epstein-Barr virus replication by phophonoformic acid. Virology 114:52-59, 1981.

36. Benz, W.C., P.J. Siegel, and J. Baer. Effects of adenine arabinoside on lymphocytes infected with Epstein-Barr virus. J. Virol. 27:475-482, 1978.

37. Colby, B.M., J.E. Shaw, G.B. Elion, and J.S. Pagano. Effect of Acyclovir [9-(2-hydroxyethoxymethyl)guanine] on Epstein-Barr virus DNA replication. J. Virol. 34:560-568, 1980.

38. Lin, J.-C., M.C. Smith, and J.S. Pagano. Prolonged inhibitory effects of 9-(1,3-Dihydroxy-2-propoxymethyl)guanine against replication of Epstein-Barr virus. J. Virol. 50:50-55, 1984.

39. Lin, J.-C., D.J. Nelson, C.U. Lambe, and J.S. Pagano. Effects of nucleoside analogs in inhibition of Epstein-Barr virus. Proc. International Virology Post-Congress Symposium on Pharmacological and Clinical Approaches to Herpesviruses and Virus Chemotherapy, Oiso, Japan. In press, 1985.

40. Lin, J.-C., M.C. Smith, E.I. Choi, E. De Clercq, A. Verbruggen, and J.S. Pagano. Effect of (E)-5-(2-Bromovinyl)-2'-deoxyuridine on replication of Epstein-Barr virus in human lymphoblastoid cell lines. Proceedings of the First International TNO Conference on Antiviral Research, Rotterdam, 30 April-3 May 1985, Elsevier, Amsterdam. Antiviral Research Suppl. 1 1: 121-126, 1985.

41. Lin, J.-C., M.C. Smith, Y.C. Cheng, and J.S. Pagano. Epstein-Barr virus: inhibition of replication by three new drugs. Science 221:578-579, 1983.

42. Krenitsky, T.W., W.W. Hall, P. De Miranda, L.M. Beauchamp, H.J. Schaeffer, and P.D. Whiteman. 6-Deoxyacyclovir: A xanthine oxidase-activated prodrug of acyclovir. Proc. Natl. Acad. Sci. USA 81:3209-13, 1984.

43. Spector, T., D.R. Averett, D.J. Nelson, C.U. Lambe, R.W. Morrison Jr., M.H. St. Clair, and P.A. Furman. Potentiation of antiherpetic activity of acyclovir by ribonucleotide reductase inhibition. Proc. Natl. Acad. Sci., 82:4254-4257, 1985.

44. Furman, P.A., J.A. Fyfe, M. H. St. Clair, K. Weinhold, J.L. Rideout, G.A. Freeman, S.N. Lehrman, D.P. Bolognesi, S. Broder, H. Mitsuya, and D.W. Barry. Phosphorylation of 3'-azido-deoxythymidine and selective interaction of the 5'-triphosphate with human immunodeficiency virus reverse transcriptase. Proc. Natl. Acad. Sci. USA, 83:8333-8337, 1986.

45. Lin, J.-C., Z.-X. Zhang, M.C. Smith, K. Biron, and J.S. Pagano. Anti-human immunodeficiency virus agent 3'-Azido-3'-Deoxythymidine inhibits replication of Epstein-Barr virus. Submitted, 1987.

46. Elion, G.B., P.A. Furman, and J.A. Fyfe. Selectivity of action of an antiherpetic agent, 9-(2-hydroxyethoxymethyl)guanine. Proc. Natl. Acad. Sci. 74:5716-5720, 1977.

47. Colby, B.M., P.A. Furman, J.E. Shaw, G.B. Elion, and J.S. Pagano. Phosphorylation of Acyclovir [9-(2-hydroxyethoxymethyl)guanine] in Epstein-Barr virus infected lymphoblastoid cell lines. J. Virol. 38:606-611, 1981.

48. Stinchcombe, T., and W. Clough. Epstein-Barr virus induces a unique pyrimidine deoxynulceotide kinase activity in superinfected and virus-producer B-cell lines. Biochem. 24:2027-2033, 1984.

49. Littler, E., J. Zeuthen, A. A. McBride, et al. Identification of an Epstein-Barr virus-coded thymidine kinase. The EMBO Journal 5:1959-1966, 1986.

50a. Chiou, J.-F., and Y.-C. Cheng. Interaction of Epstein-Barr virus DNA polymerase and 5'-triphosphates of several antiviral nucleoside analogs. Antimicr. Agents and Chemother. 27:416-418, 1985.

50b. Chiou, J.-F., J.K.K. Li, and Y.-C. Cheng. Demonstration of a stimulatory protein for virus-specified DNA polymerase in phorbol ester-treated Epstein-Barr virus-carrying cells. Proc. Natl. Acad. Sci. USA 82:5728-5731, 1985.

50. Pagano, J.S. and A.K. Datta. Perspectives on interactions of Acyclovir with Epstein-Barr and other herpes viruses. Amer. J. of Med. (Acyclovir Symposium), 18-26, 1982.

51. Furman, P.A., M.H. St. Clair, and T. Spector. Acyclovir triphosphate is a suicide inactivator of the herpes simplex virus DNA polymerase. J. Biol. Chem. 259:9575-9579, 1984.

52. Lin, J.-C., E. DeClercq, and J.S. Pagano. Novel Acyclic adenosine analogs inhibit Epstein-Barr virus replication. Antimicrob. Agents and Chemother. In Press, 1987.

53. Lin, J.-C.. D.J. Nelson, C.U. Lambe, and E. I. Choi. Metabolic activation of 9-(1,3-dihydroxy-2-propoxymethyl) guanine in human lymphoblastoid cell lines infected with Epstein-Barr virus. J. Virol. Submitted.

54. Pagano, J.S., J.W. Sixbey, and J.-C. Lin. Acyclovir and Epstein-Barr virus infection. J. Antimicrob. Chemother. 12:113-121, 1983.

MOLECULAR TARGETS FOR SELECTIVE ANTIVIRAL CHEMOTHERAPY

Erik De Clercq

Rega Institute for Medical Research
Katholieke Universiteit Leuven
B-3000 Leuven, Belgium

INTRODUCTION

Only few antiviral substances have been licensed and/or are widely available for medical use. These include, at present, amantadine and rimantadine for the prophylaxis and early treatment of influenza A virus infections, idoxuridine, trifluridine, vidarabine and acyclovir for the topical treatment of herpetic keratitis, acyclovir for the systemic (intravenous or peroral) treatment of herpes simplex virus (HSV) and varicella-zoster virus (VZV) infections, vidarabine for the systemic (intravenous) treatment of herpetic encephalitis (although acyclovir is superior to vidarabine in the treatment of herpetic encephalitis), ribavirin for the topical (aerosol) treatment of respiratory syncytial virus infection in infants, and retrovir (azidothymidine) for the systemic (intravenous or peroral) treatment of AIDS and AIDS-related complex (ARC).[1]

There are a number of other compounds which have been introduced in clinical practice, albeit on a compassionate basis : i.e. phosphonoformate and dihydroxypropoxymethylguanine for the treatment of cytomegalovirus (CMV) infections, and bromovinyldeoxyuridine (BVDU) for the treatment of HSV-1 and VZV infections, in immunosuppressed patients. Moreover, ribavirin has proven efficacious in the treatment of some hemorrhagic fever virus infections (i.e. Lassa); several new compounds have been described that show activity against a broad spectrum of DNA viruses, and others that are specifically inhibitory to rhinoviruses, and the number of compounds assumed to hold promise for the treatment of AIDS and ARC is increasing at a pace which is almost catching up on the spread of the disease itself.

Most of the antiviral agents which have been developed or are current-ly under development are targeted at one or another step in nucleic acid (RNA or DNA) biosynthesis. Notable exceptions are amantadine,[2] which inter-acts directly with the viral membrane protein M_2 of influenza A, and the isoxazole-oxazoline derivatives,[3] which bind specifically to the structural viral protein VP1 of rhinoviruses and thereby block the uncoating process. For ribavirin the mechanism of action appears to be multipronged.[4] At least three possible mechanisms of action have been proposed : a decrease in GTP synthesis due to competitive inhibition of IMP dehydrogenase, inhibition of 5'-cap formation of viral mRNAs and inhibition of initiation and elongation of viral mRNAs. To what extent any of these (or yet other) mechanisms may be entangled in the action of ribavirin may depend on the nature of the vi-rus infection.

The present review will focus on a number of nucleoside analogues which have been studied in our own laboratory, and for which the antiviral activity spectrum has been defined, the intracellular metabolism elucidated and the molecular target for their antiviral action (except for the third class of compounds) identified. The compounds fall into five categories : (I), nucleoside analogues which depend for their antiviral activity on a specific phosphorylation by the virus-induced dThd kinase (TK); (II), nu-cleoside analogues which are not dependent on this enzyme and, moreover, specifically inhibitory to thymidylate synthase; (III), phosphonylmethoxy-alkylpurines and -pyrimidines which are endowed with a broad-spectrum anti-DNA virus activity; (IV), acyclic and carbocyclic adenosine analogues which owe their broad-spectrum anti-RNA and -DNA virus activity to an inhibition of S-adenosylhomocysteine (SAH) hydrolase; and (V), 2',3'-dideoxynucleoside analogues effective against human immunodeficiency virus (HIV), the etiolo-gic agent of AIDS.

NUCLEOSIDE ANALOGUES DEPENDING FOR ANTIVIRAL ACTIVITY ON PHOSPHORYLATION BY VIRUS-ENCODED THYMIDINE KINASE

The nucleoside analogues that depend for their antiviral effects on a specific phosphorylation by the virus-encoded TK may quite schematically be divided into two groups : (i) acyclic guanosine analogues with, as the pro-totypes ACV, DHPG and DHBG (Fig. 1A), and (ii) 5-substituted 2'-deoxyuridi-nes, with EDU, CEDU, BVDU, C-BVDU and BVaraU as representative examples (Fig. 1B). For recent reviews on the chemistry and antiviral properties of these compounds, see references 5-7. Obviously, these compounds can be ex-pected to be active only against those viruses that encode a virus-specific

TK, as do HSV and VZV. This explains the selectivity of the compounds for HSV and VZV. Viruses, such as CMV, which do not encode their own TK, or viruses such as TK⁻HSV-1 variants, which have lost this propensity, are (virtually) insensitive to ACV, BVDU and their congeners (Table 1). An exception to the rule is DHPG which is a rather strong inhibitor of CMV.[8-10] Apparently, DHPG is readily converted to its active (triphosphate) form in the CMV-infected cells.[11] It is also noteworthy that EBV is quite sensitive to ACV, BVDU and DHPG. Yet, it is still uncertain whether EBV is encoding a virus-specific thymidine kinase.

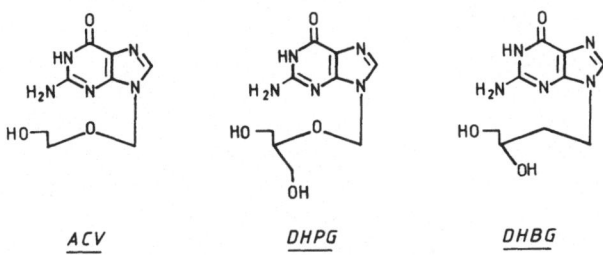

ACV *DHPG* *DHBG*

Fig. 1A. Acyclic nucleoside analogues which depend on phosphorylation by virus-induced dThd kinase to express their antiviral activity : ACV [acyclovir, 9-(2-hydroxyethoxymethyl)guanine], DHPG [9-(1,3-dihydroxy-2-propoxymethyl)guanine] and DHBG [9-(3,4-dihydroxybutyl)guanine]. The latter occurs in two enantiomeric forms [(R)-DHBG and (S)-DHBG] of which (R)-DHBG corresponds to BCV (buciclovir).

Once phosphorylated to their monophosphate by the virus-induced TK, the acyclic guanosine analogues will be subsequently converted to their di- and triphosphates by cellular enzymes. BVDU, however, may be converted to the 5'-diphosphate by the virus-induced TK, or at least some virus-induced TKs such as those specified by HSV-1, VZV, SHV-1 and BHV-1 (Fig. 2; for abbreviations of the viruses, see Table 1). These TKs thus have a dual enzymatic activity. They successively act as dThd kinase and dTMP kinase. The TKs derived from HSV-2 and EHV-1 are able to phosphorylate BVDU to its 5'-monophosphate (BVDUMP) but not further onto its 5'-diphosphate (BVDUDP) (Fig. 2).[12-15] This inefficient phosphorylation of BVDUMP to BVDUDP most probably explains why the replication of HSV-2 and EHV-1 is rather insensitive to BVDU.

The active form of all nucleoside analogues which depend for their antiviral activity on the virus-induced TK corresponds to the triphosphate.[16] In this form they may act as competitive inhibitors (with respect to the natural substrates) of the viral DNA polymerase without being incorporated into DNA via an internucleotide linkage, as has been shown for ACV,[17] DHBG[18] and BVaraU;[19] at best they may be incorporated at the 3'-terminal of the DNA chain (as is the case for ACV[17]) and thereby act as chain termina-

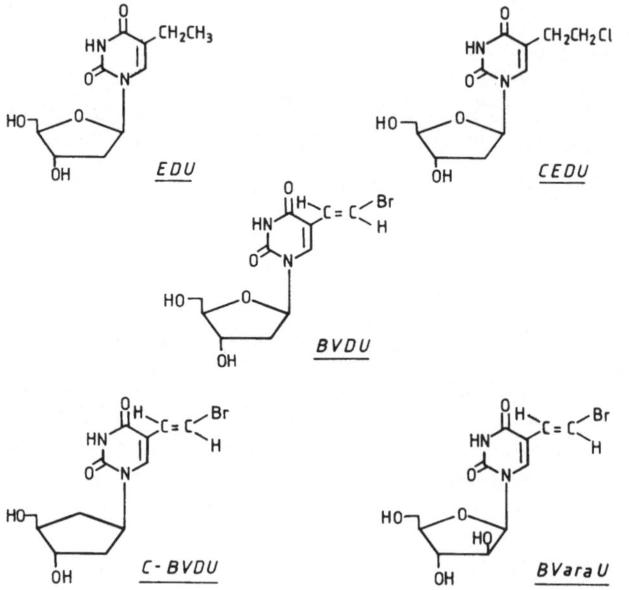

Fig. 1B. 5-Substituted 2'-deoxyuridines which depend on the phosphorylation by virus-induced dThd kinase to express their antiviral activity : EDU (5-ethyl-2'-deoxyuridine), CEDU [5-(2-chloroethyl)-2'-deoxyuridine], BVDU [(E)-5-(2-bromovinyl)-2'-deoxyuridine], C-BVDU (carbocyclic BVDU) and BVaraU [(E)-5-(2-bromovinyl)-1-β-D-arabinofuranosyluracil].

tors. In their interaction with the DNA polymerase, EDU, BVDU and C-BVDU clearly behave differently from ACV, DHBG and BVaraU. The 5-substituted 2'-deoxyuridines EDU,[20] BVDU[21] and C-BVDU[21,22] are incorporated internally into the DNA strands, that is via an internucleotide linkage, although the extent of incorporation of the latter is low (3.6 % at the most).[22] Also DHPG seems to be incorporated internally into the DNA strand.[23]

Table 1. Antiviral Activity Spectrum of BVDU and Related Compounds [Depending on phosphorylation by virus-induced dThd-dTMP kinase (TK)]

Highly susceptible
- Herpes simplex virus type 1 (HSV-1)
- Varicella-zoster virus (VZV)
- Epstein-Barr virus (EBV)
- Suid herpesvirus type 1 (SHV-1)
- Bovid herpesvirus type 1 (BHV-1)
- Simian varicella virus (SVV)

Not (or only poorly) susceptible
- Herpes simplex virus type 2 (HSV-2)
- Cytomegalovirus (CMV)
- Equid herpesvirus type 1 (EHV-1)
- Adeno- and poxviruses
- TK⁻ variants of HSV-1 and VZV
- RNA viruses

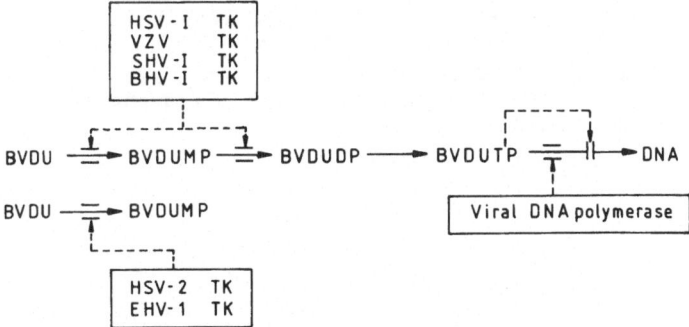

Fig. 2. Differential metabolism of BVDU in cells infected by viruses (HSV-1, VSV, SHV-1, BHV-1) which are susceptible to the antiviral action of BVDU and cells infected by viruses (HSV-2, EHV-1) which are not susceptible to the antiviral action of the compound. The susceptible viruses encode a dThd kinase (TK) which is also endowed with dTMP kinase activity. The latter ensures that BVDUMP is phosphorylated onto BVDUDP.

The consequences of the incorporation into DNA have been further studied with $[^{125}I]$IVDU and C-$[^{125}I]$IVDU (Fig. 3).[24] At concentrations achieving an equivalent antiviral effect, $[^{125}I]$IVDU is incorporated into HSV-1 DNA to a much greater extent than C-$[^{125}I]$IVDU. Furthermore, the $[^{125}I]$IVDU-substituted DNA becomes much more heterogeneous than the C-$[^{125}I]$IVDU-labeled DNA. From these data one may infer that the incorporation of $[^{125}I]$-IVDU, but not of C-$[^{125}I]$IVDU, into DNA leads to an increased susceptibi-

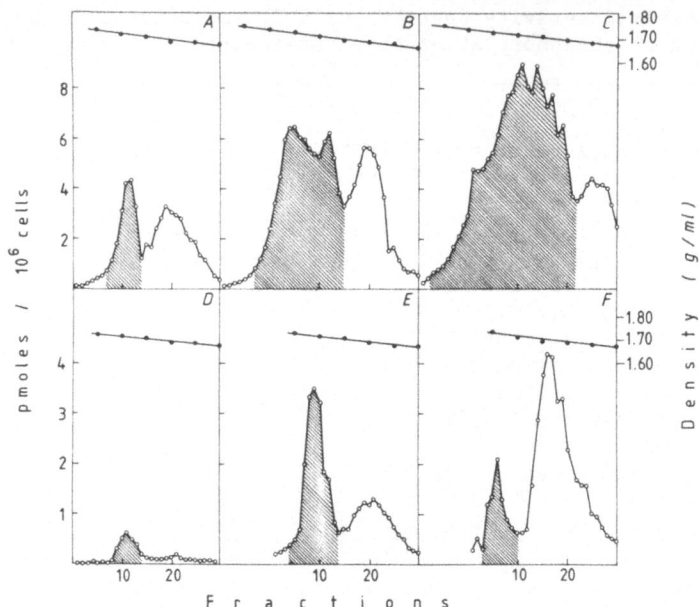

Fig. 3. Incorporation of [^{125}I]IVDU and C-[^{125}I]IVDU into DNA of HSV-1 (KOS)-infected Vero cells. Following infection the cells were incubated for 24 hours in the presence of 0.2 μM [^{125}I]IVDU (panel A), 0.5 μM [^{125}I]IVDU (panel B), 2 μM [^{125}I]IVDU (panel C), 0.5 μM C-[^{125}I]IVDU (panel D), 5 μM C-[^{125}I]IVDU (panel E) or 20 μM C-[^{125}I]IVDU (panel F). Viral and cellular DNAs were analyzed by CsCl density gradient ultracentrifugation. On the graphs viral DNA is shadowed.

lity to strand breakage. This, together with the reduced functioning of the substituted DNAs as templates for DNA and RNA polymerases may account for the compounds' ability to block virus replication.

NUCLEOSIDE ANALOGUES NOT DEPENDING ON VIRUS-ENCODED THYMIDINE KINASE AND TARGETED AT THE THYMIDYLATE SYNTHASE

Some 5-substituted 2'-deoxyuridines, i.e. those that contain a small and electronegative C-5 substituent, such as a fluorine, formyl, nitro, acetylene, cyano, thiocyano or aldoxime group, have been recognized as potent inhibitors of thymidylate (dTMP) synthase based on a number of criteria such as (i) a much greater inhibitory effect on dUrd than dThd incorporation into DNA,[25,26] (ii) inability to sustain the growth of dTMP synthase-deficient murine mammary carcinoma (FM3A/TS⁻) cells,[27,28] (iii) inhibition of tritium release from [5-^{3}H]dCyd,[29] and (iv) demonstration of inhibitory action of the 5-substituted 2'-dUMPs on the isolated dTMP synthase.[30] FDU, TFT and NDU (Fig. 4) rank among the most potent inhibitors of dTMP synthase, with K_i/K_m values[30] (for their 5'-monophosphates) in the range of 0.001 to 0.01.

That dTMP synthase may be a target for antiviral agents was originally demonstrated for NDU with regard to its inhibitory effect on vaccinia virus replication.[25] It has recently become evident that the activity spectrum of the dTMP synthase inhibitors extends to TK⁻ virus strains, i.e. those TK⁻ HSV-1 mutants that may arise during therapy with ACV or any other viral TK-dependent drug and that are insensitive to these drugs.[31] Wild-type TK⁺HSV strains are less sensitive to the dTMP synthase inhibitors than are their TK⁻ counterparts (Table 2).[31]

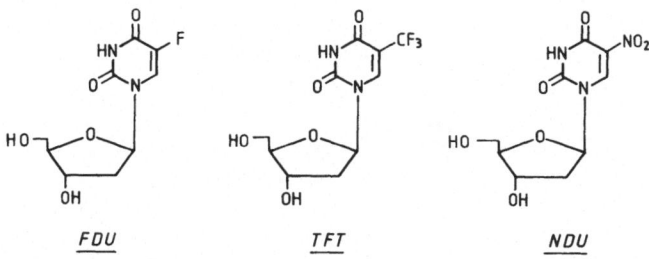

$$FDU \qquad TFT \qquad NDU$$

Fig. 4. 5-Substituted 2'-deoxyuridines which do not depend on virus-induced dThd kinase for their phosphorylation and which act as potent inhibitors of thymidylate synthase.

Table 2. Antiviral Activity Spectrum of FDU and Related Compounds (Acting as Potent Inhibitors of Thymidylate Synthase)

Highly susceptible :
- TK⁻ variants of HSV-1
- Vaccinia virus

Less susceptible :
- wild-type HSV-1, HSV-2, VZV

Not susceptible :
- RNA viruses

Apparently, TK⁻HSV mutants cannot rely on the dThd salvage pathway to ensure the necessary supply of dTMP, dTDP and dTTP for their DNA synthesis (Fig. 5), which makes them dependent on the de novo biosynthetic pathway of dTTP starting from N-carbamoyl aspartate and going through the dTMP synthase step. The latter can be considered as the bottleneck in the de novo biosynthesis of dTTP. TK⁻HSV mutants should be particularly vulnerable to any disturbance in the conversion of dUMP to dTMP as they cannot compensate for the loss in dTMP through salvaging dThd because they lack the necessary kinase for it.

Further evidence for the role of dTMP synthase in the inhibitory effect of FDU on TK⁻HSV-1 replication stems from reversal experiments (Fig. 6), which indicate that when added exogenously, dThd completely reverses the antiviral activity of FDU at concentrations as low as 1 µg/ml, whereas dUrd fails to show any reversing effect even at a concentration as high as 100 µg/ml. According to our scheme (Fig. 5), such a marked difference in the reversing effects of dUrd and dThd can only be accounted for by an inhibitory action of FDU at the dTMP synthase level.

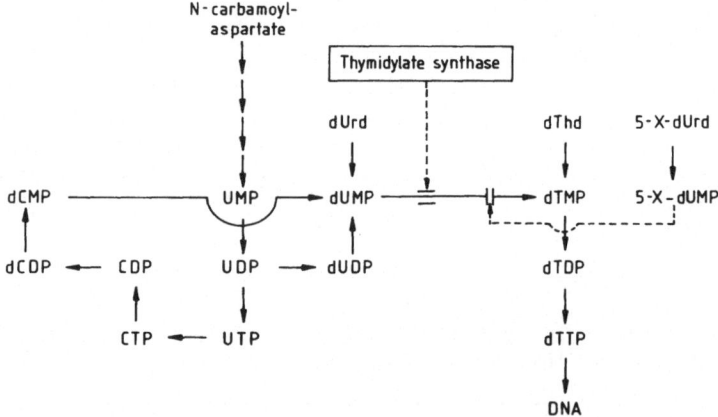

Fig. 5. Inhibitory action of 5-substituted 2'-deoxyuridines (5-X-dUrd), such as FDU, TFT and NDU, on thymidylate synthase results in a marked reduction in the de novo biosynthesis of dTTP starting from N-carbamoylaspartate.

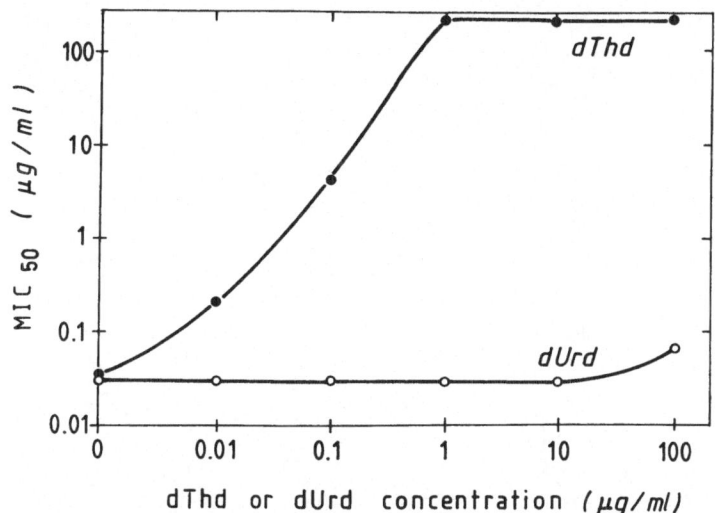

Fig. 6. Reversal of antiviral activity of FDU against TK⁻HSV-1 (B-2006) by dThd or dUrd in primary rabbit kidney cell cultures. MIC_{50} : minimum inhibitory concentration of FDU required to reduce virus-induced cytopathogenicity by 50 %.

PHOSPHONYLMETHOXYALKYLPURINES AND -PYRIMIDINES AS BROAD-SPECTRUM ANTI-DNA VIRUS AGENTS

(S)-HPMPA, PMEA and their congeners (Fig. 7) have recently been recognized as a novel class of potent and selective antiviral agents.[32] The activity spectrum of these compounds is remarkable in that it encompasses adenoviruses,[33] herpesviruses [including HSV-1, HSV-2, VZV, CMV, EBV, SHV-1, BHV-1, EHV-1, TK⁻HSV-1, TK⁻VZV as well as the seal (phocid) herpesvirus[34]], poxviruses (vaccinia virus[32]), iridoviruses (African swine fever virus[35]) and retroviruses [i.e. Moloney murine sarcoma virus (MSV) and human immunodeficiency virus (HIV)] (Table 3). (S)-HPMPA inhibits the replication of VZV in cell culture at a mean 50 % inhibitory concentration of 1.8 ng/ml (concentration range : 0.63-5.7 ng/ml, as demonstrated with a number of clinical VZV strains[36]), while not being toxic for the host cells at a concentration up to 52 µg/ml, thus achieving a selectivity index of approximately 29,000. (S)-HPMPA is also active against TK⁻HSV-1 strains[32] and this activity extends to the in vivo situation where (S)-HPMPA has been found effective against TK⁻HSV-1 keratitis in rabbits.[37]

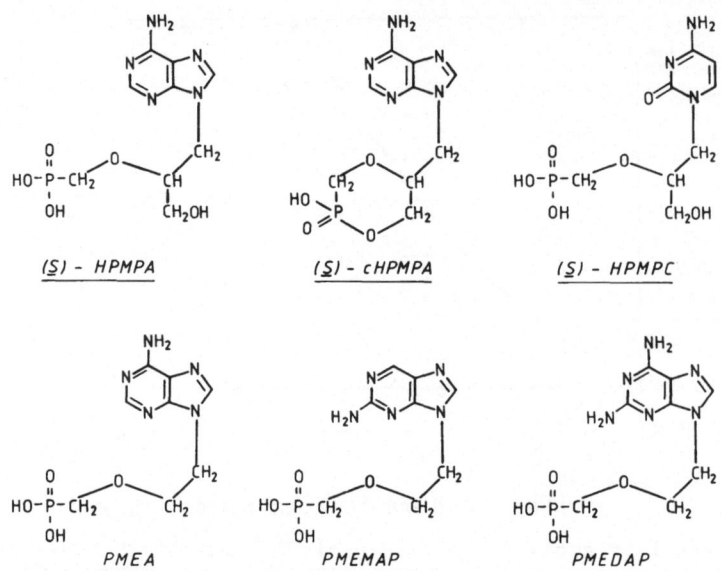

Fig. 7. Phosphonylmethoxyalkylpurine and -pyrimidine derivatives : (S)-HPMPA [(S)-9-(3-hydroxy-2-phosphonylmethoxypropyl)adenine], (S)-cHPMPA [cyclic phosphonate of (S)-HPMPA], (S)-HPMPC [(S)-1-(3-hydroxy-2-phosphonylmethoxypropyl)cytosine], PMEA [9-(2-phosphonylmethoxyethyl)adenine], PMEMAP [9-(2-phosphonylmethoxyethyl)-2-monoaminopurine] and PMEDAP [9-(2-phosphonylmethoxyethyl)-2,6-diaminopurine].

Table 3. Antiviral Activity Spectrum of (S)-HPMPA and Related Compounds (Phosphonylmethoxyalkyl Derivatives of Purines and Pyrimidines)

Moderately to highly susceptible
- Adenoviruses
- Herpesviruses (HSV-1, HSV-2, VZV, CMV, EBV, SHV-1, BHV-1, EHV-1, TK⁻ HSV-1, TK⁻ VZV, ...)
- Poxviruses (vaccinia virus)
- Iridoviruses (African swine fever virus)
- Retroviruses [human immunodeficiency virus (HIV)]

Not susceptible
- RNA viruses (except for retroviruses)

Among the different phosphonylmethoxyalkylpurines and -pyrimidines there exist rather striking differences in antiviral activity depending on the nature of the virus infection [E. De Clercq, M. Baba, T. Sakuma, R. Pauwels, J. Balzarini, R. Snoeck, I. Rosenberg and A. Holý: unpublished data (1987)]. Most active against adenovirus are (S)-HPMPA and (S)-cHPMPA, followed by (S)-HPMPDAP and (S)-HPMPC, whereas PMEA, PMEMAP and PMEDAP do not show much activity against adenovirus. The same pattern of activity is observed with respect to vaccinia virus and varicella-zoster virus (in order of decreasing activity : (S)-HPMPA ∿ (S)-cHPMPA > (S)-HPMPDAP (S)-HPMPC > PMEDAP ∿ PMEA > PMEMAP). Towards HSV-1 and HSV-2, however, (S)-HPMPA, (S)-cHPMPA, (S)-HPMPC, PMEA and PMEDAP all appear about equally active, PMEMAP being definitely less active. With a 50 %-inhibitory concentration less than 0.1 μg/ml and a selectivity index greater than 1000, (S)-HPMPC is the most potent and selective inhibitor of CMV replication described so far, whereas in terms of anti-HIV activity the order of (decreasing) activity is as follows : PMEDAP > PMEA > PMEMAP.

Using radiolabeled (S)-[U-^{14}C-adenine]HPMPA it has been demonstrated that the compound is as such taken up by the cells and subsequently converted to its monophosphoryl and diphosphoryl derivatives. This phosphorylation must be achieved by (hitherto unidentified) cellular kinases as it occurs equally well in virus-infected and mock-infected cells (Fig. 8).[38] Thus, virus-specified phosphorylation, i.e. by the virus-encoded dThd(dTMP) kinase, is not required for the antiviral action of (S)-HPMPA, a conclusion that could already be deduced from its activity against TK⁻ strains of HSV-1 and VZV.

Within the virus-infected cell, (S)-HPMPA selectively inhibits viral DNA synthesis without affecting cellular DNA synthesis. This has been clearly demonstrated in Vero cells infected with HSV-1 (Fig. 9). Only at a concentration of 500 μM (150 μg/ml) (S)-HPMPA effects a 50 % reduction in cellular DNA synthesis (in both mock-infected and virus-infected cells), while a 50 % inhibition of viral DNA synthesis is achieved at a concentration of 5 μM (1.5 μg/ml). The differential effect of (S)-HPMPA on viral and cellular DNA synthesis is even more pronounced in HEL cells infected with HSV-1, where (S)-HPMPA shuts off viral DNA synthesis at a concentration as low as 0.05 μM (0.015 μg/ml), while leaving cellular DNA synthesis unaffected at a concentration up to 50 μM (15 μg/ml).[38] The exact mechanism by which (S)-HPMPA interferes with viral DNA synthesis needs further investigation, and, hence, the exact target for its action remains to be identified. Whatever target (S)-HPMPA is directed at, it is clear that its selectivity as an antiviral agent resides in a specific inhibitory effect on viral DNA synthesis.

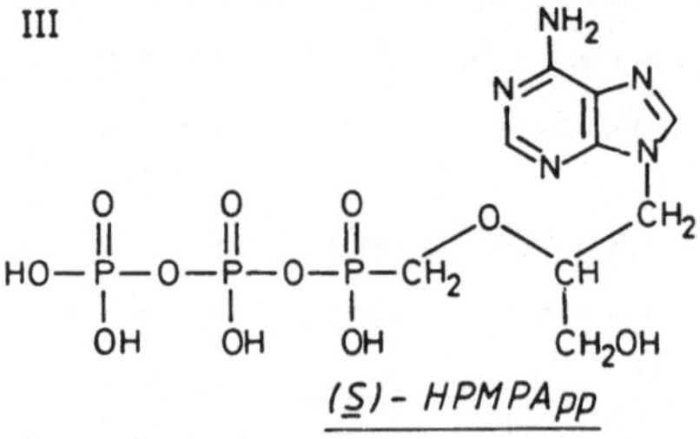

Fig. 8. Metabolites of (S)-HPMPA formed in both uninfected and virus-infected cells : I, the original compound; II, the monophosphoryl derivative of (S)-HPMPA; and III, the diphosphoryl derivative of (S)-HPMPA.

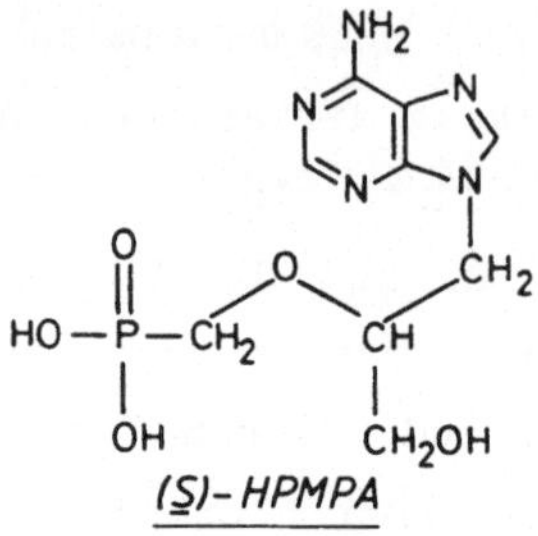

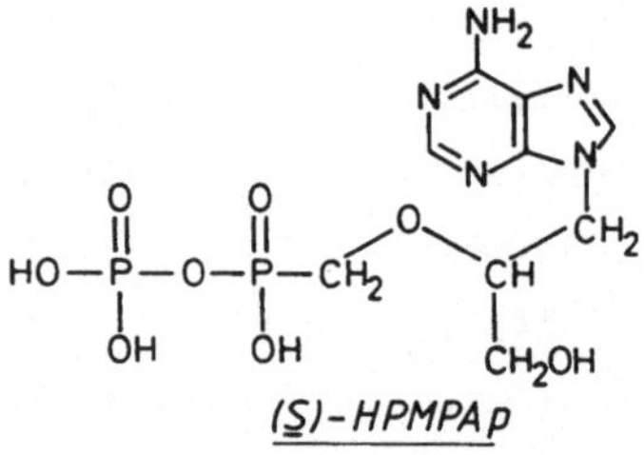

ACYCLIC AND CARBOCYCLIC ADENOSINE ANALOGUES TARGETED AT THE S-ADENOSYLHOMO-CYSTEINE HYDROLASE

The adenosine analogues which have been recognized as S-adenosylhomo-cysteine (SAH) hydrolase inhibitors fall, according to their chemical structure in two categories : acyclic adenosine analogues, i.e. (S)-DHPA and (RS)-AHPA, and carbocyclic adenosine analogues, i.e. C-c[3]Ado and neplanocin A (Fig. 10). For recent papers on the subject, see references 39-43.

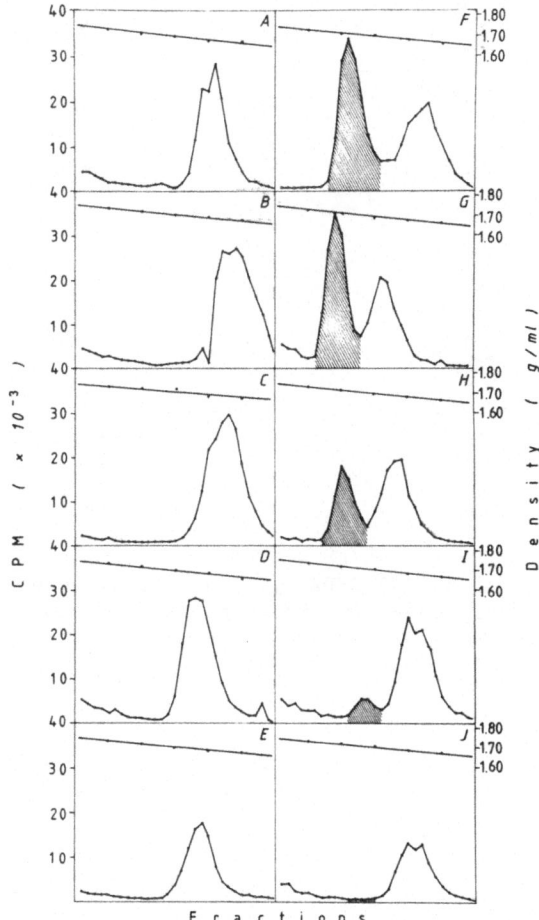

Fig. 9. Inhibitory effects of (S)-HPMPA on viral and cellular DNA synthesis in Vero cells. Panels A, B, C, D and E : mock-infected Vero cells. Panels F, G, H, I and J : HSV-1(KOS)-infected Vero cells. The cells were incubated for 24 hours in the presence of 25 µCi [^{32}P]ortophosphate together with varying concentrations of (S)-HPMPA : 0 µM (panels A and F), 0.5 µM (panels B and G), 5 µM (panels C and H), 50 µM (panels D and I) or 500 µM (panels E and J). Viral and cellular DNAs were analyzed by CsCl density gradient ultracentrifugation. On the graphs viral DNA is shadowed. After Votruba et al.[38]

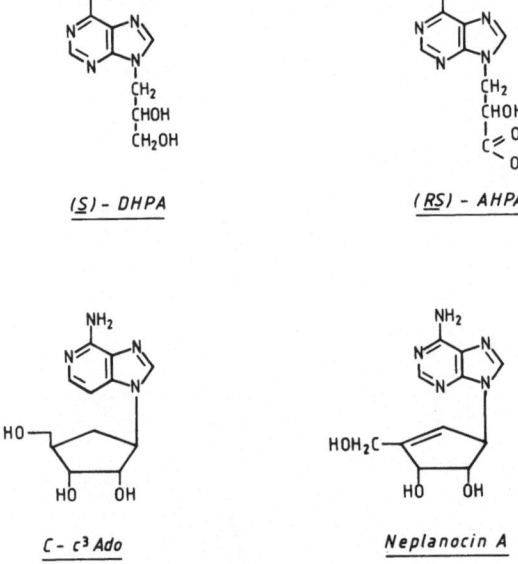

Fig. 10. Adenosine analogues which act as inhibitors of S-adenosylhomocysteine (SAH) hydrolase : (S)-DHPA [(S)-9-(2,3-dihydroxypropyl)adenine], (RS)-AHPA [(RS)-3-adenin-9-yl-2-hydroxypropanoic acid] alkyl esters, C-c3 Ado (carbocyclic 3-deazaadenosine) and neplanocin A (a cyclopentenyl derivative of adenine).

SAH hydrolase has since long been recognized as a target for inhibitors of S-adenosylmethionine (SAM)-dependent methylation reactions.[44] Inhibitors of SAH hydrolase may be expected to lead to an accumulation of SAH, itself a product inhibitor of SAM-dependent methylations. If SAH accumulates, methyltransfers will be suppressed, and as such methyltransfers are required for the maturation, i.e. 5'-capping, of viral mRNAs, the result of an inhibition at the SAH hydrolase level may be reflected by an inhibitory effect on virus replication and progeny formation.

The antiviral activity spectrum of (S)-DHPA, (RS)-AHPA, C-c3Ado and neplanocin A includes both DNA and RNA viruses : i.e. poxviruses (vaccinia virus), iridoviruses (African swine fever virus), paramyxoviruses (parainfluenza virus, measles virus), rhabdoviruses (vesicular stomatitis virus, rabies virus and infectious hematopoietic necrosis virus of fish), reoviruses (reovirus, rotavirus and infectious pancreatic necrosis virus of fish) and several plant viruses [potex (potato virus X), poty (virus isolated from Solanum palinacanthum), tymo (eggplant mosaic virus, belladonna mottle virus) and tobamo (tobacco mosaic virus)] (Table 4).[45] It is not quite clear why these viruses are included and others are excluded from the activity spectrum of the SAH hydrolase inhibitors. Possibly, viruses differ from one another in their methylation requirements, which then results in a differential susceptibility to methylation inhibitors.

The adenosine analogues [(S)-DHPA, (RS)-AHPA, C-c^{3}Ado, neplanocin A] that are assumed to interact with SAH hydrolase (Fig. 11) do not require phosphorylation and, in fact, should not be phosphorylated at all if they are to act at the SAH hydrolase level. Using a paired adenosine kinase-deficient (AK$^-$)/adenosine kinase-positive (AK$^+$) cell system,[46] we have been able to distinguish those adenosine analogues that do not need to be phosphorylated to exert their antiviral activity [i.e. (S)-DHPA, (RS)-AHPA, C-c^{3}Ado, neplanocin A] from those adenosine analogues that need such phosphorylation (i.e. tubercidin, toyocamycin, sangivamycin). The latter were much more active in AK$^+$ than in AK$^-$ cells, whereas the former were about equally, or only slightly more, active in AK$^+$ than AK$^-$ cells.

The exact target for the antiviral action of tubercidin, toyocamycin, sangivamycin and their derivatives (i.e. xylotubercidin)[47,48] has not yet been identified, but it may well correspond to the nucleic acid (RNA or DNA) polymerization process.

Table 4. Antiviral Activity Spectrum of (S)-DHPA and Related Compounds (Acyclic or Carbocyclic Adenosine Analogues Acting as SAH Hydrolase Inhibitors)

Moderately to highly susceptible
- Poxviruses (vaccinia virus)
- Iridoviruses (African swine fever virus)
- Paramyxoviruses (parainfluenza virus, measles virus)
- Rhabdoviruses (vesicular stomatitis virus, rabies virus)
- Reoviruses (reovirus, rotavirus)
- Plant viruses (potex-, poty-, tymo- and tobamoviruses)

Not or less susceptible
- Adenoviruses
- Herpesviruses (HSV-1, HSV-2, VZV, ...)
- Picornaviruses (polio virus, Coxsackie B virus, rhinovirus)
- Togaviruses (Sindbis virus, Semliki forest virus)

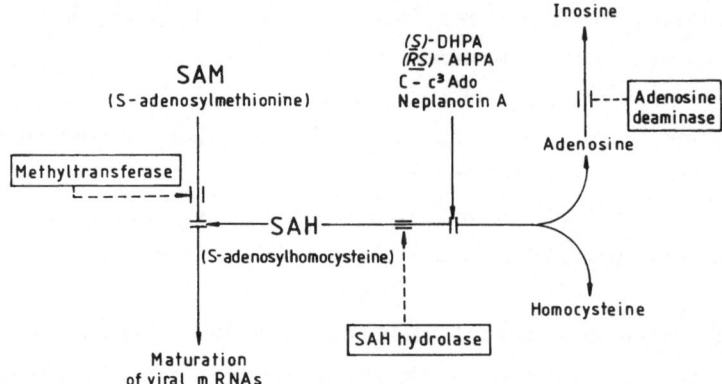

Fig. 11. Inhibitory effect of (S)-DHPA, (RS)-AHPA, C-c^{3}Ado and neplanocin A at the level of SAH hydrolase leads to an accumulation of SAH, which is in turn inhibitory to the SAM-dependent methylation reactions, i.e. those involved in the maturation of viral mRNAs.

How would an inhibitory effect on SAH hydrolase ensure a specific activity towards certain viruses ? This issue has not been resolved yet and should be addressed by directly examining SAH hydrolase activity, i.e. SAM/SAH levels, in virus-infected cells exposed to the SAH hydrolase inhibitors. That SAH hydrolase is the probable target for the antiviral action of (S)-DHPA, (RS)-AHPA, C-c^{3}Ado and neplanocin A stems from the close correlation (r = 0.986) that has been found between the antiviral potency of these compounds (against vesicular stomatitis virus) and their inhibitory effects (K$_I$/K$_m$) on the SAH hydrolase from beef liver.[49] More recently, an even closer correlation (r = 0.993) was found between the antiviral potency of the compounds (against vaccinia virus) and their inhibitory effects on SAH hydrolase extracted from the same cells (murine L929 cells) as used for the antiviral assays (Fig. 12) [M. Cools and E. De Clercq : unpublished data (1987)].

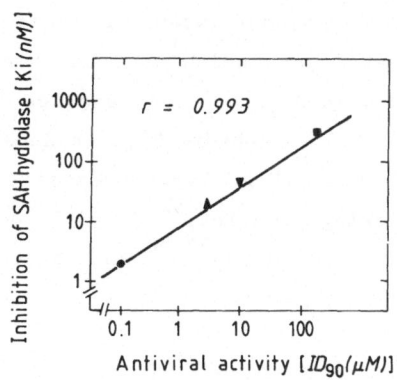

Fig. 12. Correlation between the antiviral activity of four adenosine analogues [(S)-DHPA (■), (RS)-AHPA isobutyl ester (▼), C-c^{3}Ado (▲) and neplanocin A (●)] and their inhibitory effect on SAH hydrolase. The inhibitory
effect on SAH hydrolase is expressed as K_i for the activity of SAH hydrolase isolated from murine L929 cells and measured in the direction of SAH
synthesis. The antiviral activity is expressed as ID_{90} (90 %-inhibitory
dose) or dose required to inhibit vaccinia virus growth (yield) in L929
cells by 90 %.

2',3'-DIDEOXYNUCLEOSIDE ANALOGUES TARGETED AT THE REVERSE TRANSCRIPTASE AND
EFFECTIVE AGAINST THE HUMAN IMMUNODEFICIENCY VIRUS

Most of the compounds which have been described in the past few years
as effective inhibitors of the AIDS virus (HIV) are presumed to exert their
anti-HIV activity through inhibition of the HIV-associated reverse transcriptase.[50,51] This is certainly the case for suramin, Evans blue, aurintricarboxylic acid, phosphonoformate (foscarnet), ammonium 21-tungsto-9-
antimoniate (HPA-23), and, most probably, for D-penicillamine and rifabu-
tine as well. Also, the 2',3'-dideoxynucleoside analogues, with azidothymidine (retrovir) as the prototype, are postulated to act at the reverse
transcriptase level. This requires, however, that the 2',3'-dideoxynucleo-

side analogues are first phosphorylated to their 5'-triphosphates. The 2',3'-dideoxynucleoside analogues which have been found to selectively inhibit HIV could be divided into three subcategories : 3'-azido-2',3'-dideoxy-, 2',3'-dideoxy- and 2',3'-didehydro-2',3'-dideoxyribosides. The latter are also designated as "enes" (Fig. 13). Foremost in anti-HIV potency and selectivity are the following: AzddThd,[52] ddCyd,[53] ddeCyd,[54,55] ddThd,[53,56] ddeThd,[56] ddAdo,[53,57] ddDAPR,[57] ddGuo[53,58] and AzddGuo.[58]

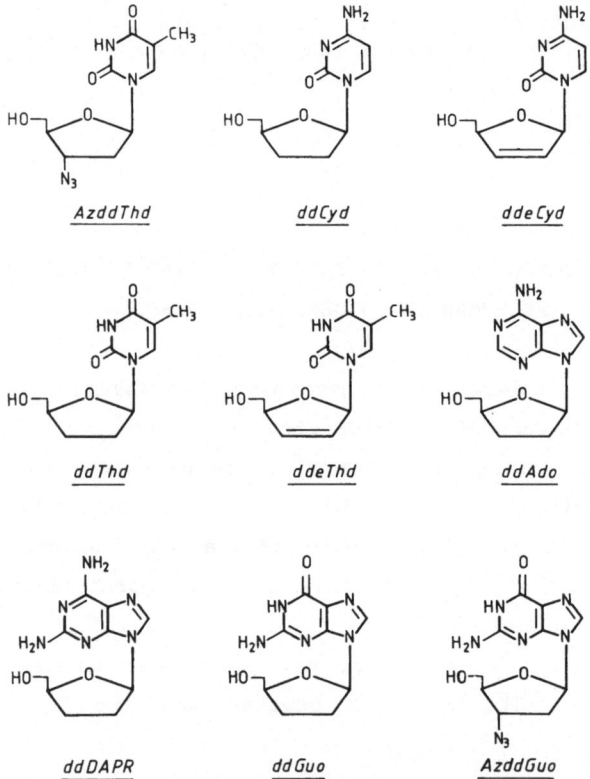

Fig. 13. 2',3'-Dideoxynucleoside analogues : AzddThd (3'-azido-2',3'-dideoxythymidine), ddCyd (2',3'-dideoxycytidine), ddeCyd (2',3'-dideoxycytidinene), ddThd (2',3'-dideoxythymidine), ddeThd (2',3'-dideoxythymidinene), ddAdo (2',3'-dideoxyadenosine), ddDAPR (2',3'-dideoxy-2,6-diaminopurineriboside), ddGuo (2',3'-dideoxyguanosine and AzddGuo (3'-azido-2',3'-dideoxyguanosine).

The antiviral activity spectrum of the 2',3'-dideoxynucleoside analogues is confined to retroviruses (Table 5). Other viruses (i.e. HSV-1, HSV-2, vaccinia virus, etc.) were evaluated for their sensitivity to AzddThd, ddCyd, ddeCyd and their congeners, but no activity was detected.[59,60] Among the retroviruses, both oncovirinae (MSV) and lentivirinae (HIV) have been found sensitive to the compounds, and it is likely that their activity extends to other oncovirinae (i.e. murine and feline leukemia and sarcoma viruses, human T-cell leukemia virus types I and II) and lentivirinae (i.e. visna and maedi viruses, caprine encephalitis virus and equine infectious anemia virus).

To exert their inhibitory effect on the HIV reverse transcriptase, the 2',3'-dideoxynucleoside analogues have to be phosphorylated (Fig. 14). As a rule, 2',3'-dideoxynucleosides (ddNs) are less efficiently phosphorylated by the cellular kinases than 2'-deoxynucleosides (dNs).[61] Furthermore, this phosphorylation process is highly dependent on the cell species : i.e., in murine cells AzddThd is readily converted to the 5'-triphosphate (AzddTTP), whereas in human and caprine cells AzddThd accumulates as the 5'-monophosphate (AzddTMP); in contrast, ddCyd is more extensively phosphorylated to its 5'-triphosphate (ddCTP) in human cells than in murine cells, so that much higher ddCTP than AzddTTP levels are achieved in human cells, an observation that may be particularly relevant from a therapeutic viewpoint.[62]

Table 5. Antiviral Activity Spectrum of 2',3'-Dideoxynucleoside Analogues

---

Moderately to highly susceptible
    Retroviruses
    - Oncovirinae (leukemia and sarcoma viruses)
    - Lentivirinae (HIV and presumably other lentiviruses as well)

Not susceptible
    - Herpesviruses (HSV-1, HSV-2, ...)
    - Poxviruses (vaccinia virus)
    - Picornaviruses (polio virus, Coxsackie B virus)
    - Togaviruses (Sindbis virus, Semliki forest virus)
    - Paramyxoviruses (parainfluenza virus)
    - Rhabdoviruses (vesicular stomatitis virus)
    - Reoviruses (reovirus)

At the level of the reverse transcriptase, the 2',3'-dideoxynucleoside 5'-triphosphates (ddNTPs) may either act as inhibitors, by direct competition with the natural substrates (dNTPs) for the same binding site at the polymerase, or be incorporated at the 3'-end of the DNA chain, thereby pre-

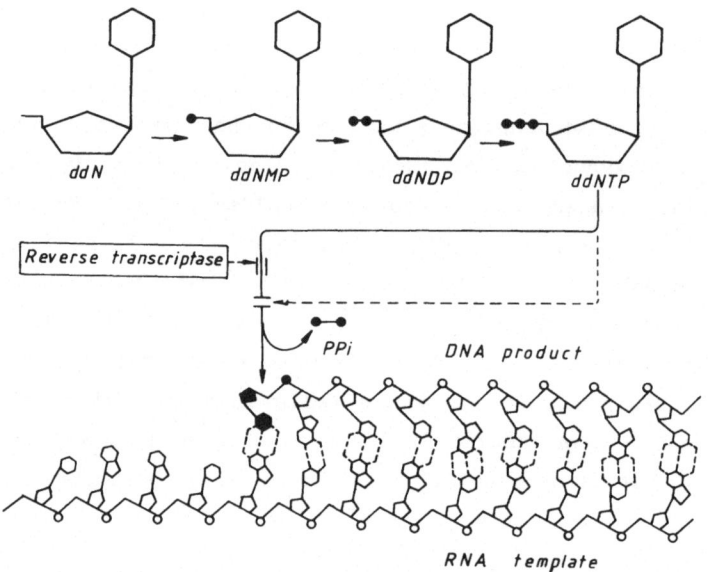

Fig. 14. Following successive phosphorylation of the 2',3'-dideoxynucleoside analogue (ddN) by cellular kinases to its 5'-monophosphate (ddNMP), 5'-diphosphate (ddNDP) and 5'-triphosphate (ddNTP), the latter interferes with the reverse transcriptase reaction. It can thereby act as either a competitive inhibitor (with respect ot the natural substrate) or an alternate substrate (and, hence, be incorporated into the DNA product). If incorporated, ddNMP will function as a chain terminator.

venting further growth of the chain (Fig. 14). Moreover, the interaction of the ddNTP with the reverse transcriptase may be facilitated if the intracellular pool levels of the competing substrate (dNTP) are reduced, for example by an inhibitory effect of the ddNs on one or another step in the anabolism of the dNs, as has been shown for AzddThd which severely reduces intracellular dTTP levels due to inhibition of dTMP kinase by AzddTMP.[63]

Thus, several factors may contribute to the potent and selective inhibition of HIV replication by 2',3'-dideoxynucleoside analogues : (i) the rate and extent by which the ddNs are phosphorylated intracellularly to their corresponding 5'-triphosphates, (ii) their interaction with the anabolism of the dNs, through a direct interference with the kinases involved in this anabolism or indirectly via feedback regulatory mechanisms, and (iii) the specificity of the ddNTPs as inhibitors of the HIV reverse transcriptase (which is 100-fold more sensitive to AzddTTP than is the cellular DNA polymerase α).[63] The interplay of these factors may then determine the relative potencies and selectivities of the 2',3'-dideoxynucleoside analogues as anti-HIV agents (Fig. 15). In this respect, it should be stressed that the 2',3'-dideoxynucleoside analogues are far more potent and selective inhibitors of HIV than other compounds which have been accredited with virustatic activity against the AIDS virus (i.e. ribavirin, HPA-23, foscarnet and suramin).

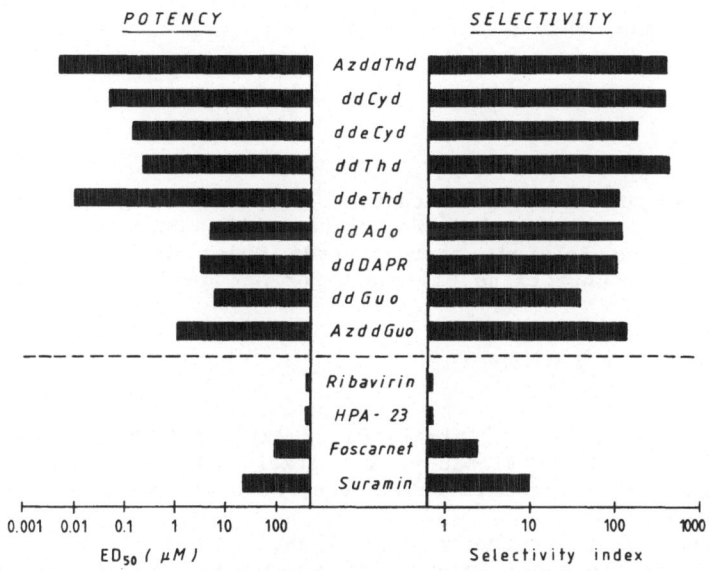

Fig. 15. Comparative potency and selectivity of 2',3'-dideoxynucleoside analogues and some reference compounds as anti-HIV agents in vitro (ATH8 or MT4 cells). The potency is expressed as ED_{50} (50 %-effective dose) or dose required to achieve 50 % protection of the cells against the cytopathic effect of HIV. The selectivity is expressed as selectivity index or ratio of the 50 %-cytotoxic dose (required to reduce the viability of the uninfected host cells by 50 %) to the ED_{50}.

CONCLUSION

Several strategies could be pursued in the development of selective antiviral agents, and this has yielded the following five groups of compounds :

- I. Compounds that are preferentially phosphorylated by the viral TK and targeted at viral DNA polymerase or viral DNA itself. These agents are specifically active against HSV and VZV.

- II. Compounds that are nonspecifically phosphorylated by the cellular TK and targeted at dTMP synthase. These agents are particularly active against TK⁻ HSV strains.

- III. Compounds that are nonspecifically phosphorylated by cellular kinases, although they selectively block viral DNA synthesis. Their exact target of action has not yet been identified. These agents are active against a broad range of DNA viruses, including adenoviruses, herpesviruses (HSV, VZV, CMV, EBV), poxviruses, and retroviruses (HIV).

- IV. Compounds that do not require phosphorylation and are targeted at SAH hydrolase. These agents are active against a wide variety of RNA and DNA viruses, in particular rhabdo-, reo-, paramyxo- and poxviruses.

- V. Compounds that are nonspecifically phosphorylated by cellular kinases and targeted at the reverse transcriptase. These agents are effective against retroviruses (HIV).

ACKNOWLEDGMENTS

The investigations of the author are supported by grants from the Belgian Fonds voor Geneeskundig Wetenschappelijk Onderzoek (Project no. 3.0040.83) and the Belgian Geconcerteerde Onderzoeksacties (Project no. 85/90-79). I thank Christiane Callebaut for her dedicated editorial assistance.

REFERENCES

1. E. De Clercq (ed.), "Clinical Use of Antiviral Drugs" (Series: Developments in Medical Virology, Y. Becker, ed.), Martinus Nijhoff Publishing, Kluwer Academic Publishers, Norwell, Massachusetts, USA, in press (1987).
2. A.J. Hay, A.J. Wolstenholme, J.J. Skehel, and M.H. Smith, The molecular basis of the specific anti-influenza action of amantadine, EMBO J. 4:3021 (1985).
3. T.J. Smith, M.J. Kremer, M. Luo, G. Vriend, E. Arnold, G. Kamer, M.G. Rossmann, M.A. McKinlay, G.D. Diana, and M.J. Otto, The site of attachment in human rhinovirus 14 for antiviral agents that inhibit uncoating, Science 233:1286 (1986).
4. B.E. Gilbert, and V. Knight, Biochemistry and clinical applications of ribavirin, Antimicrob. Agents Chemother. 30:201 (1986).

5. E. De Clercq, and R.T. Walker, Synthesis and antiviral properties of 5-vinylpyrimidine nucleoside analogs, Pharmac. Ther. 26:1 (1984).

6. E. De Clercq, and R.T. Walker, Chemotherapeutic agents for herpesvirus infections, in: "Progress in Medicinal Chemistry", vol. 23, G.P. Ellis, and G.B. West, eds., Elsevier Sci. Publ., Amsterdam, p. 187 (1986).

7. C.K. Chu, and S.J. Cutler, Chemistry and antiviral activities of acyclonucleosides, J. Heterocyclic Chem. 23:289 (1986).

8. E.-C. Mar, Y.-C. Cheng, and E.-S. Huang, Effect of 9-(1,3-dihydroxy-2-propoxymethyl)guanine on human cytomegalovirus replication in vitro, Antimicrob. Agents Chemother. 24:518 (1983).

9. M.J. Tocci, T.J. Livelli, H.C. Perry, C.S. Crumpacker, and A.K. Field, Effects of the nucleoside analog 2'-nor-2'-deoxyguanosine on human cytomegalovirus replication, Antimicrob. Agents Chemother. 25:247 (1984).

10. V.R. Freitas, D.F. Smee, M. Chernow, R. Boehme, and T.R. Matthews, Activity of 9-(1,3-dihydroxy-2-propoxymethyl)guanine compared with that of acyclovir against human, monkey, and rodent cytomegaloviruses, Antimicrob. Agents Chemother. 28:240 (1985).

11. K.K. Biron, J.A. Fyfe, S.C. Stanat, L.K. Leslie, J.B. Sorrell, C.U. Lambe, and D.M. Coen, A human cytomegalovirus mutant resistant to the nucleoside analog 9-{[2-hydroxy-1-(hydroxymethyl)ethoxy]methyl}guanine (BW B759U) induces reduced levels of BW B759U triphosphate, Proc. Natl. Acad. Sci. USA 83:8769 (1986).

12. E. De Clercq, Towards a selective chemotherapy of virus infections. Development of bromovinyldeoxyuridine as a highly potent and selective antiherpetic drug, Verh. K. Acad. Geneeskd. Belg. 48:261 (1986).

13. N.K. Ayisi, E. De Clercq, R.A. Wall, H. Hughes, and S.L. Sacks, Metabolic fate of (E)-5-(2-bromovinyl)-2'-deoxyuridine in herpes simplex virus- and mock-infected cells, Antimicrob. Agents Chemother. 26:762 (1984).

14. S. Kit, H. Ichimura, and E. De Clercq, Phosphorylation of nucleoside analogs by equine herpesvirus type 1 pyrimidine deoxyribonucleoside kinase, Antiviral Res. 7:53 (1987).

15. S. Kit, H. Ichimura, and E. De Clercq, Differential metabolism of (E)-5-(2-iodovinyl)-2'-deoxyuridine (IVDU) by equine herpesvirus type 1- and herpes simplex virus-infected cells, Antiviral Res. 8:41 (1987).

16. E. De Clercq, Biochemical aspects of the selective antiherpes activity of nucleoside analogues, Biochem. Pharmacol. 33:2159 (1984).

17. D. Derse, Y.-C. Cheng, P.A. Furman, M.H. St. Clair, and G.B. Elion, Inhibition of purified human and herpes simplex virus-induced DNA polymerases by 9-(2-hydroxyethoxymethyl)guanine triphosphate, J. Biol. Chem. 256: 11447 (1981).

18. K. Stenberg, A. Larsson, and R. Datema, Metabolism and mode of action of (R)-9-(3,4-dihydroxybutyl)guanine in herpes simplex virus-infected Vero cells, J. Biol. Chem. 261: 2134 (1986).

19. J. Descamps, R.K. Sehgal, E. De Clercq, and H.S. Allaudeen, Inhibitory effect of E-5-(2-bromovinyl)-1-β-D-arabinofuranosyluracil on herpes simplex virus replication and DNA synthesis. J. Virol. 43:332 (1982).

20. E. De Clercq, and R. Bernaerts, Specific phosphorylation of 5-ethyl-2'-deoxyuridine by herpes simplex virus-infected cells and incorporation into viral DNA, J. Biol. Chem., in press (1987).

21. E. De Clercq, R. Bernaerts, J. Balzarini, P. Herdewijn, and A. Verbruggen, Metabolism of the carbocyclic analogue of (E)-5-(2-iodovinyl)-2'-deoxyuridine in herpes simplex virus-infected cells, J. Biol. Chem. 260:10621 (1985).

22. J. Sági, E. De Clercq, A. Szemzö, A.H. Csárnyi, T. Kovács, and L. Ötvös, Incorporation of the carbocyclic analogue of (E)-5-(2-bromovinyl)-2'-deoxyuridine 5'-triphosphate into a synthetic DNA, Biochem. Biophys. Res. Commun., in press (1987).

23. Y.-c. Cheng, S.P. Grill, G.E. Dutschman, K. Nakayama, and K.F. Bastow, Metabolism of 9-(1,3-dihydroxy-2-propoxymethyl)guanine, a new anti-herpes virus compound, in herpes simplex virus-infected cells, J. Biol. Chem. 258:12460 (1983).

24. R. Bernaerts, A. Verbruggen, and E. De Clercq, Mechanism of antiviral action of 5-substituted 2'-deoxyuridines: (E)-5-(2-iodovinyl)-2'-deoxyuridine (IVDU) as compared to its carbocyclic analogue (C-IVDU), in "Frontiers in Microbiology. From Antibiotics to AIDS" (E. De Clercq, ed.), Martinus Nijhoff Publishers, Dordrecht, p. 289 (1987).

25. E. De Clercq, J. Descamps, G.-F. Huang, and P.F. Torrence, 5-Nitro-2'-deoxyuridine and 5-nitro-2'-deoxyuridine 5'-monophosphate: antiviral activity and inhibition of thymidylate synthetase in vivo, Mol. Pharmacol. 14:422 (1978).

26. E. De Clercq, J. Balzarini, P.F. Torrence, M.P. Mertes, C.L. Schmidt, D. Shugar, P.J. Barr, A.S. Jones, G. Verhelst, and R.T. Walker, Thymidylate synthetase as target enzyme for the inhibitory activity of 5-substituted 2'-deoxyuridines on mouse leukemia L1210 cell growth, Mol. Pharmacol. 19:321 (1981).

27. J. Balzarini, E. De Clercq, D. Ayusawa, and T. Seno, Thymidylate synthetase-deficient mouse FM3A mammary carcinoma cell line as a tool for studying the thymidine salvage pathway and the incorporation of thymidine analogues into host cell DNA, Biochem. J. 217:245 (1984).

28. J. Balzarini, E. De Clercq, D. Ayusawa, and T. Seno, Thymidylate synthetase-positive and -negative murine mammary FM3A carcinoma cells as a useful system for detecting thymidylate synthetase inhibitors, FEBS Lett. 173:227 (1984).

29. J. Balzarini, and E. De Clercq, Strategies for the measurement of the inhibitory effects of thymidine analogs on the activity of thymidylate synthase in intact murine leukemia L1210 cells, Biochim. Biophys. Acta 785:36 (1984).

30. J. Balzarini, E. De Clercq, M.P. Mertes, D. Shugar, and P.F. Torrence, 5-Substituted 2'-deoxyuridines: correlation between inhibition of tumor cell growth and inhibition of thymidine kinase and thymidylate synthetase, Biochem. Pharmacol. 31:3673 (1982).

31. E. De Clercq, J. Béres, and W.G. Bentrude, Potent activity of 5-fluoro-2'-deoxyuridine and related compounds against thymidine kinase-deficient (TK⁻) herpes simplex virus: targeted at thymidylate synthase, Mol. Pharmacol. 32:286 (1987).

32. E. De Clercq, A. Holý, I. Rosenberg, T. Sakuma, J. Balzarini, and P.C. Maudgal, A novel selective broad-spectrum anti-DNA virus agent, Nature 323:464 (1986).

33. M. Baba, S. Mori, S. Shigeta, and E. De Clercq, Selective inhibitory effect of (S)-9-(3-hydroxy-2-phosphonylmethoxypropyl)adenine and 2'-nor-cyclic GMP on adenovirus replication in vitro, Antimicrob. Agents Chemother. 31:337 (1987).

34. A.D.M.E. Osterhaus, J. Groen, and E. De Clercq, Selective inhibitory effects of (S)-9-(3-hydroxy-2-phosphonyl-methoxypropyl)adenine and 1-(2-deoxy-2-fluoro-β-D-arabinofuranosyl)-5-iodouracil on seal herpesvirus (phocid herpesvirus 1) infection in vitro, Antiviral Res. 7:221 (1987).

35. C. Gil-Fernandez, and E. De Clercq, Comparative efficacy of broad-spectrum antiviral agents as inhibitors of African swine fever virus replication in vitro, Antiviral Res 7:151 (1987).

36. M. Baba, K. Konno, S. Shigeta, and E. De Clercq, In vitro activity of (S)-9-(hydroxy-2-phosphonylmethoxypropyl)adenine against newly isolated clinical varicella-zoster virus strains, Eur. J. Clin. Microbiol.: 6:158 (1987).

37. P.C. Maudgal, E. De Clercq, and P. Huyghe, Efficacy of (S)-HPMPA against thymidine kinase-deficient herpes simplex virus-keratitis, Invest. Ophthalmol. Vis. Sci. 28:243 (1987).

38. I. Votruba, R. Bernaerts, T. Sakuma, E. De Clercq, A. Merta, I. Rosenberg, and A. Holý, Intracellular phosphorylation of broad-spectrum anti-DNA virus agent (S)-9-(3-hydroxy-2-phosphonylmethoxypropyl)adenine and inhibition of viral DNA synthesis, Mol. Pharmacol., in press (1987).

39. E. De Clercq, and J.A. Montgomery, Broad-spectrum antiviral activity of the carbocyclic analog of 3-deazaadenosine, Antiviral Res. 3:17 (1983).

40. E. De Clercq, and A. Holý, Alkyl esters of 3-adenin-9-yl-2-hydroxypropanoic acid: a new class of broad-spectrum antiviral agents, J. Med. Chem. 28:282 (1985).

41. E. De Clercq, Antiviral and antimetabolic activities of neplanocins, Antimicrob. Agents Chemother. 28:84 (1985).

42. R.T. Borchardt, B.T. Keller, and U. Patel-Thombre, Neplanocin A. A potent inhibitor of S-adenosylhomocysteine hydrolase and of vaccinia virus multiplication in mouse L929 cells, J. Biol. Chem. 259:4353 (1984).

43. R.I. Glazer, K.D. Hartman, M.C. Knode, M.M. Richard, P.K. Chiang, C.K.H. Tseng, and V.E. Marquez, 3-Deazaneplanocin: a new and potent inhibitor of S-adenosylhomocysteine hydrolase and its effects on human promyelocytic leukemia cell line HL-60, Biochem. Biophys. Res. Commun. 135:688 (1986).

44. R.T. Borchardt, S-Adenosyl-L-methionine-dependent macromolecule methyltransferases: potential targets for the design of chemotherapeutic agents, J. Med. Chem. 23:347 (1980).

45. E. De Clercq, S-Adenosylhomocysteine hydrolase inhibitors as broad-spectrum antiviral agents, Biochem. Pharmacol. 16:2567 (1987).

46. M. Cools, E. De Clercq, and J.C. Drach, Role of adenosine kinase in the biological (antiviral and anticellular) activities of adenosine analogues, Nucleosides & Nucleotides 6:423 (1987).

47. E. De Clercq, and M.J. Robins, Xylotubercidin against herpes simplex virus type 2 in mice, Antimicrob. Agents Chemother. 30:719 (1986).

48. E. De Clercq, J. Balzarini, D. Madej, F. Hansske, and M.J. Robins, Nucleic acid related compounds. 51. Synthesis and biological properties of sugar-modified analogues of the nucleoside antibiotics tubercidin, toyocamycin, sangivamycin and formycin, J. Med. Chem. 30:481 (1987).

49. E. De Clercq, and M. Cools, Antiviral potency of adenosine analogues: correlation with inhibition of S-adenosylhomocysteine hydrolase, Biochem. Biophys. Res. Commun. 129:306 (1985).

50. E. De Clercq, Chemotherapeutic approaches to the treatment of the acquired immune deficiency syndrome (AIDS), J. Med. Chem. 29:1561 (1986).

51. E. De Clercq, New selective antiviral agents active against the AIDS virus, Trends in Pharmacological Sciences (TIPS) 8:339 (1987).

52. H. Mitsuya, K.J. Weinhold, P.A. Furman, M.H. St. Clair, S. Nusinoff Lehrman, R.C. Gallo, D. Bolognesi, D.W. Barry, and S. Broder, 3'-Azido-3'-deoxythymidine (BW A509U): an antiviral agent that inhibits the infectivity and cytopathic effect of human T-lymphotropic virus type III/lymphadenopathy-associated virus in vitro. Proc. Natl. Acad. Sci. USA 82:7096 (1985).

53. H. Mitsuya, and S. Broder, Inhibition of the in vitro infectivity and cytopathic effect of human T-lymphotrophic virus type III/lymphadenopathy-associated virus (HTLV-III/LAV) by 2',3'-dideoxynucleosides, Proc. Natl. Acad. Sci. USA 83:1911 (1986).

54. J. Balzarini, R. Pauwels, P. Herdewijn, E. De Clercq, D.A. Cooney, G.-J. Kang, M. Dalal, D.G. Johns, and S. Broder, Potent and selective anti-HTLV-III/LAV activity of 2',3'-dideoxycytidinene, the 2',3'-unsaturated derivative of 2',3'-dideoxycytidine, Biochem. Biophys. Res. Commun. 140:735 (1986).

55. T.-S. Lin, R.F. Schinazi, M.S. Chen, E. Kinney-Thomas, and W.H. Prusoff, Antiviral activity of 2',3'-dideoxycytidin-2'-ene (2',3'-dideoxy-2',3'-didehydrocytidine) against human immunodeficiency virus in vitro, Biochem. Pharmacol. 36:311 (1987).

56. M. Baba, R. Pauwels, P. Herdewijn, E. De Clercq, J. Desmyter, and M. Vandeputte, Both 2',3'-dideoxythymidine and its 2',3'-unsaturated derivative (2',3'-dideoxythymidinene) are potent and selective inhibitors of human immunodeficiency virus replication in vitro, Biochem. Biophys. Res. Commun. 142:128 (1987).

57. J. Balzarini, R. Pauwels, M. Baba, M.J. Robins, R. Zou, P. Herdewijn, and E. De Clercq, The 2',3'-dideoxyriboside of 2,6-diaminopurine selectively inhibits human immunodeficiency virus (HIV) replication in vitro, Biochem. Biophys. Res. Commun. 145:269 (1987).

58. M. Baba, R. Pauwels, J. Balzarini, P. Herdewijn, and E. De Clercq, Selective inhibition of human immunodeficiency virus (HIV) by 3'-azido-2',3'-dideoxyguanosine in vitro, Biochem. Biophys. Res. Commun. 145:1080 (1987).

59. E. De Clercq, J. Balzarini, J. Descamps, and F. Eckstein, Antiviral, antimetabolic and antineoplastic activities of 2'- or 3'-amino or -azido-substituted deoxyribonucleosides, Biochem. Pharmacol. 29:1849 (1980).

60. J. Balzarini, G.-J. Kang, M. Dalal, P. Herdewijn, E. De Clercq, S. Broder, and D. G. Johns, The anti-HTLV-III (anti-HIV) and cytotoxic activity of 2',3'-didehydro-2',3'-dideoxyribonucleosides. A comparison with their parental 2',3'-dideoxyribonucleosides, Mol. Pharmacol. 32:162 (1987).

61. M.A. Waqar, M.J. Evans, K.F. Manly, R.G. Hughes, and J.A. Huberman, Effects of 2',3'-dideoxynucleosides on mammalian cells and viruses, J. Cell. Physiol. 121:402 (1984).

62. J. Balzarini, R. Pauwels, M. Baba, P. Herdewijn, E. De Clercq, S. Broder, and D.G. Johns, The in vitro and in vivo anti-retrovirus activity, and intracellular metabolism of 3'-azido-2',3'-dideoxythymidine and 2',3'-dideoxycytidine are highly dependent on the cell species, Biochem. Pharmacol., in press (1987).

63. P.A. Furman, J.A. Fyfe, M.H. St. Clair, K. Weinhold, J.L. Rideout, G.A. Freeman, S. Nusinoff Lehrman, D.P. Bolognesi, S. Broder, H. Mitsuya, and D.W. Barry, Phosphorylation of 3'-azido-3'-deoxythymidine and selective interaction of the 5'-triphosphate with human immunodeficiency virus reverse transcriptase, Proc. Natl. Acad. Sci. USA 83:8333 (1986).

INHIBITION OF S-ADENOSYLMETHIONINE-DEPENDENT TRANSMETHYLATION AS AN APPROACH TO THE DEVELOPMENT OF ANTIVIRAL AGENTS.

Bradley T. Keller and Ronald T. Borchardt

Department of Pharmaceutical Chemistry, University of Kansas

Lawrence, Kansas 66045 U.S.A.

The enzymatic transfer of methyl groups from S-adenosylmethionine (AdoMet) to an acceptor molecule, i.e. biological transmethylation, is widely recognized as a ubiquitous set of reactions involved in a diverse array of physiological processes (Usdin et al., 1982; Borchardt et al., 1986). In addition to their established role in the metabolism of a variety of small molecules such as histamine, catecholamines and phospholipids, it is now apparent that methylation of proteins and nucleic acids is equally significant as a mechanism for regulating the biochemical activity of these macromolecules. Regardless of the type of substrate, however, one of the most important and unifying features of virtually all AdoMet-dependent methyltransferases studied to date is that they follow a reaction scheme (Figure 1) which results in formation of the product, S-adenosylhomocysteine (AdoHcy). Owing to the fact that AdoHcy is a potent competitive inhibitor of these AdoMet-dependent methyltransferases, the rate of cellular methylation is regulated by the existing intracellular ratio of AdoHcy/AdoMet (Cantoni and Chiang, 1980; Chiang and Cantoni, 1979). Consequently, AdoHcy must be continuously degraded or eliminated in order to maintain some potential for methylation to proceed.

In eukaryotic cells, the only route for metabolism of AdoHcy is via hydrolysis by AdoHcy hydrolase (EC 3.3.1.1) to adenosine and homocysteine (de la Haba and Cantoni, 1959). This enzyme has more recently been identified in some prokaryotic cell lines as well (Shimizu et al., 1984). In either case, the hydrolysis of AdoHcy by this enzyme is a reversible reaction (Figure 1), with the equilibrium lying strongly in favor of the synthetic direction. In vivo, however, the rapid metabolic elimination of both adenosine and homocysteine drive the reaction in the hydrolytic direction, maintaining relatively low intracellular levels of AdoHcy and permitting methylation to continue at the required rate.

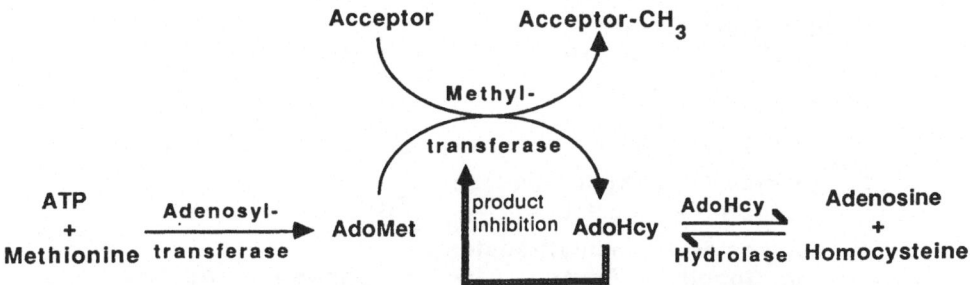

FIGURE 1. *General Reaction Scheme for AdoMet-dependent Transmethylation.*

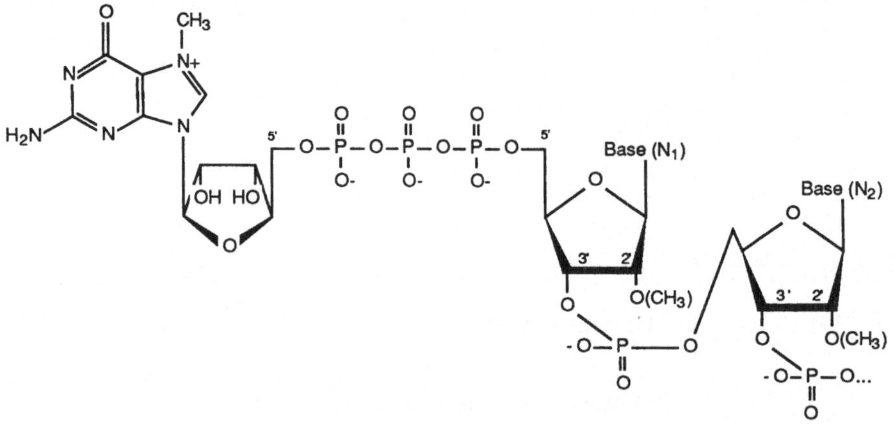

FIGURE 2. Structure of mRNA 5'-terminal Cap.

One specific biological process in which AdoMet-dependent transmethylation is known to play an important role is in the capping of eukaryotic and viral mRNAs. Since the presence of such blocked and methylated cap structures on the 5'-terminus of mRNA was first reported in 1974 (Reddy et al., 1974; Perry and Kelly, 1974), much has been learned about their structure, mechanism of synthesis and functional role (Banerjee, 1980). Capping has been shown to protect mRNA against nuclease digestion from the 5'-end of the molecule, thereby enhancing its stability in the cytoplasm (Furuichi et al., 1977). In addition, methylation of the blocking nucleoside (below) increases the affinity for ribosome binding to the 5'-end of the mRNA during formation of the translational initiation complex (Both et al., 1975b). As a result, translational efficiency of capped mRNAs is significantly increased over that of uncapped mRNAs (Both et al., 1975a). Because capping is observed to occur during the early stages of precursor heterogeneous nuclear RNA (hnRNA) synthesis, it has also been proposed that these structures are intimately involved in, and perhaps required for, biosynthesis of complete precursor transcripts (Darnell, 1979). In several studies of viral transcriptional systems, there appears to be a direct relationship between the presence of 5'-cap structures and completed viral mRNA transcripts and thus, viral replication (Banerjee et al., 1977; Krug et al., 1979).

The general structure of the 5'-cap, as illustrated in Figure 2, includes a blocking guanosine nucleoside bound to the penultimate residue of the mRNA through a unique, 5'-5' inverted linkage. While the capping moiety is, with few exceptions, 7-methylguanosine, the penultimate (N1) and adjacent nucleoside (N2) can be either purine or pyrimidine bases. Moreover, the methylation of these latter two nucleosides can vary giving rise to three distinct cap structures: ^{7}mG(5')ppp(5')N1pN2p...(cap 0), ^{7}mG(5')ppp(5')N1^{m}pN2p... (cap 1), and ^{7}mG(5')ppp(5')N1^{m}pN2^{m}p...(cap 2). At least three specific enzymes are responsible for the biosynthesis of capped and fully methylated mRNAs (Figure 3). mRNA guanylyltransferase, in the presence of GTP, catalyzes the first step of these reactions in which the capping nucleoside is attached to a 5'-nucleotide chain

1. **RNA Polymerase:**
   ATP + CTP + GTP + UTP ---► pppN- + PPi

2. **guanylyltransferase:**

   $\overset{\gamma\,\beta\,\alpha}{pppG}$ + $\overset{\gamma\,\beta\,\alpha}{pppN\text{-}}$ ----------► $\overset{\alpha\,\beta\,\alpha}{GpppN\text{-}}$ + $\overset{\gamma}{Pi}$ + $\overset{\gamma\,\beta}{PPi}$

3. **guanine-7 methyltransferase:**
   GpppN- + AdoMet -------► m^{7}GpppN-´ + AdoHcy

4. **nucleoside-2' methyltransferase:**
   m^{7}GpppN- + AdoMet ----► m^{7}GpppNm- + AdoHcy

FIGURE 3. Enzymes Involved in the Biosynthesis of mRNA Cap Structures.

(primer) through the 5'-5' inverted linkage (Figure 2). Subsequent methylation of this capped structure is accomplished by two specific AdoMet-dependent methyltransferases. The 7 position of the capping guanosine residue serves as a substrate for mRNA (guanine-7-)methyltransferase and, as noted, this position is virtually always observed to be methylated in capped mRNAs *in vivo*. In part this is likely due to the fact that methylation of the 2'-O-ribose moieties in nucleotides N1 and N2 by (nucleoside-2')methyltransferase (to generate cap 1 or cap 2 structures) is strictly dependent on prior methylation of the 7 position of the 5'-terminal guanosine residue.

Since capping and methylation of mRNAs were first identified, it has become apparent that cap structures are present on virtually all eukaryotic mRNA. Similarly, mRNAs from most plant and animal viruses have been shown to contain caps (see Banerjee, 1980 for review). These encompass a broad spectrum of viruses including double-stranded DNA viruses (poxviruses, e.g. vaccinia), double-stranded ($\pm$) RNA viruses (reoviruses, e.g. rotavirus; orbivirus, e.g. cytoplasmic polyhedrosis virus) and both single-stranded (-) RNA viruses (rhabdovirus, e.g. vesicular stomatitis; paramyxovirus, e.g. influenza, Newcastle disease) and single-stranded (+) RNA viruses (togavirus, e.g. Sindbis; oncovirus, e.g. Rous sarcoma). Some representative examples of viral cap structures are presented in Figure 4. Only picornavirus (e.g. polio) and the plant viruses cowpea mosaic virus (CMV), satellite tobacco nerosis virus and its helper virus have been found to lack cap structures. However, in place of a cap the first two of this group (polio and CMV) have a protein covalently linked to the 5'-end of their RNA genome. With the exception of these few uncapped mRNAs, viral cap structures contain exclusively m^7G as the capping base. Moreover, only the mRNA strand of the viral genome (i.e., the positive strand) is capped implying that this modification is unique to mRNAs. The 5'-penultimate base of viral messages is always a purine which may be methylated, further suggesting that initiation of RNA synthesis or processing is purine-specific. Also, when the penultimate nucleoside is Ado it is usually found to be dimethylated (N^6mA^m). While plant viruses usually contain cap 0 structures, these are rarely found in animal viruses which generally exhibit cap 1 (*in vitro*) or cap 2 (*in vivo*) structures.

In light of the evidence that capped and methylated viral mRNAs are required for efficient translation into proteins which are essential for viral replication, these post-transcriptional modifications have become an attractive target for the design of antiviral agents. Based on the depth of knowledge currently available with respect to the regulation of AdoMet-dependent transmethylation in general and our present understanding of the virus-specific methyltransferases responsible for carrying out mRNA cap methylations, our laboratory, along with several others, has focused on this step of viral mRNA maturation as one approach to the design of antiviral agents. In the present paper, the recent progress that has been made in developing inhibitors of AdoMet-dependent trans-methylation as antiviral agents is considered.

Class (Type)	Virion	In vitro mRNA	Cytoplasmic mRNA
Poxvirus (ds DNA) e.g., vaccinia	(not applicable)	$^7mGpppA^m$ p $^7mGpppG^m$ p	$^7mGpppm^6A^m$ p $^7mGpppG^m$ p
Reovirus ($\pm$ dsRNA) e.g., reovirus	$^7mGpppG^m$ (+) ppG (-)	$^7mGpppG^m$ p	$^7mGpppG^m$ pC^m p
Rhabdovirus (- ssRNA) e.g., vesicular stomatitis	(p)ppA	$^7mGpppA^m$ p	$^7mGpppm^6A^m$ pA^m p
Paramyxovirus (- ssRNA) e.g., parainfluenza e.g., Newcastle disease	pppA (undetermined)	No cap $^7mGpppNp$	$\Big\langle$ $^7mGpppm^6A^m$ p $^7mGpppG^m$ p $^7mGpppNp$
Orbivirus ($\pm$ dsRNA) e.g., cytoplasmic polyhedrosis	$^7mGpppA^m$ p (+) ppA (-)	$^7mGpppA^m$ p	(undetermined)

[Reference: A. K. Banerjee, *Microbiological Reviews*, 44(2): 175-205, (1980)]

FIGURE 4.   *Examples of Viruses Which Contain 5'-terminal Caps.*

Regarding the reaction scheme presented in Figure 1, three distinct approaches to designing inhibitors of AdoMet-dependent methylation have been investigated based on the three types of enzymes involved (Borchardt, 1980). One approach has been to attempt to decrease the intracellular levels of the substrate, AdoMet, by altering the activity of adenosyltransferase. Although a few potent L-methionine analog inhibitors of this enzyme have been shown to decrease AdoMet in several tissues following administration, most have demonstrated poor inhibitory activity due to the high structural specificity of adenosyl-transferase (Lombardini and Talalay, 1973; Sufrin et al., 1979). In addition, another major disadvantage of this approach is that altering AdoMet levels will result in a general inhibitory effect on all AdoMet-dependent methyltransferases as well as a direct effect on the production of decarboxylated AdoMet and thus, on the biosynthesis of polyamines.

As a second approach, a substantial amount of effort has also been exerted toward designing inhibitors of the various AdoMet-dependent methyltransferases, themselves. Numerous structural analogs of both AdoMet and AdoHcy have been prepared and examined with the aim of increasing their specificity and eliminating potential metabolic side effects (Zappia et al., 1969; Schlenk and Dainko, 1974; Borchardt et al., 1976; Schlenk et al., 1978). In general, AdoMet analogs have exhibited poor substrate and/or inhibitor properties, similarly reflecting the high degree of structural specificity required by the methyltransferase enzymes. The relative instability of many of these compounds has also been an important concern. Two AdoMet analogs, however, Sinefungin and A9145C which are antifungal antibiotics, were found to elicit potent inhibitory activity toward a variety of methyltransferases (Vedel et al., 1978; Fuller and Nagarajan, 1978; Borchardt et al., 1979), including the virion mRNA(guanine-7-)methyltransferase from both vaccinia and Newcastle disease viruses and the mRNA(nucleoside-2')-methyltransferase from vaccinia virus (Pugh et al., 1978). Although these compounds were also observed to inhibit vaccinia virus plaque formation in monolayer cultures of mouse L929 cells, relatively high concentrations were required to produce these effects (Borchardt and Pugh, 1979) reflecting their low cellular permeability. This appears to be a major problem limiting the *in vivo* efficacy of AdoMet analogs which exhibit good *in vitro* activity.

Similarly, systematic analysis of base, amino acid or sugar modified analogs of AdoHcy on a variety of AdoMet-dependent methyltransferases has been carried out in an attempt to determine the structure-activity relationships for binding of specific inhibitors to particular methyltransferases. Using purified viral mRNA methyltransferases, this approach has delineated significant differences in the structural requirements not only between Newcastle disease (NDV) and vaccinia virus mRNA(guanine-7-)methyltrans-ferase, but also between this enzyme and the accompanying mRNA(nucleoside-2')-methyltransferase within the vaccinia virion itself (Borchardt and Pugh, 1979), suggesting a possibility for the design of site specific agents. For example, the base modified analogs S-tubercidinyl-L-homocysteine (TubHcy) and 8-aza-AdoHcy were found to have potent inhibitory activity toward NDV mRNA(guanine-7-)methyltransferase but were essentially inactive toward the vaccinia virion enzyme, although TubHcy did exhibit potent activity toward the vaccinia mRNA(nucleoside-2')methyltransferase. 3-Deaza-AdoHcy, however, was more selective for both the vaccinia enzymes as compared to that from NDV. In contrast to TubHcy which, like AdoHcy, has been observed to be a more general inhibitor of AdoMet-dependent methylation, other AdoHcy analogs such as AdoHcy sulfone, AdoHcy sulfoxide and N^6-methyl AdoHcy are reported to be more selective inhibitors for virion mRNA(guanine-7-)methyltransferases (Pugh et al., 1977). Despite the promising activity that some of these AdoHcy analogs have exhibited *in vitro*, few have been found to elicit significant *in vivo* activity. Those with most notable activity are TubHcy which has been shown to inhibit internal N^6-adenosine methylation in mRNA (as high as 80% at 500 µM) in phytohemagglutinin-stimulated rat lymphocytes, Novikoff heptoma cells and HeLa cells (Chang and Coward, 1975; Kaehler et al., 1979; Camper et al., 1984) and 5'-deoxy-5'-(isobutylthio)adenosine (SIBA) which has been observed to inhibit the replication of several viruses and virus-induced transformation of chick embryo cells at relatively high (500 µM) concentrations (Robert-Géro et al., 1975; Raies et al., 1976; Jacquemont and Huppert, 1977). Thus, like AdoMet analogs, the low cellular permeability of AdoHcy analogs appears to be a major factor limiting their *in vivo* effectiveness.

In recent years, investigators have recognized that a more promising approach to the development of methylation inhibitors as antiviral agents is to focus on AdoHcy hydrolase as a primary target (De Clercq, 1982,1985b). Being the key enzyme responsible for AdoHcy metabolism (Figure 1), inhibition of the hydrolase will result in a direct increase in intracellular levels of the endogenous product inhibitor of AdoMet-dependent methyltransferase reactions. Based on earlier reports that 5'-deoxy-5'-(isobutylthio)-adenosine elicited antiviral effects (Robert-Géro et al., 1975; Raies et al., 1976) apparently related to the inhibition of viral RNA methylation (Jacquemont and Huppert, 1977), Cantoni and his colleagues clearly established the efficacy of AdoHcy hydrolase as an antiviral target with their findings that two inhibitors of this enzyme, 3-deaza-adenosine and 5'-deoxy-5'-(isobutylthio)-3-deazaadenosine prevented the replication of Rous sarcoma and other viruses as well as viral transformation of chick embryo cells (Bader et al., 1978; Chiang et al., 1978; Bodner et al., 1981).

As a result of the increased understanding of the biochemical properties and mechanism of catalysis of AdoHcy hydrolase obtained since these initial studies, a wide range of adenosine analogs have been synthesized and evaluated as hydrolase inhibitors with antiviral activity. Among those currently observed to elicit significant activity are (S)-9-(2,3-dihydroxypropyl)adenine [(S)-DHPA] (De Clercq et al., 1978; De Clercq and Holý, 1979), (D)eritadenine (Holý et al., 1982; De Clercq et al., 1984), (R,S)-3-adenin-9-yl-2-hydroxypropanoic acid [(RS)-AHPA] alkyl esters (De Clercq and Holý, 1985), 3-deazaadenosine (Bader et al. ,1978; Bodner et al., 1981; De Clercq et al., 1984), aristeromycin (De Clercq et al., 1984), 3-deazaaristeromycin (Montgomery et al., 1982; De Clercq and Montgomery, 1983; De Clercq et al., 1984), 7-deaza-aristeromycin (De Clercq et al., 1984; Secrist et al., 1984), neplanocin A (Borchardt et al., 1984; De Clercq, 1985a), adenosine dialdehyde (Keller and Borchardt, 1987), 9-(trans-2', trans-3'-dihydroxycyclopent-4'-enyl)-adenine and 9-(trans-2', trans-3'-dihydroxycyclopent-4'-enyl)-3-deazaadenine (Borcherding et al., 1987; Hasobe et al., 1987a,b) (Figure 5). 3-Deazaneplanocin A has recently been synthesized and shown to be another potent inhibitor of AdoHcy hydrolase (Glazer et al., 1986 a,b), although the antiviral properties of this compound have yet to be reported.

A striking feature of all these AdoHcy hydrolase inhibitors is the similarity in their broad-spectrum antiviral activity indicating a common mechanism of action. These compounds are observed to exhibit more selective activity toward DNA poxviruses (i.e., vaccinia), double-stranded (±) RNA viruses (i.e., reo- and rotavirus) and single-stranded (-) RNA viruses (i.e., vesicular stomatitis, rabies, parainfluenza and measles). Based on the knowledge that these viruses not only require methylated 5'-cap structures on their mRNA but also contain virus-specific AdoMet-dependent mRNA methyltransferases associated with their virion, it is presumed that the mechanism of antiviral action of AdoHcy hydrolase inhibitors involves an inhibition of viral mRNA cap methylation. As a result, the translation of viral proteins would also be inhibited thereby suppressing viral replication. (It should be noted that inhibition of mRNA cap methylation has similarly been implicated in the antiviral action of interferon on vesicular stomatitis virus-infected HeLa cells (de Ferra and Baglioni, 1981,1983)). In support of such a mechanism, DeClercq and Cools (1985) have demonstrated that a direct correlation exists between the antiviral potency of several of these adenosine analogs and their relative inhibitory effects on AdoHcy hydrolase. Further evidence has been provided by work from our laboratory with adenosine dialdehyde (Keller and Borchardt, 1987). We have shown that, as an effective inhibitor of vaccinia virus replication in mouse L929 cells, this compound is also a potent inhibitor of the cellular AdoHcy hydrolase resulting in a 10-fold increase in the intracellular ratio of AdoHcy/AdoMet. Moreover, the dialdehyde was found to be most effective when administered within the first eight hours after infection (Figure 6) implying that early events in viral replication are the site of action. These findings are in agreement with the observation that the synthesis of early virus-specific proteins (i.e., incorporation of [^{35}S]methionine) is dramatically inhibited by a 12 hour exposure to the dialdehyde (to elevate AdoHcy/AdoMet) prior to infection (Figure 7). It is interesting that this treatment was not observed to inhibit cellular protein synthesis nor did it block the characteristic shut-down in host translation related to the vaccinia infection (Rice and Roberts, 1983). In addition, we also determined that the incorporation of [^{3}H]-methyl groups into RNA isolated from vaccinia virus-infected, dialdehyde-

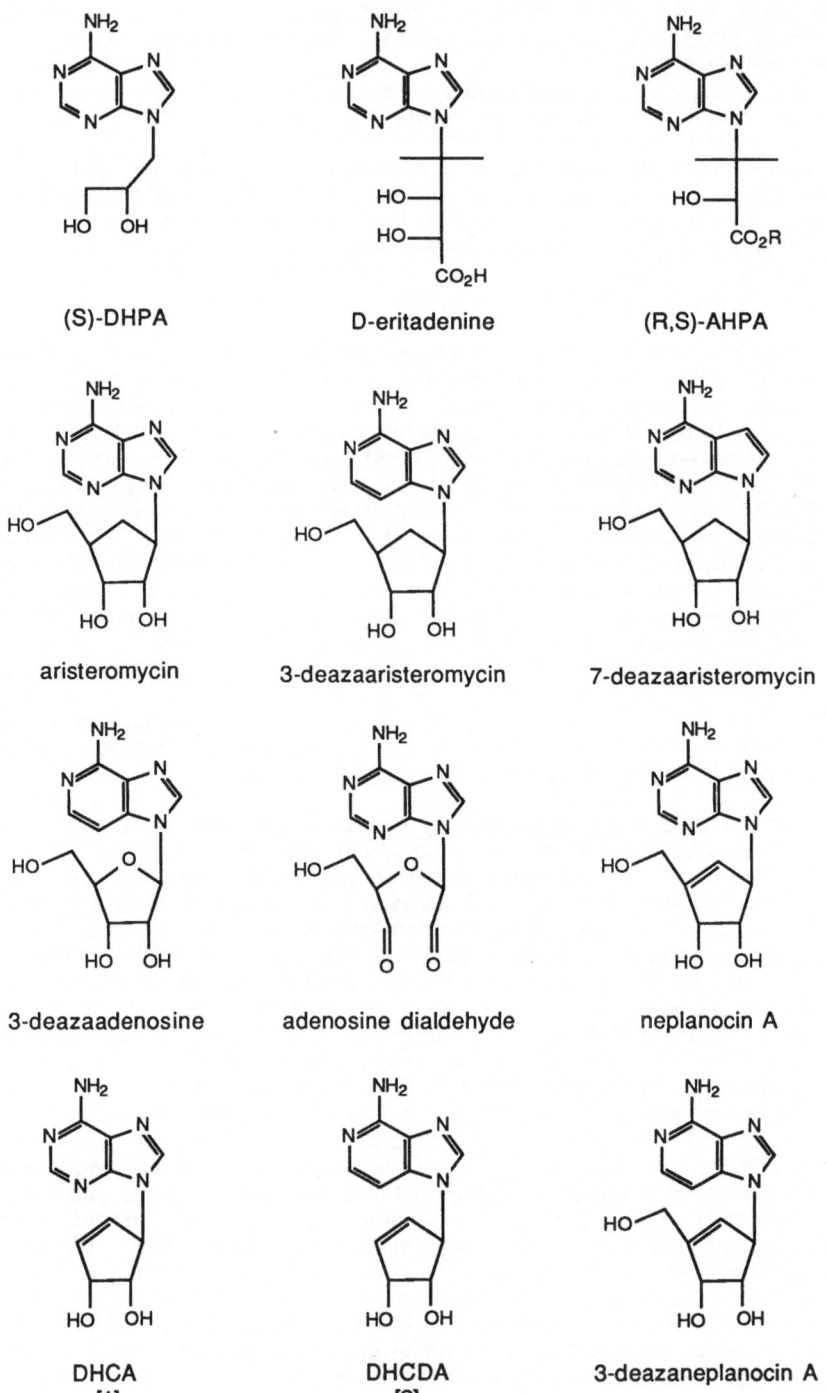

FIGURE 5. Structures of AdoHcy Hydrolase Inhibitors. *(S)-DHPA, (S)-9-*
(2,3-dihydroxypropyl)adenine; (R,S)-AHPA, (R,S)-3-adenin-9-yl-2-hydroxypropanoic acid
(alkyl ester); DHCA [1], 9-(trans-2', trans-3'-dihydroxycyclopent-4'-enyl)-adenine; DHCDA
[2], 9-(trans-2', trans-3'-dihydroxycyclopent-4'-enyl)-3-deazaadenine.

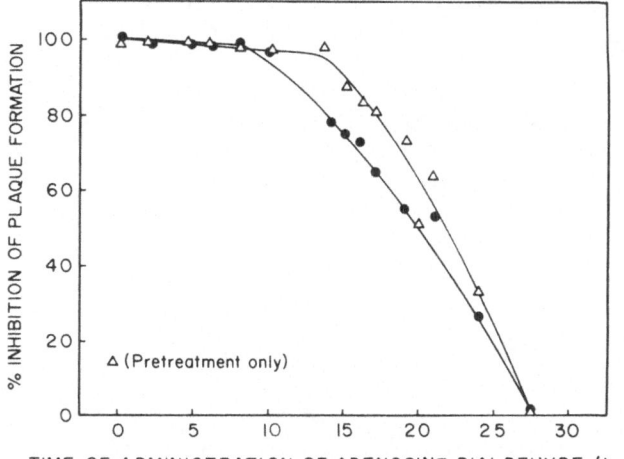

FIGURE 6. **The Effect of Time of Administration of Adenosine Dialdehyde on Vaccinia Plaque Formation.** *Monolayer cultures of mouse L929 cells were infected with vaccinia virus for 60 min then refed with fresh growth medium. At the indicated times thereafter, the medium was also supplemented with 5 µM adenosine dialdehyde. After a total of 72 hrs incubation (37°) post infection, the dishes were stained and the viral plaques counted. Untreated controls averaged 220 plaques/dish. △, cultures were pretreated for 60 min prior to infection with 5 µM adenosine dialdehyde;   non-pretreated cultures.*

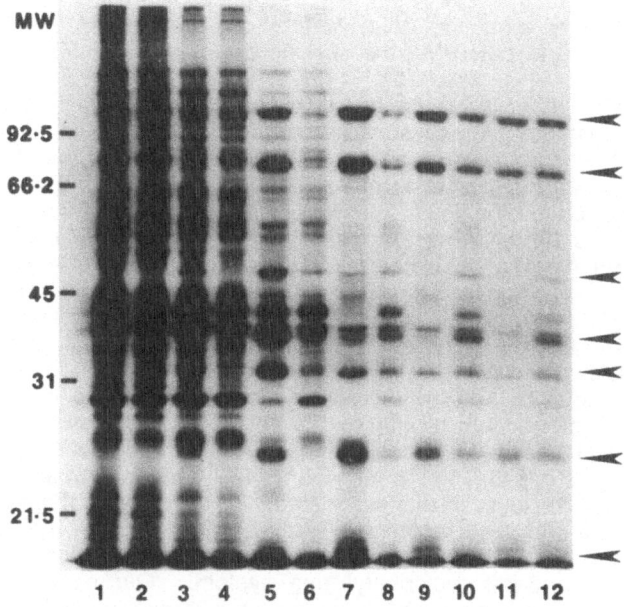

FIGURE 7. **Autoradiogram of [^{35}S]-labeled Proteins from Vaccinia Virus-infected Mouse L929 Cells Treated With or Without Adenosine Dialdehyde.** *Following a 12 hr treatment with or without 5 µM adenosine dialdehyde, monolayer cultures were infected with virus (30 pfu/cell) for 45 min then pulse-labeled for 15 min with 25 µCi of [^{35}S]methionine at the indicated times thereafter. The cells were solubilized and the proteins resolved by SDS-polyacrylamide gel electrophoresis and autoradiograpy. Time of labeling: lanes 1,2 - 0 hr; lanes 3,4 - 1 hr; lanes 5,6 - 2 hr; lanes 7,8 - 4 hr; lanes 9,10 - 6 hr; lanes 11,12 - 8 hr. Lanes 1,3,5,7,9,11 - untreated controls; lanes 2,4,6,8,10,12 - treated with 5 µM adenosine dialdehyde. (◄) indicates viral protein bands.*

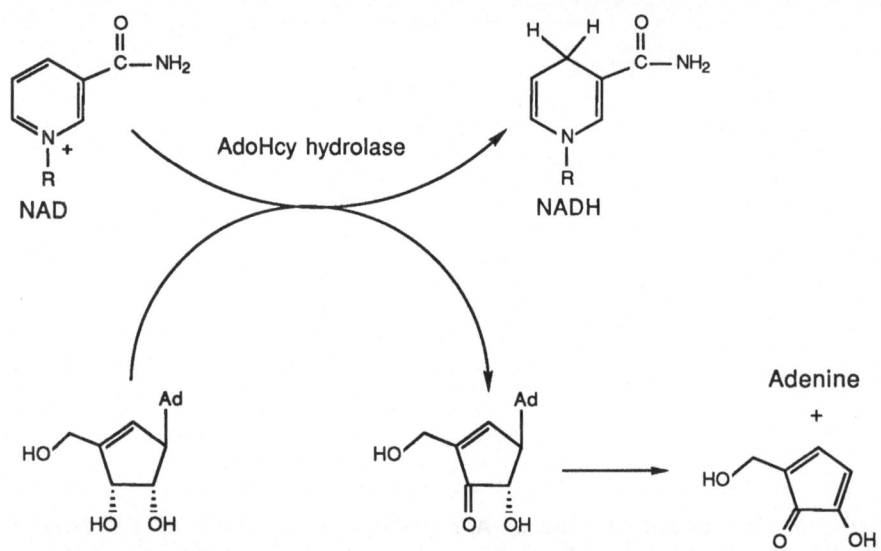

FIGURE 8. *The Role of Enzyme-bound NAD⁺ in the Inactivation of AdoHcy Hydrolase by Neplanocin A.*

treated cells was inhibited by approximately 30% in the cytoplasmic fraction and approximately 15% in the poly A⁺-mRNA fraction as compared to untreated controls providing preliminary evidence of an effect on viral mRNA methylation. Work is currently in progress to determine the specific site(s) of this inhibition.

Current studies in our laboratory have focused on both the antiviral and the metabolic properties of neplanocin A and some newly synthesized analogs of this carbocyclic nucleoside. Originally isolated as an antitumor antibiotic from the microorganism *Ampullariella regularis* (Hayashi et al., 1981; Yaginuma et al., 1980,1981), neplanocin A has been observed to be a potent inhibitor of AdoHcy hydrolase from both eukaryotic and prokaryotic sources (Borchardt et al., 1984; Matuszewska and Borchardt, 1987a). This inhibition is both time- and concentration-dependent, yielding a K_i of 4.0 nM for the purified bovine liver enzyme. In addition, the stoichiometry of the reaction between neplanocin A and AdoHcy hydrolase has been determined to be one molecule of inhibitor per molecule (tetramer) of the enzyme. The catalytic mechanism of AdoHcy hydrolase, as described by Palmer and Abeles (1976,1979), involves an NAD⁺- dependent oxidation of the 3'-hydroxyl group of AdoHcy (hydrolytic direction) resulting in the generation of a series of 3'-keto-intermediates prior to the release of adenosine and homocysteine. Inhibition of the hydrolase by neplanocin A has been shown to be a K_{cat} type of mechanism in which the enzyme-bound NAD⁺ is reduced to NADH (Wolfson et al., 1986; Matuszewska and Borchardt, 1987b), but remains associated with the enzyme (Figure 8). Although the release of adenine was also reported by Wolfson et al. (1986), it was unclear whether this was related to inactivation of the enzyme or rather a consequence of the formation of a chemically unstable 3-ketocyclopentyl ring system. Our observation that *in vitro* enzymatic activity can be recovered by incubation with excess NAD⁺ (Matuszewska and Borchardt, 1987b) implies that it is the reduction of NAD⁺ to NADH which is critical to the loss of hydrolase activity.

Neplanocin A, as mentioned above, has been shown to exhibit potent antiviral activity at concentrations of 1 μM and lower (Borchardt et al., 1984; De Clercq, 1985) which appears to be related to its inhibition of cellular AdoHcy hydrolase. Studies similar to those described for adenosine dialdehyde have confirmed that this compound likewise inhibits early virus-specific protein synthesis and methylation of poly A⁺-mRNA isolated from vaccinia virus-infected L929 cells (Keller and Borchardt, 1986). Based on the promising results obtained with neplanocin A, we have also investigated the antiviral activity and the inhibition of AdoHcy hydrolase by a series of several synthetic analogs of

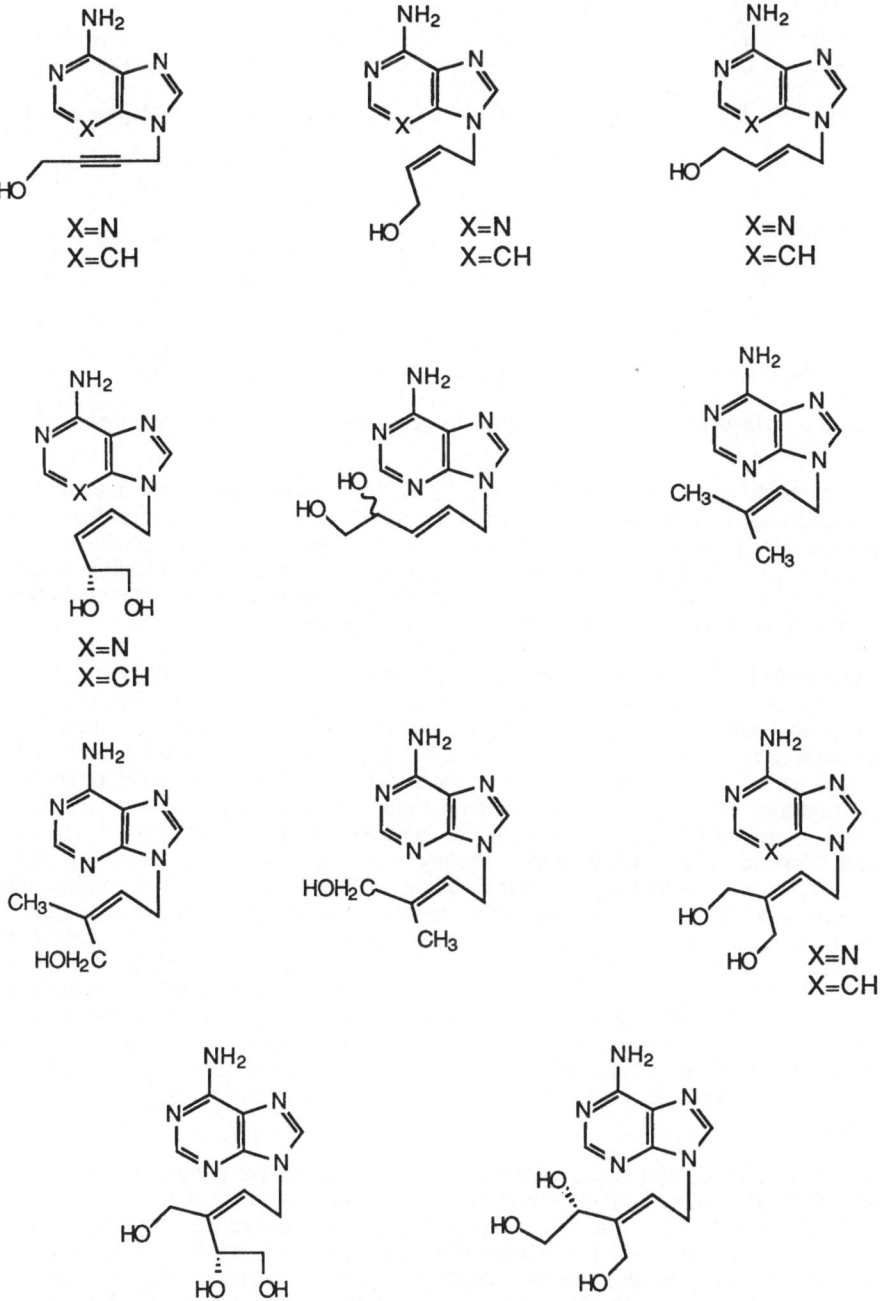

FIGURE 9. *Acyclic Derivatives of Neplanocin A.*

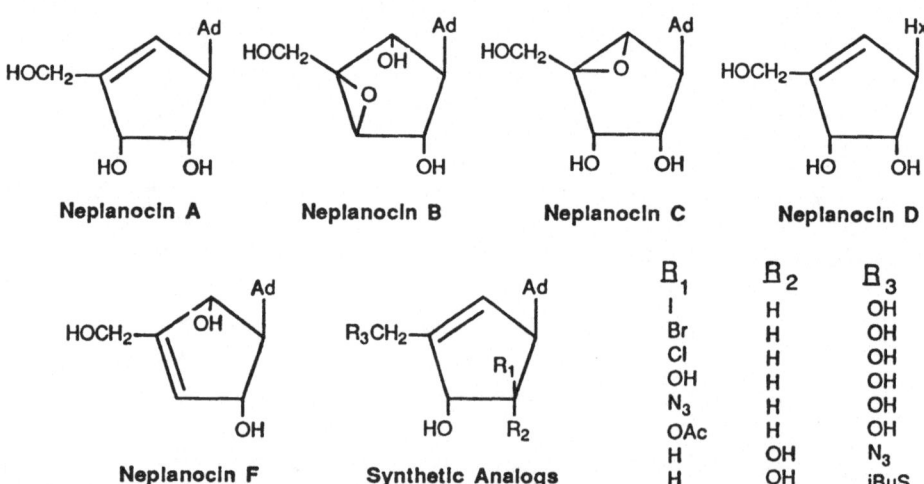

FIGURE 10. *Naturally Occurring and Synthetic Analogs of Neplanocin A.*

neplanocin A (both carbocyclic and acyclic) as well as some other naturally occurring analogs of neplanocin A (Figures 9,10; Keller and Borchardt, 1986; Borcherding, 1987). These compounds, however, were observed to have substantially less activity as either antiviral agents or inhibitors of the hydrolase, consistent with previous studies on some of these compounds (De Clercq, 1985) and further supporting the correlation between these two activities reported by De Clercq and Cools (1985).

Neplanocin A has additionally been shown to inhibit AdoHcy hydrolase in variety of cultured cells lines (Borchardt et al., 1984; Aarbakke et al., 1985; Keller and Borchardt, 1986; Whaun et al., 1986; Ramakrishnan and Borchardt, 1987) resulting in an accumulation of AdoHcy and a dramatic increase (10-12-fold) in the intracellular ratio of AdoHcy/AdoMet. These changes are subsequently reflected in a decrease in AdoMet-dependent methylation in the treated cells. Neplanocin A, however, exhibits multifunctional activity (Figure 11) since it can also serve as a substrate for adenosine deaminase (Tsujino et al., 1980) which rapidly converts it to the biologcially inactive neplanocin D, and for adenosine kinase leading to the formation of the triphosphate derivative which is subsequently utilized as a substrate for AdoMet synthetase to generate the corresponding AdoMet derivative, S-neplanocylmethionine (NpcMet) (Keller and Borchardt, 1984; Glazer and Knode, 1984; Keller et al., 1985; Saunders et al., 1985; Inaba et al., 1986). Although deamination appears to be of little significance in studies with cultured cells, (i.e., inhibitors of adenosine deaminase do not enhance the activity of neplanocin A), the latter metabolic route has been proposed to be responsible for the observed suppression of cellular RNA synthesis and the significant cytotoxic action of neplanocin A. The cytotoxicity of neplanocin A is the major factor limiting its utility as an antiviral agent.

In an attempt to design more specific monofuntional antiviral agents (i.e., AdoHcy hydrolase inhibitors) with minimal other metabolic effects (i.e., adenosine deaminase, adenosine kinase) two analogs, 9-(trans-2'trans-3'-dihydroxycyclopent-4'enyl)-adenine (DHCA [1]) and -3-deazaadenine (DHCDA [2]) have been synthesized in our laboratory (Borcherding et al., 1987a) which lack the 4' hydroxymethyl group of the corresponding parent molecules (Figure 5). These analogs, like neplanocin A, show time- and concentration-dependent inhibition of AdoHcy hydrolase (Borcherding et al., 1987b) while lacking substrate activity with either the deaminase or the kinase, confirming their monofunctional design (Figure 12). In mouse L929 cells, the concentrations required to give 95% inhibition of AdoHcy hydrolase activity were 0.2 μM, 0.5 μM and 0.5 μM for neplanocin A, 1 and 2, repectively (Hasobe et al.,1987b). Moreover, these analogs exhibited inhibition of cellular AdoHcy hydrolase which persisted throughout 72 hours whereas that of neplanocin A began to recover after 48 hours. Similarly, the elevation of

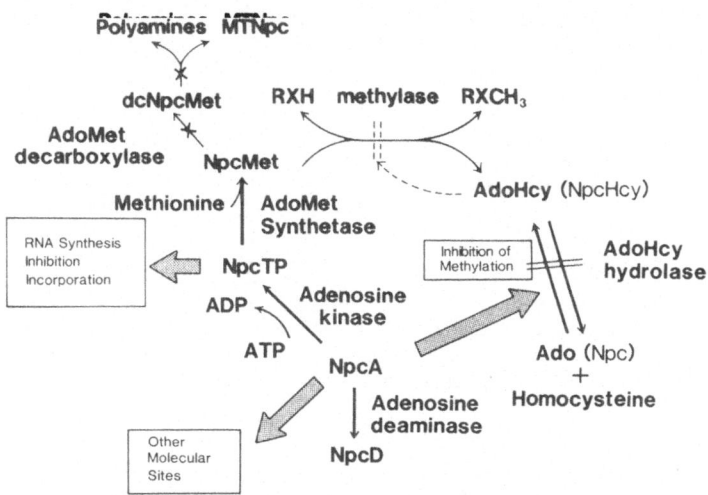

FIGURE 11. **Multifunctional Metabolic Potential of Neplanocin A.** *NpcA, neplanocin A; NpcD, neplanocin D; NpcTP, neplanocin A triphosphate; NpcMet, S-neplanocyl-methionine; dcNpcMet, decarboxylated neplanocylmethionine; MTNpc, methylthioneplanocin; NpcHcy, neplanocylhomocysteine; Ado, adenosine; RXH, methyl acceptor molecule; RXCH₃, methylated acceptor molecule.*

intracellular levels of AdoHcy by 1 and 2 was maintained throughout 72 hours in comparison to that of neplanocin A-treated cells which had decreased to near control levels by this time. These compounds were also observed to have no effect on either cellular RNA or DNA synthesis at concentrations up to 10 μM.

Both compounds 1 and 2 were determined to have potent antiviral activity against vaccinia virus-infected L929 cells, exhibiting IC_{50} values (concentration required to produce 50% inhibition of plaque formation) of 0.28 μM and 0.95 μM, respectively, as compared to 0.08 μM for neplanocin A (Hasobe et al., 1987a). More striking, however, was the reduced cytotoxicity observed with these analogs: ID_{50} (concentration required to produce 50% inhibition of cell growth) was determined to be 0.5 μM, 17 μM and 56 μM, respectively, for neplanocin A, 1 and 2. As a consequence of the reduced cytotoxicity of compounds 1 and 2, their antiviral effectiveness (ID_{50}/IC_{50}), 61 and 59, respectively, was found to be significantly higher than that of neplanocin A (i.e., 6).

In summary, significant progress has been made in the development of inhibitors of AdoMet-dependent transmethylation as antiviral agents. The most successful approach to date has been to utilize AdoHcy hydrolase as the primary target for the design of new inhibitors, with the objective of elevating intracellular levels of AdoHcy. Although this product of transmethylation reactions is recognized as a universal inhibitor of AdoMet-dependent methyltransferases, our recent studies indicating that the intracellular ratio of AdoHcy/AdoMet required for antiviral activity is significantly lower than that which elicits cytotoxic effects suggests that the viral methyltransferases may be inherently more sensitive to changes in this ratio than the cellular enzymes. Presently, the design of such inhibitors has been directed at developing more specific monofunctional (i.e., AdoHcy hydrolase-specific) agents which are devoid of other metabolic activities (e.g., adenosine deaminase, adenosine kinase) in an attempt to increase their stability and prolong their effects, thereby lowering the concentrations required to produce and main-

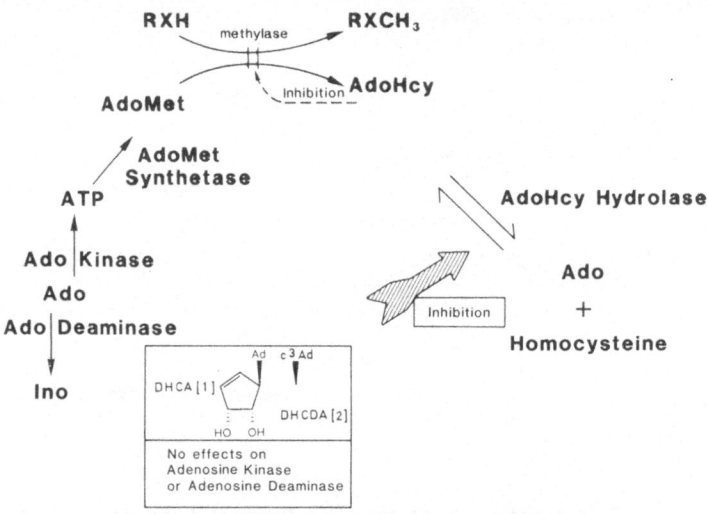

FIGURE 12. *Monofunctional Metabolic Potential of New Synthetic Neplanocin A Analogs.* Ado, adenosine; Ino, inosine; RXH, methyl acceptor molecule; RXCH₃, methylated acceptor molecule.

tain effective antiviral activity. Based on the results obtained to date, the future prospects of this approach to developing antiviral agents appear to hold substantial promise.

Acknowledgements - This work was supported by United States Public Health Service Grant (GM-29332). Neplanocin A and its natural and synthetic analogs were kindly provided by the Toyo Jozo Co., Ltd., Japan. Compounds 1 and 2 and the acyclic neplanocin analogs and were synthesized by Dr. David R. Borcherding and Steven A. Scholtz.

References

Aarbakke, J., Gordon, R.K., Cross, A.S., Miura, G.A. and Chiang, P.K., 1985, Correlation between DNA hypomethylation and differentiation of HL-60 promyelocyte cells induced by neplanocin A and 3-deaza nucleosides., Fed. Proc., 44:464, Abstract #313.

Bader, J.P., Brown, N.R., Chiang, P.K. and Cantoni, G.L., 1978, 3-Deazaadenosine, an inhibitor of adenosylhomocysteine hydrolase, inhibits reproduction of Rous sarcoma virus and transformation of chick embryo cells. Virology, 89:495-505.

Banerjee, A.K., 1980, 5'-terminal cap structure in eukaryotic messenger ribonucleic acid., Microbiol. Rev., 44:175-205.

Banerjee, A.K., Abraham, G. and Colonno, R.J., 1977, Vesicular stomatitis virus: mode of transcription. J. Gen. Virol., 34:1-8.

Bodner, A.J., Cantoni, G.L. and Chiang, P.K., 1981, Antiviral activity of 3-deazaadenosine and 5'-deoxy-5'-isobutylthio-3-deazaadenosine (3-deaza-SIBA)., Biochem. Biophys. Res. Commun., 98:476-481.

Borchardt. R.T., 1980, S-Adenosyl-L-methionine-dependent macromolecular methyltransferases: Potential targets for the design of chemotherapeutic agents., J. Med. Chem., 23:347-357.

Borchardt, R.T. and Pugh, C.S.G., 1979, Analogues of S-adenosyl-L-homocysteine as inhibitors of viral mRNA methyltransferases., in: "Transmethylation," Usdin, E., Borchardt, R.T. and Creveling, C.R., eds., pp.197-206, Elsevier/North-Holland, New York, Amsterdam, Oxford.

Borchardt, R.T., Wu, Y.S., Huber, J.A. and Wycpalek, A.F., 1976, Potential inhibitors of S-adenosylmethionine-dependent methyltransferases. 6. Structural modifications of S-adenosylmethionine. J. Med. Chem., 19:1104-1110.

Borchardt, R.T., Eiden, L.E., Wu, B.S. and Rutledge, C.O., 1979, Sinefungin, a potent inhibitor of S-adenosylmethionine:protein O-methyltransferase., Biochem. Biophys. Res. Commun., 89:919-924.

Borchardt, R.T., Keller, B.T. and Patel-Thombre, U., 1984, Neplanocin A. A potent inhibitor of S-adenosylhomocysteine hydrolase and of vaccinia virus multiplication in mouse L929 cells., J. Biol. Chem., 259:4353-4358.

Borchardt, R.T., Creveling, C.R. and Ueland, P.M. (eds.), 1986, "Biological Methylation and Drug Design," The Humana Press, Clifton, NJ.

Borcherding, D.R., 1987, The Synthesis and Biological Activity of Acyclic and Carbocyclic Analogs of Neplanocin A., Ph.D. Thesis, The University of Kansas.

Borcherding, D.R., Scholtz, S.A. and Borchardt, R.T., 1987a, The synthesis of analogs of neplanocin A: Utilization of optically active dihydroxycyclopentenones derived from carbohydrates., J. Amer. Chem. Soc., submitted.

Borcherding, D.R., Scholtz, S.A., Narayanan, S.R., Keller, B.T. and Borchardt, R.T., 1987b, 9-(trans-2', trans-3'-dihydroxycyclopent-4'-enyl)-adenine and -3-deaza-adenine: potent inhibitors of bovine liver S-adenosylhomocysteine hydrolase. J. Biol. Chem., submitted.

Both, G.W., Banerjee, A.K. and Shatkin, A.J., 1975a, Methylation-dependent translation of viral messenger RNAs in vitro., Proc. Natl. Acad. Sci. U.S.A., 72:1189-1193.

Both, G.W., Furuichi, Y., Muthukrishnan, S. and Shatkin, A.J., 1975b, Ribosome binding to reovirus mRNA in protein synthesis requires 5'-terminal 7-methylguanosine., Cell, 6:185-195.

Camper, S.A., Albers, R.J., Coward, J.K. and Rottman, F.M., 1984, Effect of undermethylation of mRNA on cytoplasmic appearance and half-life., Mol. Cell. Biol., 4:538-543.

Cantoni, G.L. and Chiang, P.K., 1980, The role of S-adenosylhomocysteine and S-adenosylhomocysteine hydrolase in the control of biological methylations., in: "Natural Sulfur Compounds: Novel Biochemical and Structural Aspects," Cavallini, D., Gaull, G.E. and Zappia, V., eds., pp. 67-80, Plenum Publishing Corp., New York and London.

Chiang, P.K. and Cantoni, G.L., 1979, Perturbation of biochemical transmethylation by 3-deazaadenosine in vivo., Biochem. Pharmacol., 28:1897-1902.

Chang, C. D. and Coward, J.K., 1975, Effect of S-adenosylhomocysteine and S-tubercidinylhomocysteine on transfer ribonucleic acid methylation in phytohemagglutinin-stimulated lymphocytes. Mol. Pharmacol., 11:701-707.

Chiang, P.K., Cantoni, G.L., Bader, J.P., Shannon, W.M., Thomas, H.J. and Montgomery, J.A., 1978, Adenosylhomocysteine hydrolase inhibitors: synthesis of 5'-deoxy-5'-(isobutylthio)-3-deazaadenosine and its effect on Rous sarcoma virus and Gross murine leukemia virus., Biochem. Biophys. Res. Commun., 82:417-423.

Darnell, J.E., Jr., 1979, Transcription units for mRNA production in eukaryotic cells and their DNA viruses. Prog. Nucleic Acid Res. Mol. Biol., 22:327-353.

De Clercq, E., 1982, Specific targets for antiviral drugs., Biochem. J., 205:1-13.

De Clercq, E., 1985a, Antiviral and antimetabolic activities of neplanocins., Antimicrob. Agents Chemother. 28:84-89.

De Clercq, E., 1985b, Recent trends and development in antiviral chemotherapy., Antiviral Res., Suppl. 1:11-19.

De Clercq, E. and Cools, M., 1985, Antiviral potency of adenosine analogues: correlation with inhibition of S-adenosylhomocysteine hydrolase., Biochem. Biophys. Res. Commun., 129:306-311.

De Clercq, E. and Holý, A., 1979, Antiviral activity of aliphatic nucleoside analogues: structure-function relationship., J.Med Chem., 22:510-513.

De Clercq, E. and Holý, A., 1985, Alkyl esters of 3-adenin-9-yl-2-hydroxypropanoic acid: A new class of broad-spectrum antiviral agents., J. Med. Chem., 28:282-287.

De Clercq, E. and Montgomery, J.A., 1983, Broad-spectrum antiviral activity of the carbocyclic analog of 3-deazaadenosine., Antiviral Res., 3:17-24.

De Clercq, E., Descamps, J., De Somer, P. and Holy, A., 1978, (S)-9-(2,3-dihydroxy-propyl)adenine: an aliphatic nucleoside analog with broad spectrum antiviral activity., Science (Wash. D.C.), 200:563-565.

De Clercq, E., Bergstrom, D.E., Holý, A. and Montgomery, J.A., 1984, Broad-spectrum antiviral activity of adenosine analogues., Antiviral Res., 4:119-133.

de Ferra, F. and Baglioni, C., 1981, Viral messenger RNA unmethylated in the 5'-terminal guanosine in interferon-treated HeLa cells infected with vesicular stomatitis virus., Virology, 112:426-435.

de Ferra, F. and Baglioni, C., 1983, Increase in S-adenosylhomocysteine concentration in interferon-treated HeLa cells and inhibition of methylation of vesicular stomatitis virus mRNA., J. Biol. Chem., 258:2118-2121.

de la Haba, G. and Cantoni, G.L., 1959, The enzymatic synthesis of S-adenosyl-L-homo-cysteine from adenosine and homocysteine., J. Biol. Chem., 234:603-608.

Fuller, R.W. and Nagarajan, R., 1978, Inhibition of methyltransferases by some new analogs of S-adenosylhomocysteine. Biochem. Pharmacol., 27:1981-1983.

Furuichi, Y., LaFinandra, A. and Shatkin, A.J., 1977, 5'-terminal structure and mRNA stability. Nature (London), 266:235-239.

Glazer, R.I. and Knode, M.C., 1984, Neplanocin A. A cyclopentenyl analog of adenosine with specificity for inhibiting RNA methylation. J. Biol. Chem., 259:12964-12969.

Glazer, R.I., Hartman, K.D., Knode, M.C., Richard, M.M., Chiang, P.K., Tseng, C.K.H. and Marquez, V.E., 1986a, 3-Deazaneplanocin: a new potent inhibitor of S-adenosylhomo-cysteine hydrolase and its effects on human promyelocytic leukemia cell line HL-60., Biochem. Biophys. Res. Commun., 135:688-694.

Glazer, R.I., Knode, M.C., Tseng, C.K.H., Haines, D.R. and Marquez, V.E., 1986b, 3-Deaza-neplanocin A: a new inhibitor of S-adenosylhomocysteine synthesis and its effects in human colon carcinoma cells., Biochem. Pharmacol., 35:4523-4527.

Hasobe, M., McKee, J.G., Borcherding, D.R. and Borchardt, R.T., 1987a, 9-(trans-2', trans-3'-dihydroxycyclopent-4'-enyl)-adenine and -3-deazaadenine: analogs of neplanocin A which retain potent antiviral activity but exhibit reduced cytotoxicity., Antimicrob. Agents Chemother., submitted.

Hasobe, M., McKee, J.G., Borcherding, D.R., Keller, B.T. and Borchardt, R.T., 1987b, Effects of 9-(trans-2', trans-3'-dihydroxycyclopent-4'-enyl)-adenine and -3-deazaadenine on the metabolism of S-adenosylhomocysteine hydrolase in mouse L929 cells., Mol. Pharmacol., submitted.

Hayashi, M., Yaginuma, S., Yoshioka, H. and Nakatsu, K., 1981, Studies on neplanocin A, new antitumor antibiotic. II. Structure determination., J. Antibiot., 34:675-680.

Holý, A., Votruba, I. and De Clercq, E., 1982, Synthesis and antiviral activity of stereoisomeric eritadenines., Coll. Czech. Chem. Commun., 47:1392-1407.

Inaba, M., Nagashima, K., Tukagoshi, S. and Sakurai, Y., 1986, Biochemical mode of cyto-toxic action of neplanocin A in L1210 leukemic cells., Cancer Res., 46:1063-1067.

Jacquemont, B. and Huppert, J., 1977, Inhibition of viral RNA methylation in Herpes simplex virus type 1-infected cells by 5' S-isobutyl-adenosine., J. Virol., 22:160-167.

Kaehler, M., Coward, J. and Rottman, F., 1979, Cytoplasmic location of undermethylated messenger RNA in Novikoff cells., Nucleic Acids Res., 6:1161-1175.

Keller, B.T. and Borchardt, R.T., 1986, Metabolism and mechanism of action of neplanocin A - A potent inhibitor of S-adenosylhomocysteine hydrolase., in: "Biological Methylation and Drug Design," Borchardt, R.T., Creveling, C.R. and Ueland, P.M., eds., pp. 385-396, The Human Press, Clifton, NJ.

Keller, B.T. and Borchardt, R.T., 1987, Adenosine dialdehyde: a potent inhibitor of vaccinia virus multiplication in mouse L929 cells., Mol. Pharmacol., 31: (in press).

Keller, B.T., Clark, R.S., Pegg, A.E. and Borchardt, R.T., 1985, Purification and charac-terization of some metabolic effects of S-neplanocylmethionine., Mol. Pharmacol., 28:364-370.

Krug, R.M., Broni, B.A. and Bouloy, M., 1979, Are the 5' ends of influenza viral mRNAs synthesized in vivo or donated by host mRNAs?, Cell, 18:329-334.

Lombardini, J.B. and Talalay, P., 1973, Effects of inhibitors of adenosine triphosphate: L-methionine S-adenosyltransferase on levels of S-adenosyl-L-methionine and L-methionine in normal and malignant mammalian tissues., Mol. Pharmacol., 9:542-560.

Matuszewska, B. and Borchardt, R.T., 1987a, Inhibition of S-adenosylhomocysteine hydrolase from Alcaligenes faecalis by neplanocin A., Arch. Biochem. Biophys., submitted.

Matuszewska, B. and Borchardt, R.T., 1987b, The role of nicotinamide adenine dinucleo-tide in the inhibition of bovine liver S-adenosylhomocysteine hydrolase by neplanocin A., J. Biol. Chem., 262:265-268.

Montgomery, J.A., Clayton, S.J., Thomas, H.J., Shannon, W.M., Arnett, G., Bodner, A.J., Kim, I.-K., Cantoni, G.L. and Chiang, P.K., 1982, Carbocyclic analogue of 3-deazaadeno-sine: a novel antiviral agent using S-adenosylhomocysteine hydrolase as a pharmaco-logical target., J. Med. Chem., 25:626-629.

Palmer, J.L. and Abeles, R.H., 1976, Mechanism for enzymatic thioether formation. Mechanism of action of S-adenosylhomocysteinase., J. Biol. Chem., 5817-5819.

Palmer, J.L. and Abeles, R.H., 1979, The mechanism of action of S-adenosylhomocystei-nase., J. Biol. Chem., 254:1217-1226.

Perry, R.P. and Kelley, D.E., 1974, Existence of methylated messenger RNA in mouse L cells., Cell, 1:37-42.

Pugh, C.S.G., Borchardt, R.T. and Stone, H.O., 1977, Inhibition of Newcastle disease virion messenger RNA (guanine-7-)methyltransferase by analogues of S-adenosylhomocys-teine., Biochemistry, 16:3928-3932.

Pugh, C.S.G., Borchardt, R.T. and Stone, H.O., 1978, Sinefungin, a potent inhibitor of virion mRNA(guanine-7-)methyltransferase, mRNA(nucleoside-2'-)methyltransfer-ase, and viral multiplication., J. Biol. Chem. 253:4075-4077.

Raies, A., Lawrence, F., Robert-Géro, M., Loche, M. and Cramer, R., 1976, Effect of 5'-deoxy-5'-S-isobutyl adenosine on polyoma virus replication., FEBS Letts. 72:48-52.

Ramakrishnan, V. and Borchardt, R.T., 1987, Adenosine dialdehyde and neplanocin A: Potent inhibitors of S-adenosylhomocysteine hydrolase in neuroblastoma N2a cells., Nerochem. Int., in press.

Reddy, R., Ro-Choi, T.S., Henning, D. and Busch, H., 1974, Primary sequence of U-1 nuclear ribonucleic acid of Novikoff hepatoma ascites cells., J. Biol. Chem., 249: 6486-6494.

Rice, A.P. and Roberts, B.E., 1983, Vaccinia virus induces cellular mRNA degradation. J. Virol., 47:529-539.

Robert-Gero, M., Lawrence, F., Farrugia, G., Berneman, A., Blanchard, P., Vigier, P. and Lederer, E., 1975, Inhibition of virus-induced cell transformation by synthetic analogues of S-adenosylhomocysteine., Biochem. Biophys. Res. Commun., 65: 1242-1249.

Saunders, P.P., Tan, M.-T. and Robins, R.K., 1985, Metabolism and action of neplanocin A in chinese hamster ovary cells., Biochem. Pharmacol., 34:2749-2754.

Schlenk, F. and Dainko, J.L., 1974, The S-n-propyl analogue of S-adenosylmethionine., Biochem. Biophys. Acta, 385:312-323.

Schlenk, F., Hannum, C.H. and Ferro, A.J., 1978, Biosynthesis of adenosyl-D-methionine and adenosyl-2-methylmethionine by Candida utilis., Arch. Biochem. Biophys., 187:191-196.

Secrist III, J.A., Clayton, S.J. and Montgomery, J.A., 1984, (±)-3-(4-Amino-1H-pyrrolo[2,3-d]pyrimidin-1-yl)-5-(hydroxymethyl)-(1α,2α,3β,5β)-1,2-cyclo-pentanesdiol, the carbocyclic analogue of tubercidin., J. Med. Chem., 27:534-536.

Shimizu, S., Shiozaki, S., Ohshiro, T. and Yamada, H., 1984, Occurence of S-adenosyl-homocysteine hydrolase in prokaryotic cells. Characterization of the enzyme from Alcaligenes faecalis and role of the enzyme in the activated methyl cycle., Eur. J. Biochem., 141:385-392.

Sufrin, J.R., Coulter, A.W. and Talalay, P., 1979, Structural and conformational analogues of L-methionine as inhibitors of the enzymatic synthesis of S-adenosyl-L-methionine. IV. Further mono-, bi-, and tricyclic amino acids., Mol. Pharmacol., 15: 661-677.

Tsujino, M., Yaginuma, S., Fujii, T., Hayano, K., Matsuda, T., Watanabe, T. and Abe, J., 1980, Neplanocins, new antitumor agents: Biological activities., in: "Current Chemo-therapy and Infectious Disease," Nelson, J.D. and Grassi, C., eds., Vol. 2, pp. 1559-1561, The American Society for Microbiology, Washington, DC.

Usdin E., Borchardt, R.T. and Creveling, C.R. (eds.), 1982, "Biochemistry of S-Adenosyl-methionine and Related Compounds," Macmillan Press Ltd., London.

Vedel, M., Lawrence, F., Robert-Géro, M. and Lederer, E., 1978, The antifungal antibiotic sinefungin as a very active inhibitor of methyltransferases and of the transformation of chick embryo fibroblasts by Rous sarcoma virus., Biochem. Biophys. Res. Commun., 85:371-376.

Whaun, J.M., Miura, G.A., Brown, N.D., Gordon, R.K. and Chiang, P.K., 1986, Anti-malarial activity of neplanocin A with perturbations in the metabolism of purines, polyamines and S-adenosylmethionine., J. Pharmacol. Exp. Therap., 236: 277-293.

Wolfson, G., Chisholm, J., Tashjian, A.H., Jr., Fish, S. and Abeles, R.H., 1986, Neplanocin A: Actions on S-adenosylhomocysteine hydrolase and on hormone synthesis by GH_4C_1 cells., J. Biol. Chem., 261:4492-4498.

Yaginuma, S., Tsujino, M., Muto, N., Otani, M., Hayashi, M., Ishimura, F., Fujii, T., Watanabe, S., Matsuda, T., Watanabe, T. and Abe, J., 1980, Neplanocins, new antitumor agents: Discovery, isolation, and structure., in: "Current Chemotherapy and Infectious Disease," Nelson, J.D. and Grassi, C., eds., Vol. 2, pp. 1558-1559, The American Society for Microbiology, Washington, DC.

Yaginuma, S., Muto, N., Tsujino, N., Sudate, M., Hayashi, M. and Otani, M., 1981, Studies on neplanocin A, new antitumor antibiotic. I. Producing organism, isolation and characterization., J. Antibiot., 34:359-366.

Zappia, V., Zydek-Cwick, C.R. and Schlenk, F., (1969), The specificity of S-adenosyl-methionine derivatives in methyl transfer reactions. J. Biol. Chem., 244:4499-4509.

ASSAY SYSTEMS: TESTING OF ANTIVIRAL DRUGS IN CELL CULTURE

(IN VITRO)

Hans J. Eggers

Institut für Virologie der Universität zu Köln
Fürst Pückler Str. 56
D-5000 Köln 41

INTRODUCTION

Testing for virus-inhibitory activity of chemicals in cell culture has sometimes furnished important leads in the development of antiviral drugs. The shortcomings of this so-called in vitro testing are obvious: substances which lack an effect on the viral replication cycle but, e.g., enhance antibody formation, improve nonspecific resistance, strengthen the interferon systems will not be detected by this method. Likewise, although so far only rarely described, some compounds affecting the viral replication cycle might reveal their antiviral effects only in an animal model, since metabolic activities of the organism, absent in cell culture, are required to convert the drug into its effective form. Most of the time, however, the reverse is true, viz. more compounds do appear active in a cell culture system than will later be found useful in the organism (Buthala, 1965; Eggers, 1977b). I shall refer to this problem in a later section.

When search for antiviral substances began, not too many virus-cell culture systems were yet available. Thus, e.g., the at the time well-established chick-embryo techniques served as welcome substitutes and permitted, e.g., screening for antiviral activity against myxoviruses. Indeed, availability of a good in vitro system determined antiviral testing to a large extent, i.e., even viruses were included which appeared to be of very little if any medical importance.

I should hasten to add, however, that in the fifties and sixties several of the few investigators who believed at all in the possibility of antiviral chemotherapy hoped for a broad-spectrum antiviral agent like a "lapis philosophorum" of medieval alchemy which could heal more or less all virus diseases. Consequently, the screening system used would not matter too much. When finally some virus-selective chemical inhibitors emerged such as methisazone against poxviruses (Bauer and Sadler, 1960), 2-(α-hydroxybenzyl)-benzimidazole against enteroviruses (Eggers and Tamm, 1961) guanidine against polioviruses (and some other enteroviruses) (Rightsel et al., 1961; Crowther and Melnick, 1961; Loddo et al., 1962), and 1-aminoadamantane against

influenza A viruses (Davies et al., 1964) often regret was expressed as to the narrow and limited range of activity, in contrast to the wonder drug interferon discovered some years earlier
in 1957. It was only when some insight was gained into the biochemical basis of selective inhibition of virus replication
(Eggers and Tamm, 1961; Eggers and Tamm, 1962) that virus group
specificity of an antiviral agent was realized as a logical consequence of virus-selective action.

Today, deliberations of this kind are no longer in the center
of discussions: for most viruses a good cell culture system is
now available with notable exceptions such as hepatitis B virus
and the human papilloma viruses. Accepting the "group specificity"
of antiviral agents the search is concentrated on virus diseases
in particular need for treatment. In passing it may also be
pointed out, that commercial interests will naturally be of considerable significance in choosing a system to be evaluated.

ANTIVIRAL SCREENING IN CELL CULTURES MAINTAINED IN SOLID MEDIA

A few years after introduction of the plaque technique into
animal virology (Dulbecco, 1952) De Somer and Prinzie (1957) in
Leuven used filter paper discs to measure neutralizing antibody
titers for poliomyelitis virus. Monkey kidney cell sheets were infected with an appropriate number of poliovirus plaque units and
covered with agar. After solidification of the agar, paper discs
soaked with serum to be tested for neutralizing antibodies were
placed on the surface of the agar. After 2 to 3 days plaques
appeared. However, if a serum contained neutralizing antibodies,
a circular area of the cell sheet was protected from virus infection. The authors demonstrated a relationship between the diameter
of the protected area and neutralizing antibody concentration in
the serum. They also pointed out that this method had been found
useful to screen substances for their activity against viruses,
analogous to the experiences with assays for antibiotics in
bacteriology.

The paper disc-agar-technique has been exploited by Herrmann
(Herrmann et al., 1960; Herrmann, 1961) and others. When the disc
contained an effective antiviral agent, a halo of living cells,
free of plaques, was produced around the disc. This zone is
stainable with dyes taken up by vital cells. Toxicity of the
compound produces a halo of dead, unstained cells around the
disc. If - as to be expected - the material is toxic at higher
concentrations but exhibits antiviral activity at lower concentrations, an inner zone of unstained dead cells and an outer zone
of stained, living, plaque-free cells will be observed. An
advantage of this method consisted of the possibility to process
large numbers of substances per week, and that a first quantitative estimate of antiviral and toxic activity of a compound could
be made (Herrmann, 1961; Buthala, 1965). A very similar procedure
has been described by Rada et al. (1960) except that the substance to be tested was put into a cylinder which was placed
into the agar plate.

These types of test are certainly limited to viruses that
produce plaques under agar. For many viruses, however, nowadays plaque tests are available. A more serious drawback of
the method appears to be its comparatively low sensitivity to
detect both the cytotoxic and antiviral levels of test compounds
(Bucknall, 1973).

In passing it should be mentioned that substances to be test for antiviral activity may also be incorporated directly into t agar overlay or into a more fluid overlay containing methyl-cellulose. The quantitation of virus inhibition in this type test is not easy. Certainly, the reduction of plaque number in treated as compared to the untreated cultures can easily determined, but the regularly observed reduction of plaque size is difficult to assess (Tamm and Eggers, 1963), though it is obviously an important parameter.

TESTS IN FLUID MEDIA PRINCIPLES

With the development of microsystems (microtiter plates and complementary mechanical and electronic equipment) tests for antiviral substances in fluid media gained more and more genera acceptance. It might be stated, however, that the principles of these assays for antiviral activity of chemical or biological substances had been developed earlier in macrosystems. How then does one determine antiviral and toxic effects of a com-pound in such a virus-cell culture system?

A still widely employed and reliable test is to utilize vira and toxic cytopathogenicity as indicator. Formerly this was per formed in cell culture tubes (Eggers and Tamm, 1961). Appropria cell cultures were infected with a low dose of the virus to be tested. The development of viral cytopathic effects was observe in untreated control cultures as compared to cultures treated with various concentrations of the test compound. Since a low virus dose had been applied, several viral replicative cycles h to occur before development of visible and finally complete vir cytopathic effects, i.e., any step of virus multiplicatio potentially susceptible to an antiviral compound might be detec-ted using this procedure. The inhibitory effect of the different concentrations of a compound might also provide a quantitative measure of the antiviral activity.

In the same test toxicity of the compound can be evaluated in treated uninfected cultures (Eggers and Tamm, 1961). For the experienced observer, this system is reliable, fast, and economical (see also Bucknall, 1973).

We and others (Bucknall, 1973), however, have noted certain deficiences of this method to assess compound toxicity. As indicated, toxicity is measured in already established cell cultures. Consequently, it may be difficult to observe toxic effects of compounds affecting cell growth and multiplication. Thus, inhibitors of nucleic acid synthesis, of protein synthesis, or of uncouplers of oxidative phosphorylation may give misleading results. In a resting, confluent cell monolayer very little DNA or protein synthesis may be needed, and likewise, little energy. Con-sequently, in confluent monolayers it may take some time before morphological changes will appear. Obviously, however, such compounds may seriously affect virus replication which requires extensive nucleid acid and protein synthesis.

We, therefore, also looked for RNA and protein synthesis as well as energy-yielding processes in uninfected cells treated with virus-inhibitory concentrations of the compounds to be tested (Eggers and Tamm, 1961). Additionally, we measured cell multip-lication under these conditions to detect affects of a compound on cellular activities in a most comprehensive and sensitive manner (Eggers and Tamm, 1961; 1962). Certainly, subtle changes such as mutagenic effects, though of paramount importance, will not be uncovered by this method.

SOME FREQUENTLY USED ANTIVIRAL ASSAYS

In different ways the same principles have subsequently been applied to the current micromethods, with individual modifications introduced almost regularly by the different laboratories to meet special needs. E.g., uninfected cells not forming monolayers are exposed in wells of microtiter plates to different concentrations of the prospective antiviral compounds, and, at various times, cell counts are being performed, often applying a dye exclusion method to measure viable cells (e.g., Mitsuya and Broder, 1986; Pauwels et al., 1987). Alternatively, in the presence of representative concentrations of the compound, adherence of cells and formation of a monolayer may be examined by microscopy or after staining by gross inspection (e.g., Griengl et al., 1985). In essence then, as originally proposed by Eggers and Tamm (1961), growing cells represent a more sensitive test for the evaluation of cytotoxicity of antiviral compounds than preformed cell sheets. Additionally, cell adherence can also be measured by this procedure.

At the same time inhibition of virus multiplication may be monitored by microscopic examination of cell sheets for viral cytopathic effects (e.g., De Clercq et al., 1980). An alternative possibility consists of fixation and staining of the cell sheets in the microtiter plates, which allows an estimation of the extent of virus-induced cytopathic effects in infected controls and drug-treated wells (e.g., Griengl et al., 1985).

Viral cytopathic effects may also be evaluated by measuring survival and growth of virus-infected cells in the presence of various dilutions of a test compound (e.g.,Mitsuya and Broder, 1986; Pauwels et al., 1987). The test conditions, however, have to be well defined. It is easy to see, e.g., that under conditions of a low infecting virus dose and applying an effective antiviral compound, cell multiplication may proceed seemingly undisturbed, since most of the cells in the culture will remain uninfected in contrast to an untreated, infected control culture in which virus infection will spread undisturbed. This effect, however, may be obscured, when a high multiplicity of infection is being used: despite the presence of a potentially effective antiviral inhibitor, the infected cells may be destroyed by the virus although virus replication might have been strongly inhibited. Obviously, the results depend on the step in the replication cycle being affected by an antiviral substance (Bablanian et al., 1965; Eggers, 1977a; Rosenwirth and Eggers, 1977).

Experiments of the sort described may also yield estimates of the selectivity of a compound: the virus-inhibitory concentration of a substance as measured by the number of viable cells after treatment and virus infection is compared to the number of viable cells in uninfected cultures treated with serial concentrations of the compound in question (e.g., Mitsuya and Broder, 1986; Pauwels et al., 1987).

Inhibition of virus multiplication can certainly be determined in various other ways. One possibility concerns measurement of viral antigen expression by immunofluorescence microscopy (e.g., Pauwels et al., 1987). The data may be monitored by laser flow cytofluorography: the intensity of fluorescence of each of a large number of cells may be measured and printed in form of a histogram (Pauwels et al., 1987).

Miller et al. (1970) in an attempt to develop a rapid, automated procedure used inhibition of viral RNA synthesis for assaying antiviral activity in the case of RNA viruses. This method is applicable to those RNA viruses which are insusceptible to actinomycin D, a compound suppressing cellular RNA synthesis. The effect of a compound on host cell RNA synthesis can likewise be easily measured.

Radioactive tracers can certainly also be used to quantitate proliferation of cells by measuring ^{3}H-thymidine uptake (Wendler et al., 1987). In this way inhibition of human immunodeficiency virus (HIV)-induced cytopathogenicity by nucleoside derivatives can be determined. Similarly, the toxicity of a compound for un-infected cells may be assayed (Hartmann et al., 1987).

Finally, another test for compound- and virus-induced cellular toxicity may be mentioned. Aliquots of cells, at the end of treat-ment with various concentrations of a potentially antiviral compound are stained with neutral red. After extensive washing, the stained cells are subsequently dissolved in ethanol, and the optical density of the solution is measured at 546 nm. The absorbance has been found to be proportional to the number and vitality of the cells (Griengl et al., 1985). This method has originally been devised by Finter (1970) who showed that viral cytopathic or cytotoxic effects could be quantitated by measuring the reduction in the uptake of neutral red by cells.

Rightsel et al. (1956) have proposed earlier an analogous system based on the fact that damaged cells (damaged either by virus infection or toxic effects of a chemical) are deficient in metabolism and will not change the pH of their medium as shown by an indicator.

SOME REMARKS ON QUANTITATION OF ANTIVIRAL ACTIVITY

A simple way to quantitate antiviral activity of a compound consists of applying serial dilutions of that compound and to de-fine a suitable endpoint, e.g., 50 or 75 per cent inhibition of viral cytopathogenicity after a defined time post infection with a low virus dose as compared to untreated control cultures (e.g., Tamm et al., 1969).

In another assay single cycles of virus replication are performed with various concentrations of the test compound and the reduction of virus yield is determined as compared to the untreated control culture (Eggers and Tamm, 1961; Eggers, 1977a).

These certainly are only basic principles which do allow numerous modifications suitable for special needs.

ALTERNATIVE ASSAYS FOR DETECTION OF ANTIVIRAL DRUGS.

The procedures described so far concentrated mainly on the

effects of an antiviral compound on the virus multiplication cycle as a whole, i.e., in this way effects on any of the various steps of virus replication might be detected.

Since virus-cell culture systems were not always easily available in the past and required a great deal of personnel, not to speak of the substantial current expenses, one was looking for shortcuts, i.e., testing was focussed on selected viral functions, easily to assay. Thus, compounds were tested for inhibition of neuraminidase activity, an enzyme involved in the replication of myxoviruses (Meindl et al., 1974). In this way also compounds inhibitory to the replication of influenza and parainfluenza viruses were discovered (Palese and Schulman, 1974).

Reverse transcriptase is an essential and unique enzyme for replication of retroviruses. It thus has long been considered an attractive target for selective agents against retroviruses (e.g., Gallo et al., 1972; De Clercq, 1979). Needless to say that with the discovery of human immunodeficiency viruses interest has again focussed on inhibitors of reverse transcriptase. This subject can not be treated here in any detail, but it is important to keep in mind that by this assay only one series of steps in the retroviral cycle, though an essential one, can be monitored. Effects of a compound, e.g., on virus maturation will obviously not be detected. Some agents may behave differently as expected. Thus interferon appears to affect retroviral maturation (Aboud and Hassan, 1983).

LIMITATIONS OF IN VITRO TESTING

As indicated in the introduction testing of chemicals for virus-inhibitory activity in cell culture suffers from obvious shortcomings: only substances affecting the viral replication cycle directly will possibly be recognized. Drugs requiring at first metabolic activities of the host which are absent in cell culture will register negative (though this appears to be a rare event). Furthermore, it turned out that, e.g., ribavirin (virazole) in some cell culture systems yielded rather disappointing results. We now presume that this host cell dependence may partly be related to different pool sizes of nucleotides in various cell types (Smith and Canonico, 1984). At any rate, ribavirin has proven effective in the therapy of Lassa fever (McCormick et al., 1986). Depending on the screening system used, ribavirin nevertheless might have passed the net unnoticed.

Largely, however, more compounds exhibiting antiviral activity will be found in cell culture as compared to the animal. In some cases, at least, this appears to be related to a rapid metabolic degradation of the drug into inactive derivatives. We have studied this problem in great detail for 2-(α-hydroxybenzyl)-benzimidazole, an antiviral compound which is far more active in cell culture than in adult mice. Its significant prophylactic and therapeutic effects in enterovirus-infected mice can be achieved only by considerable pharmacologic and pharmacokinetic adjustments (Eggers, 1982).

I am not aware of data which would support the possibility that the test virus strains selected for screening might influence the test result significantly. Potent selective inhibitors appear to affect structures and/or processes relevant for the whole virus group or the virus type, respectively.

CONCLUSIONS

Any antiviral screening program today will rely on virus-cell culture systems. For most viruses of medical importance appropriate systems are now available. The introduction of micromethods does permit screening of large numbers of compounds under economical conditions. Thus significant leads may be discovered, and extensive structure-activity studies are feasible. With the recent development of high-power x-ray crystallography even a rational design of active compounds appears in sight (Smith et al., 1986), though these possibilities are still limited.

Despite of the remarkable progress in the field of in vitro testing, animal experiments will be indispensable in the development of antiviral chemotherapy.

REFERENCES

Aboud, M., and Hassan, Y., 1983, Accumulation and breakdown of RNA-deficient intracellular virus particles in interferon-treated NIH 3T3 cells chronically producing Moloney murine leukemia virus.
J. Virol. 45:489-495.

Bablanian, R., Eggers, H.J., and Tamm, I., 1965, Studies on the mechanism of poliovirus-induced cell damage. I. The relation between poliovirus-induced metabolic and morphological alterations in cultured cells.
Virology 26:100-113.

Bauer, D.J., and Sadler, P.W., 1960, The structure-activity relationships of the antiviral chemotherapeutic activity of isatin ß-thiosemicarbazone.
Br. J. Pharmac. Chemother. 15:101-110.

Bucknall, R.A., 1973, The continuing search for antiviral drugs
Adv. Pharmacol. Chemotherapy 11:295-319.

Buthala, D.A., 1965, Experience gained in screening for synthetic antiviral compounds.
Ann. N.Y. Acad. Sci. 130:Art. 1, 17-23.

Crowther, D., and Melnick, J.L., 1961, Studies of the inhibitory action of guanidine on poliovirus multiplication in cell cultures.
Virology 15:65-74

Davies, W.L., Grunert, R.R., Haff, R.F., McGahen, J.W., Neumayer, E.M., Paulshock, M., Watts, J.C., Wood, T.R., Herrmann, E.C., and Hoffmann, C.E., 1964, Antiviral activity of 1-adamantanamine (amantadine).
Science 144:862-863.

De Clercq, E., 1979, Suramin: a potent inhibitor of the reverse transcriptase of RNA tumor viruses
Cancer Letters 8:9-22.

De Clercq, E., Descamps, J., Verhelst, G., Walker, R.T., Jones, A.S., Torrence, P.F., and Shugar, D., 1980, Comparative

efficacy of antiherpes drugs against different strains of
herpes simplex virus.
J. Inf. Dis. 141:563-574.

De Somer, P., and Prinzie, A., 1957, Poliomyelitis virus neutralizing antibodies determination by filter paper discs on soli-
dified bottle cultures.
Virology 4: 387-388.

Dulbecco, R., 1952, Production of plaques in monolayer tissue
cultures by single particles of an animal virus.
Proc. Natl. Acad. Sci. 38:747-752.

Eggers, H.J., 1977a, Selective inhibition of uncoating of echovirus 12 by rhodanine.
Virology 78:241-252.

Eggers, H.J., 1977b, Specific antiviral substances: testing
in experimental animals and in "in vitro" systems. In:
International Symposium on Experimental Animals and "in
vitro" Systems in Medical Microbiology.
WHO Collaborating Centre for Collection and Evaluation of
Data on Comparative Virology. Munich, pp. 277-282.

Eggers, H.J., 1982, Benzimidazoles. Selective inhibitors of
picornavirus replication in cell culture and in the organism.
In: Chemotherapy of Viral Infections. P.E. Came, and L.A.
Caliguiri (eds.).
Springer-Verlag Berlin, Heidelberg, New York, pp. 377-417.

Eggers, H.J., and Tamm, I., 1961, Spectrum and characteristics of
the virus inhibitory action of 2-(α-hydroxybenzyl)-benzimida-
zole.
J. Exp. Med. 113:657-682.

Eggers, H.J., and Tamm, I., 1962, On the mechanism of selective
inhibition of enterovirus multiplication by 2-(α-hydroxyben-
zyl)-benzimidazole.
Virology 18:426-438.

Finter, N.B., 1970, Methods for screening in vitro and in vivo
for agents active against myxoviruses.
Ann. N.Y. Acad. Sci. 173:Art. 1, 131-138.

Gallo, R.C., Smith, R.C., Whang-Peng, J., Ting, R.C., Yang,
S.S., and Abrell, J.W., 1972, RNA tumor viruses, DNA polymerases, and oncogenesis: some selective effects of rifampicin
derivatives.
Medicine 51:159-168.

Griengl, H., Bodenteich, M., Hayden, W., Wanek, E., Streicher,
W., Stütz, P., Bachmayer, H., Ghazzouli, I., and Rosenwirth,
B., 1985, 5-(Haloalkyl)-2'-deoxyuridines: a novel type of
potent antiviral nucleoside analogue.
J.Med. Chem. 28:1679-1684.

Hartmann, H., Hunsmann, G., and Eckstein, F., 1987, Inhibition
of HIV-induced cytopathogenicity in vitro by 3'-azido-2',3'-
dideoxyguanosine.
Lancet I:115-116.

Herrmann, E.C.,Jr., 1961, The detection, assay and evaluation
    of antiviral drugs.
    Progr. Med. Virol. 3:158-192.

Herrmann, E.C.,Jr., Gabliks, J., Engle, C., and Perlman, P.L.,
    1960, Agar diffusion method for detection and bioassay of
    antiviral antibiotics.
    Proc. Soc. Exp. Biol. Med. 103:625-628.

Loddo, B., Ferrari, W., Brotzu, G., and Spanedda, A., 1962,
    In vitro inhibition of infectivity of polio viruses by
    guanidine.
    Nature 193:97-98.

McCormick, J.B., King, I.J., Webb, P.A., Scribner, C.L.,
    Craven, R.B., Johnson, K.M., Elliott, L.H., and Belmont-
    Williams, R., 1986, Lassa fever. Effective therapy with
    ribavirin.
    N. Engl. J. Med. 314:20-26.

Meindl, P., Bodo, G., Palese, P., Schulman, J., and Tuppy, H.,
    1974, Inhibition of neuraminidase activity by derivatives of
    2-deoxy-2,3-dehydro-N-acetylneuraminic acid.
    Virology 58:457-463.

Miller, P.A., Lindsay, H.L., Cormier, M., Mayberry, B.R.,
    and Trown, P.W., 1970, Rapid semi-automated procedures for
    assaying antiviral activity.
    Ann. N.Y. Acad. Sci. 173:Art. 1, 151-159.

Mitsuya, H., and Broder, S., 1986, Inhibition of the in vitro
    infectivity and cytopathic effect of human T-lymphotrophic
    virus type III/lymphadenopathy-associated virus (HTLV-III/
    LAV) by 2',3'-dideoxynucleosides.
    Proc. Natl. Acad. Sci. 83:1911-1915.

Palese, P., and Schulman, J.L., 1974, Inhibition of influenza
    and parainfluenza virus replication in tissue culture by 2-
    deoxy-2,3-dehydro-N-trifluoroacetylneuraminic acid (FANA).
    Virology 59:490-498.

Pauwels, R., De Clercq, E., Desmyter, J., Balzarini, J., Goubau,
    P., Herdewijn, P., Vanderhaeghe, H., and Vandeputte, M., 1987,
    Sensitive and rapid assay on MT-4 cells for detection of
    antiviral compounds against the AIDS virus.
    J. Virol. Meth. in press.

Rada, B, Blaškovič, D., Šorm, F., and Škoda, J., 1960, The inhibi-
    tory effect of 6-azauracil riboside on the multiplication of
    vaccinia virus.
    Experientia 15:487-488.

Rightsel, W.A., Dice, J.R., McAlpine, R.J., Timm, E.A., Mc-
    Lean, I.W., Jr., Dixon, G.J., and Schabel, F.M., Jr.,
    1961, Antiviral effect of guanidine.
    Science 134:558-559.

Rightsel, W.A., Schultz, P., Muething, D., and McLean, I.W.,
    Jr., 1956, Use of vinyl plastic containers in tissue cultures
    for virus assays.
    J. Immunol. 76:464-474.

Rosenwirth, B., and Eggers, H.J., 1977, Echovirus 12-induced host cell shutoff is prevented by rhodanine. Nature 267:370-371.

Smith, R.A., and Canonico, P., 1984, Ribavirin. In: Antiviral Drugs and Interferon: the Molecular Basis of their Activity, Y. Becker (ed.). Martinus Nijhoff Publishing, Boston, pp. 253-269.

Smith, Th.J., Kremer, M.J., Luo, M., Vriend, G., Arnold, E., Kamer, G., Rossmann, M.G., McKinlay, M.A., Diana, G.D., and Otto, M.J., 1986, The site of attachment in human rhinovirus 14 for antiviral agents that inhibit uncoating. Science 233:1286-1293.

Tamm, I., and Eggers, H.J., 1963, Unique susceptibility of enteroviruses to inhibition by 2-(α-hydroxybenzyl)-benzimidazole and derivatives. 2nd International Symposium of Chemotherapy, H.P. Kuemmerle, P. Preziosi, P. Rentchnick (eds.). S. Karger, Basel, New York, pp. 88-118.

Tamm, I., Eggers, H.J., Bablanian, R., Wagner, A.F., and Folkers, K., 1969, Structural requirements of selective inhibition of enteroviruses by 2-(α-hydroxybenzyl)-benzimidazole and related compounds. Nature 223:785-788.

Wendler, I., Bienzle, U., and Hunsmann, G., 1987, Neutralizing antibodies and the course of HIV induced disease. AIDS Research and Human Retroviruses, in press.

ANIMAL MODELS AS ASSAY SYSTEMS FOR THE DEVELOPMENT OF ANTIVIRALS

Earl R. Kern

Division of Infectious Diseases
Department of Pediatrics
University of Utah School of Medicine
Salt Lake City, Utah 84132

INTRODUCTION

The importance of experimental viral infections in animal models for development and testing of new antiviral agents prior to their use in man should not be understated. While tissue culture systems are of great value in determining if a new drug has activity against a particular virus, these systems should not be used as indicators or predictors of activity in humans. Only where suitable animal models are not available, should a compound be taken from tissue culture directly into human trials. Although one can legitimately argue that most, if not all, animal model infections are not identical to the human disease, it can be demonstrated that a compound does in fact have activity in an in vivo system and early indications of its antiviral activity, tissue distribution, metabolic disposition, pharmacokinetics, and acute toxicity can be realized. Importantly, all of these parameters of drug pharmacodynamics can be correlated with inhibition of viral replication in target organs. Additionally, our understanding of the pathogenesis of many viral infections, the response of the host to infection and interaction between the infection, the host's response, and a therapeutic agent has been enhanced greatly through the use of animal model systems.

When selecting or developing an animal model for determining efficacy of an antiviral, one should utilize systems that best simulate the corresponding human disease. The properties of an "ideal" animal model are shown in table 1. It is obvious that there is currently no experimental viral infection that meet all of these properties. The most common animal species used are rodents and these differ from humans by

Table 1. Properties of an "Ideal" Animal Model

---

1. The most appropriate animal species should be used.
2. Use the human virus with minimum alteration through adaptation if possible.
3. The natural route of infection with similar inoculum size should be utilized when feasible.
4. The course of infection and pathogenesis are similar to that in human disease.
5. The host response is similar to that seen in humans.
6. Drug pharmacodynamics, metabolism and toxicity are similar to that observed in humans.
7. The endpoints used for evaluation of efficacy are the same as used in human trials.

---

most criteria. The use of chimpanzees or other non-human primates, however, can not be used as models for most viral diseases due to their lack of availability, cost, space requirements, or the inability to experimentally infect these animals with many human viruses. An additional problem is that many human viruses such as cytomegalovirus, hepatitis B virus, papillomavirus and human immunodeficiency virus are species specific and do not cause disease in animals. In these cases an animal virus counterpart must be used in order to do antiviral evaluations in animal models. There are also significant differences in drug metabolism and pharmacokinetics between humans and most animal species.

Although we do not have "ideal" animal models, there are numerous examples of good experimental viral infections in animals that can be used to evaluate new antiviral agents. In order to obtain the best information regarding the predictability of an animal model system for efficacy of a new compound, a number of guidelines for using animal models should be considered. These are listed in table 2. The use of as many of these guidelines as possible will result in an accumulation of data that should allow one to make predictions regarding the efficacy of a compound in man. Once human trials are completed, the data obtained should then be compared with those acquired in the animal system to help establish the predictability of the animal model systems.

It is not the intent of this communication to review all the animal model infections, but to focus on those used most commonly, to highlight those that have demonstrated predictability for human disease and to discuss briefly those diseases where new animal models are needed.

Table 2. Guidelines for Animal Model Testing

Drug should be evaluated in more than one species.
Drug administration must be varied - route, dose and time
    of initiation, so that a maximum tolerated dose and a
    minimum effective dose can be determined.
Virus should be varied - route, inoculum, size: an
    infectivity or mortality rate of 80-95% should be used.
Age of animal should be considered.
Human infection should be closely simulated.
Sensitivity of virus to drug should be known in cells from
    same species.
Drug metabolism and toxicity should be known.
Pathogenesis of virus and effect of drug should be defined.
Effect of drug on immune system should be determined.
Animals should be free of adventitious agents.
Results from models should be compared with human trials.

ANIMAL MODELS OF HERPESVIRUS INFECTIONS

The herpesviruses and particularly herpes simplex virus (HSV) are the etiologic agents of a wide variety of diseases in man. It is important that new antiviral agents be tested in animal models that simulate each of those disease states in humans. A summary of some of the animal model systems for HSV infections are shown in table 3. Some specific examples of these animal models utilized in our laboratory will be used to illustrate how they can be used to determine efficacy of antiviral agents. In addition, since animal models for HSV infections have been the most widely used, the predictability for most of these infections can be established. The first experimental system to be considered is a model of HSV-1 encephalitis in which weanling mice are inoculated intranasally with virus. After inoculation the virus travels from the nasopharynx via nerve tracts to the brain resulting in death from an acute encephalitis. In this model, treatment with adenine arabinoside (Ara-A) resulted in significant protection from mortality if drug was initiated within 24-48 h after infection (table 4). In clinical trials of HSV-1 encephalitis, treatment with Ara-A also resulted in significant protection against mortality.[1,2] In the animal model infection, treatment with acyclovir (ACV) was considerably more effective than Ara-A in preventing mortality. Concentrations as low as 15 mg/kg significantly reduced mortality when therapy was begun 72 h after infection (table 5), a time when virus is replicating in the CNS.[3] Based on these experimental data, we predicted that ACV would be more effective than Ara-A in treatment of HSV-1 encephalitis in humans. Subsequent clinical trials confirmed these observations in the animal model as ACV

Table 3. Animal Models for Herpes Simplex Virus Infections

Virus	Animal	Disease	Measure of efficacy
HSV type 1	mice	encephalitis	mortality pathogenesis
HSV type 1	rabbits	eye disease	lesion severity virus titers
HSV type 1	guinea pigs	skin lesions	lesion severity virus titers
HSV type 1	guinea pigs	genital herpes	lesion severity virus titers
HSV type 2	mice	disseminated neonatal herpes encephalitis	mortality pathogenesis
HSV type 2	mice guinea pigs	genital herpes	lesion severity virus titers recurrences

was shown to be significantly more effective than Ara-A.[4]

Mice inoculated intranasally with HSV-2 appear to be a good model for disseminated neonatal herpes. After viral inoculation, replication of virus is initially detectable in lung with subsequent dissemination to visceral organs. Virus is also concommitantly transmitted by neural routes from the nasopharynx to olfactory lobe, cerebellum - brain stem, cerebrum and spinal cord.[5] The effect of treatment with Ara-A on final mortality is shown in table 6. Although there was no protection against mortality, alteration of viral replication in some target organs has been demonstrated.[5,6] This is a complex infection involving both CNS and non-CNS tissues and historically has not responded well to most antiviral agents. In placebo controlled clinical trials, however, Ara-A did reduce mortality[7] but many survivors had permanent neurological sequelae.[8] Parenteral treatment with ACV in the murine model did reduce mortality (table 7) and we predicted that ACV would again be more effective in humans than Ara-A. In the recently completed clinical trials in which Ara-A and ACV were compared for efficacy against neonatal herpes, the two agents were judged to be comparable (Richard J. Whitley, personal communication). We have reported previously[3] that oral administration of ACV is considerably more effective than parenteral therapy in the mouse model but this mode of treatment has not been utilized for therapy of the

Table 4. Effect of Treatment With Ara-A on the Mortality of Mice Inoculated Intranasally With HSV-1

| Treatment[a] | Mortality | | | | |
	Number	Percent	P-value	MDD[b]	P-value
None	13/15	87	-	4.8	-
Placebo + 24 h	13/15	87	NS[c]	5.8	<0.05
Ara-A 250 mg/kg + 24 h	1/15	7	<0.001	7.0	NS
250 mg/kg + 48 h	5/15	33	<0.01	7.8	<0.05
250 mg/kg + 72 h	10/15	67	NS	7.6	NS
250 mg/kg + 96 h	12/15	80	NS	6.3	NS
Ara-A 83.3 mg/kg + 24 h	2/15	13	<0.001	4.5	NS
83.3 mg/kg + 48 h	8/15	53	NS	6.2	NS
83.3 mg/kg + 72 h	5/15	33	<0.01	6.6	NS
83.3 mg/kg + 96 h	14/15	93	NS	6.4	NS
Ara-A 27.8 mg/kg + 24 h	9/15	60	NS	6.8	NS
27.8 mg/kg + 48 h	9/15	60	NS	7.4	NS
27.8 mg/kg + 72 h	8/15	53	NS	7.4	<0.01
27.8 mg/kg + 96 h	13/15	87	NS	5.1	0.05
Ara-A 9.2 mg/kg + 24 h	13/15	87	NS	7.2	<0.05
9.2 mg/kg + 48 h	12/15	80	NS	5.8	NS
9.2 mg/kg + 72 h	11/15	73	NS	7.4	<0.05
9.2 mg/kg + 96 h	13/15	87	NS	5.5	NS

a Animals were treated i.p. twice daily for 7 days with stated concentrations and initiated at the times indicated.

b MDD = Mean Day of Death.   c NS = Not Significant.

human disease. The experimental model infection, therefore, accurately predicted efficacy for treatment of human disease but apparently failed to determine which of the two drugs is most active.

The rabbit model for herpetic eye disease was probably the first experimental animal system that demonstrated predictability for antiviral efficacy in humans. The effectiveness of iododeoxyuridine, adenine arabinoside, trifluorothymidine, and interferon was first determined in the rabbit model and subsequently confirmed in HSV-1 infections of the eye in humans.

Cutaneous infections of guinea pigs with HSV-1 appears to be a good model for skin and oral infections caused by this virus in man. When the animals are inoculated according to the procedure originally described by Hubler et. al.[9], the skin infection is characterized by rapid development of discreet 3-5 mm diameter vesicular lesions. The lesions then crust and heal completely by 10-12 days. The lesions and pattern of healing are similar to human cutaneous HSV disease.[10] The primary advantage of this model is that animals do not develop zosteriform lesions or die of

Table 5. Effect of Treatment With the Sodium Salt of ACV on the Mortality of Mice Inoculated Intranasally With HSV-1

Treatment[a]	Mortality			MDD[b]	P-value
	Number	Percent	P-value		
None	15/15	100	-	5.3	-
Placebo + 24 h	14/15	93	NS[c]	6.1	NS
ACV 60 mg/kg + 24 h	5/15	33	0.001	4.2	0.01
60 mg/kg + 48 h	3/15	20	<0.001	4.3	NS
60 mg/kg + 72 h	4/15	27	<0.001	4.2	<0.05
60 mg/kg + 96 h	9/14	64	NS	5.3	NS
ACV 30 mg/kg + 24 h	0/15	0	<0.001	-	-
30 mg/kg + 48 h	0/15	0	<0.001	-	-
30 mg/kg + 72 h	5/15	33	0.001	6.4	NS
30 mg/kg + 96 h	12/15	80	NS	6.8	NS
ACV 15 mg/kg + 24 h	3/15	20	<0.001	9.3	0.01
15 mg/kg + 48 h	2/15	13	<0.001	11.0	<0.05
15 mg/kg + 72 h	7/15	47	0.01	5.7	NS
15 mg/kg + 96 h	14/15	93	NS	6.1	NS
ACV 7.5 mg/kg + 24 h	6/15	40	<0.01	8.7	0.01
7.5 mg/kg + 48 h	5/15	33	0.001	8.2	0.01
7.5 mg/kg + 72 h	10/15	67	NS	6.9	NS
7.5 mg/kg + 96 h	14/15	93	NS	5.9	NS

a Animals were treated i.p. twice daily for 7 days with stated concentrations and initiated at the times indicated.

b MDD = Mean Day of Death.   c NS = Not Significant.

neurological disease, commonly seen in mouse models of cutaneous HSV-1 infections. One of the disadvantages of the model is that recurrent skin lesions do not develop spontaneously nor have they successfully been induced. While a large variety of antiviral compounds have been tested in this model, there are few compounds that have been shown to be efficacious in man. Comparative data in the guinea pig and humans using acyclovir continues to be accumulated, however, at this time, the predictive value of the model has not been established.[11] Another advantage of the animal model system is that it lends itself well to studying skin penetration of antiviral drugs both in vitro and in vivo, and is especially useful for determining the best formulation for topical therapy.[12]

Genital HSV infections continue to be a major health problem worldwide. Although the use of ACV has contributed greatly to the management of primary and recurrent disease, there is still a need for additional modes of therapy for the prevention or treatment of these infections. As pointed out previously, it is important that animal model infections utilized for evaluation of new antiviral compounds be selected

Table 6. Effect of Treatment With Ara-A on Mortality of Mice
Inoculated Intranasally With HSV-2

| Treatment[a] | Mortality | | | | |
	Number	Percent	P-value	MDD[b]	P-value
None	11/15	73	-	7.3	-
Placebo + 24 h	12/14	86	NS[c]	8.8	NS
Ara-A 250 mg/kg + 24 h	10/15	67	NS	9.4	NS
+ 48 h	7/15	47	NS	11.6	NS
+ 72 h	12/15	80	NS	10.7	NS
+ 96 h	10/15	67	NS	7.1	<0.05
Ara-A 83.3 mg/kg + 24 h	12/15	80	NS	9.2	NS
+ 48 h	9/15	60	NS	7.9	NS
+ 72 h	10/14	71	NS	8.0	NS
+ 96 h	10/15	67	NS	8.0	NS
Ara-A 27.8 mg/kg + 24 h	10/15	67	NS	8.3	NS
+ 48 h	12/15	80	NS	10.0	NS
+ 72 h	12/15	80	NS	8.2	NS
+ 96 h	10/15	67	NS	8.4	NS
Ara-A 9.2 mg/kg + 24 h	13/15	87	NS	8.8	NS
+ 48 h	14/15	93	NS	8.3	NS
+ 72 h	10/15	67	NS	8.8	NS
+ 96 h	13/15	87	NS	8.3	NS

a Animals were treated i.p. twice daily for 7 days with stated
concentrations and initiated at the times indicated.

b MDD = Mean Day of Death.   c NS = Not Significant.

that most closely represent the target disease in man. A number of
animal species including mice, hamsters, guinea pigs, rabbits, monkeys
and non-human primates have been utilized as model infections for genital
herpes. In our laboratory we have utilized mice and guinea pigs
inoculated intravaginally with HSV-2 or HSV-1 as models of genital
herpes. As with all experimental infections, there are advantages and
disadvantages to the two experimental systems. The size and cost of mice
make them particularly useful for use in initial screening experiments,
and to determine optimal dosage and treatment regimens. However, since
mice inoculated with HSV by the intravaginal route generally die of acute
encephalomyelitis, do not develop typical lesions on external genital
skin, and do not exhibit spontaneous recurrent lesions, the model does
not accurately reflect the disease seen in humans. The advantages of the
guinea pig as a model for genital HSV infection include: 1) animals are
susceptible to infection with HSV-2 or HSV-1; 2) initial viral
replication occurs in the vaginal tract; 3) natural external lesions
develop on the external genital skin; 4) there is low mortality; 5) all
surviving animals become latently infected; and 6) spontaneous recurrent
lesions appear on the external genital skin. The main disadvantage of

Table 7. Effect of Treatment With the Sodium Salt of ACV on the
Mortality of Mice Inoculated Intranasally with HSV-2

Treatment[a]	Mortality			MDD[b]	P-value
	Number	Percent	P-value		
None	15/15	100	-	7.7	-
Placebo + 24 h	15/15	100	NS[c]	7.1	NS
ACV 60 mg/kg + 24 h	5/15	33	<0.001	14.2	<0.001
60 mg/kg + 48 h	4/15	27	<0.001	14.2	0.001
60 mg/kg + 72 h	11/15	73	NS	8.8	NS
60 mg/kg + 96 h	14/15	93	NS	7.9	NS
ACV 30 mg/kg + 24 h	9/15	60	0.01	9.9	<0.001
30 mg/kg + 48 h	9/15	60	0.01	9.1	<0.01
30 mg/kg + 72 h	12/15	80	NS	7.3	NS
30 mg/kg + 96 h	14/15	93	NS	7.8	<0.05
ACV 15 mg/kg + 24 h	13/15	87	NS	8.8	<0.01
15 mg/kg + 48 h	13/15	87	NS	9.4	<0.001
15 mg/kg + 72 h	15/15	100	NS	7.0	NS
15 mg/kg + 96 h	15/15	100	NS	7.7	NS
ACV 7.5 mg/kg + 24 h	14/15	93	NS	8.0	<0.05
7.5 mg/kg + 48 h	14/15	93	NS	8.6	<0.01
7.5 mg/kg + 72 h	15/15	100	NS	7.5	NS
7.5 mg/kg + 96 h	14/15	93	NS	7.4	NS

a Animals were treated i.p. twice daily for 7 days with stated
concentrations and initiated at the times indicated.

b MDD = Mean Day of Death.   c NS = Not Significant.

the guinea pig include cost, space requirements, and the need for
additional care. We have utilized both the mouse and guinea pig models
for many years and have tested a large number of potential antiviral
agents in both systems. In essentially all cases efficacy in the two
models has been identical. Since genital HSV-2 infection in the guinea
pig results in the development of external lesions similar to those seen
in humans and has spontaneous recurrent lesions, the guinea pig is the
model that most closely simulates the disease in humans.

The natural history of primary and recurrent HSV infection in the
guinea pig, the measures of efficacy utilized for evaluation of
antivirals and the effectiveness of a number of antiviral agents in this
model have been reported previously[13-19] and will be reviewed only
briefly to illustrate the similarity to the human disease. Intravaginal
inoculation of guinea pigs with HSV-2 results in a primary infection
characterized by viral replication in the vaginal tract followed by the
development of vesicular lesions on the external genital skin similar to
those seen in human disease.[20] After recovery from the primary
infection, spontaneous recurrent lesions appear at the site of initial

primary lesions. Unlike recurrent genital herpes in humans, the recurrent lesions in guinea pigs remain small, do not spread, last only 1-3 days, have a frequency of about one episode per week and generally one cannot isolate virus from the lesions. At this time, however, the genital HSV-2 infection of guinea pigs appears to be the model that most closely resembles primary and recurrent disease in humans.

The ultimate test of any animal model system is its predictability for humans. Until recently there has not been enough experience in testing an antiviral agent in animal models and human trials to establish the predictability of a model system for genital HSV-2 infections. The large body of information that has been collected using ACV now allows one to make comparisons of efficacy in the genital HSV-2 infection of guinea pigs with those reported in clinical trials. These comparisons using a number of measures of efficacy are summarized in table 8.

Topical treatment (intravaginal and external) during primary infection with 5% ACV significantly reduced viral replication in the vaginal tract, prevented or reduced the severity of external lesion development, and markedly reduced systemic clinical symptoms such as paralysis and urinary retention.[14,19] Subsequent recurrence rates, however, were not altered. Similar results have been observed in patients treated topically for primary or recurrent genital HSV infection.[21,22] During primary infection, oral or parenteral administration of ACV effectively reduced lesion severity, virus shedding and clinical symptoms in both guinea pigs[14,15] and in human disease.[23,24] Subsequent recurrent rates were not altered in the guinea pig[19] and generally are unchanged in humans.[23-25] With long term follow-up there are suggestions from one study that treatment during primary infection may have an effect on the frequency of recurrent episodes,[25] but this observation has not been confirmed.[26]

When therapy with oral ACV is initiated during recurrent disease, the frequency and severity of the episodes are reduced significantly in the guinea pig infection[19], and markedly altered in human disease during the time treatment is administered.[27,28] After cessation of therapy, however, the frequency of recurrences returns to pre-treatment levels in both the guinea pig model and in humans. The data collected to date using topical or oral administration of ACV in the guinea pig and in humans indicate that the genital HSV-2 infection of guinea pigs is predictive for the outcome of antiviral therapy of genital herpes in humans.

Table 8. Predictability of the Guinea Pig Model for Treatment of
Genital HSV Infection in Humans

	Topical ACV		Oral ACV	
	Guinea Pig	Humans	Guinea Pig	Humans
Primary Infection				
Reduce lesion severity	yes	yes	yes	yes
Reduce virus shedding	yes	yes	yes	yes
Reduce clinical symptoms	yes	yes	yes	yes
Reduce recurrence rates	no	no	no	no
Recurrent Disease				
Reduce frequency during treatment	no	no	yes	yes
Reduce severity during treatment	no	no	yes	yes
Reduce frequency after treatment	no	no	no	no

ANIMAL MODELS FOR OTHER HERPESVIRUS INFECTIONS

The experimental animal systems for other herpesviruses including, cytomegalovirus (CMV), varicella or varicella zoster virus (VZV) and perhaps Epstein-Barr virus (EBV) are listed on table 9. Cytomegalovirus is the causative agent of a variety of clinical syndromes in the fetus, neonate and particularly in patients that are immunosuppressed for organ transplantation, during chemotherapy for malignancies, or as a result of another infection such as AIDS. Due to the strict species specificity of human CMV, it is not possible to test potential antiviral agents against this virus in experimental animals. There are a number of natural CMV infections in various animal species, however, the two experimental systems that have been utilized for antiviral evaluation are murine CMV (MCMV) and guinea pig CMV (GPCMV).

Inoculation of mice with MCMV provides a model infection that shares many characteristics with the human disease. Both acute lethal and chronic non-lethal MCMV infections have been used in our laboratory to determine efficacy of antivirals.[29] After i.p. inoculation of 3 week-old Swiss Webster female mice with 1×10^6 pfu of MCMV, 90-100% of animals die with a mean day of death of 5-6 days. With an inoculum of 1×10^5 pfu all animals survive. With either inoculum, high titers of virus are present in lung, liver, spleen, kidney, and blood within 24 h and in salivary gland by 48-72 h. In the non-lethal infection, persistent viral replication occurs in lung, liver, kidney and spleen for at least 20-30 days and in salivary glands for months. The MCMV infection, therefore, involves many of the same target organs as HCMV. When guinea pigs are

Table 9. Animal Models for Other Herpesvirus Infections

Virus	Animal	Disease	Measure of efficacy
Cytomegalovirus (MCMV)	mice	generalized, acute, chronic	mortality pathogenesis
Cytomegalovirus (GPCMV)	guinea pig	generalized, acute, chronic	pathogenesis histopathology
Varicella or varicella zoster	guinea pig	viremia, rash nasal shedding	virus isolation seroconversion
Simian varicella	monkeys	disseminated w/skin lesions	mortality, rash virus titers blood chemistry

inoculated with GPCMV, virus can be isolated from the blood, lung, liver, kidney, urine and salivary glands. This model infection is analogous to many of the clinical syndromes seen in humans including transplacental transmission of virus, congenital infections, mononucleosis, interstitial pneumonia, and transmission of virus by blood transfusion.[30] It has been difficult to establish which, if either, of these two animal models is predictive of antiviral efficacy for human CMV infection, since until recently there has not been a form of effective therapy in the animal models or in humans.

The acyclic nucleoside analogue of acyclovir, 9-(1,3-dihydroxy-2-propoxymethyl) guanine (DHPG, 2'NDG, BW759) is a highly effective inhibitor of human CMV replication in tissue culture cells and is approximately 50 times as active against this virus as acyclovir. To determine the predictability of the murine model for human CMV infections we determined the sensitivity of MCMV to DHPG in tissue culture cells and compared the results with those published for human CMV. The results of numerous experiments indicated that MCMV was inhibited by 3.0 - 9.0 μ M of DHPG using a 50 percent plaque-reduction assay in MEF cells. Investigators from other laboratories have reported that approximately 1-7 μ M of DHPG is required to inhibit human strains of CMV in human fibroblast cells. These data indicate that both murine and human strains of CMV are equally susceptible to DHPG and support the validity of using the murine infection as a model for human CMV infections.

The effect of treatment with DHPG on mortality due to MCMV is illustrated in table 10. Groups of 15 mice were inoculated i.p. with virus and oral treatment with 50 mg/kg of DHPG was begun 6, 24, or 48 h

Table 10. Effect of Treatment With DHPG on the Mortality of Weanling Mice Inoculated i.p. With MCMV

Treatment		Mortality Number	Percent	P-value	MDD	P-value
Oral[a]						
Placebo	+ 6 h	14/15	93	-	6.6	-
DHPG	+ 6 h	0/15	-	<0.001	-	-
	+ 24 h	1/15	7	<0.001	7.0	NS
	+ 48 h	8/15	53	<0.05	8.4	<0.001
Intraperitoneal[b]						
None	+ 24 h	13/15	87	-	5.3	-
Placebo	+ 24 h	14/15	93	NS	5.1	NS
DHPG 40 mg/kg	+ 6 h	0/15	0	<0.001	-	-
	+ 24 h	0/14	0	<0.001	-	-
	+ 48 h	8/15	53	<0.05	8.5	NS
	+ 72 h	15/15	100	NS	5.5	NS
DHPG 20 mg/kg	+ 6 h	0/15	0	<0.001	-	-
	+ 24 h	0/15	0	<0.001	-	-
	+ 48 h	12/15	80	NS	4.1	<0.05
	+ 72 h	15/15	100	NS	4.6	NS
DHPG 10 mg/kg	+ 6 h	0/15	0	<0.001	-	-
	+ 24 h	2/15	13	<0.001	6.0	<0.05
	+ 48 h	10/15	67	NS	8.0	<0.05
	+ 72 h	14/15	93	NS	5.5	NS
DHPG 5 mg/kg	+ 6 h	4/15	27	<0.001	9.5	<0.05
	+ 24 h	6/15	40	<0.01	5.7	NS
	+ 48 h	14/15	93	NS	5.9	<0.05
	+ 72 h	14/15	93	NS	5.8	<0.05
None		13/15	87	NS	5.4	NS

a Treatment with 50 mg/kg was initiated at times indicated and continued twice daily for seven days.

b Treatment administered twice daily for 5 days.

after viral inoculation. There was significant reduction in mortality when therapy was begun as late as 48 h after MCMV infection. To determine the effect of DHPG treatment by the i.p. route, mice were inoculated with MCMV and treated with 40, 20, 10, or 5 mg/kg of DHPG beginning 6, 24, 48 or 72 h after infection. Treatment with 40 mg/kg gave protection when initiated as late as 48 h after infection, whereas, at 20, 10 or 5 mg/kg a decrease in mortality occurred only when treatment was initiated at 24 h post infection. Since treatment with DHPG effectively reduced mortality in the lethal infection, we next determined the effect of therapy with DHPG on the pathogenesis of a chronic MCMV infection of mice (data not shown). Groups of mice were inoculated with a sublethal concentration of MCMV and treated with either placebo or 40 mg/kg of DHPG beginning at 24 h and continued for 14 days. There was a significant reduction in virus titers of DHPG-treated animals in blood, lung, liver, spleen and kidney. Viral clearance was also more rapid in these organs of treated animals. In salivary gland, there was a 2 day delay in virus replication, and virus titers were consistently lower for the first 15 days. On day 22, there was little difference between the 2 groups. Both MCMV and HCMV are sensitive to DHPG in vitro and the results from the animal model experiments suggested that this compound may be an excellent candidate for treatment of CMV infections in humans.

The 50% inhibitory dose of DHPG against GPCMV has been reported to be 71 μM in tissue culture cells,[31] indicating that GPCMV is considerably less susceptible to DHPG than either murine or human CMV. When guinea pigs were inoculated i.p. with GPCMV and treated twice daily for 7 days with 25 mg/kg of DHPG, virus titers in the lung were higher in drug-treated animals than in those treated with placebo. There were no significant differences in the severity of histopathologic lesions in the lung, liver and spleen between drug and placebo-treated animals, however DHPG-treated animals had fewer lesions in kidney and salivary glands than those treated with placebo. Virus titers in salivary gland were also lower in DHPG-treated animals.[31] These data indicate that DHPG had only minimal activity against GPCMV both in tissue culture and in the animal model.

Although DHPG has not been approved for general use, it is currently being used to treat human CMV infections on a compassionate plea basis. Since there has not been any double-blind, placebo-controlled studies conducted regarding the efficacy of DHPG against HCMV infections, its effectiveness has not been established. However, there are numerous reports attesting to its activity in retinitis, esophagitis, and colitis due to CMV.[32,33] The effect of DHPG treatment for CMV-induced pneumonia has been minimal. In one study designed to determine the effect of DHPG therapy on CMV replication in lung tissue, there was a dramatic 4-6 log decrease in viral titers after therapy with DHPG.[34] The results obtained in the clinical studies, particularly those that did virus cultures, correlated well with the data obtained in the murine model. There is not as good a correlation with the guinea pig model, presumably due to the lack of susceptibility of GPCMV to DHPG. Based on the limited amount of data available, it would appear that the mouse model, at least for DHPG, is more predictive of antiviral efficacy for human CMV than is the guinea pig model.

Varicella-zoster virus (VZV) infections in the immunocompromised host continues to be a problem of great magnitude. Due to the species specificity of VZV, there is currently no established animal model for evaluating antiviral drugs for their activity against this virus. It has been reported that weanling guinea pigs are susceptible to VZV that has been adapted in fetal guinea pig cells.[35] After intranasal or subcutaneous inoculation the virus replicates in the nasopharynx, viremia can be detected, specific humoral antibodies are produced and animal to animal transmission has been documented. Since this model infection has not yet been utilized for antiviral studies, the suitability of this model infection is unknown at the present time. The model should certainly aid in enhancing our understanding of the pathophysiology of this infection and help elucidate the role of the immune system in VZV infections. It may also provide a system for investigating the mechanism of latent infections.

A model for VZV that does not use the human virus is simian varicella virus infection of monkeys. After intratracheal inoculation the animals develop fever and viremia followed by a disseminated infection with skin rash.[36] A large number of antiviral agents have been tested in this animal model, including Ara-A, ACV, alpha interferon, and BVDU. These agents have also been evaluated in placebo-controlled human studies. When African green monkeys were infected with simian varicella virus and treated i.v. with 15 mg/kg of Ara-A per day for 10 days beginning 48h after infection no significant differences in viremia, rash or serum transaminase levels were observed between treated and control animals.[37] ACV treatment of African green monkeys at 10 mg/kg i.v. twice daily for 10 days also had no effect on clinical symptoms, development of rash or viremia.[38] In contrast, ACV administration at 100 mg/kg i.m. effectively prevented or reduced viremia, appearance of rash and elevated transaminases.[39] The difference in the two studies appeared to be due to achievement of sufficient plasma levels of ACV when the higher dose was given by the i.m. route. Administration of 10^6 units/kg of rIFN-αA twice daily for 8 days also inhibited simian varicella virus infection of monkeys whether treatment was begun 4 h prior to or 44 h after virus inoculation.[40] Similar efficacy has been obtained with oral, i.m., or i.v. administration of BVDU.[41]

In immunocompromised patients, both ACV and Ara-A have been reported to prevent the development of visceral disease in children with primary varicella and reduce dissemination of VZV.[42-45] In a study comparing the two drugs it was shown that ACV was significantly more effective than Ara-A treatment for VZV infections in severly immunocompromised patients.[46] In other studies, human leukocyte interferon or BVDU have also been reported to be effective in treatment of varicella or VZV infections of immunocompromised patients.[47,48] Although the experimental simian varicella infection of African green monkeys did not accurately predict efficacy of Ara-A in humans, it was predictive of efficacy for ACV, interferon and BVDU. The model is also of great value in determining drug pharmacokinetics, and toxicity prior to use in man.

Another herpesvirus for which there is no animal model available is Epstein-Barr virus (EBV). It was reported recently that guinea pigs infected with guinea pig herpes like virus (GPHLV) undergo a lymphocytosis and lymphoid hyperplasia similar to that seen in chronic EBV syndrome.[49] This model will need additional characterization, and a better understanding of the clinical syndrome is necessary before the usefulness of this model is established.

ANIMAL MODELS FOR RESPIRATORY VIRUS INFECTIONS

There have been few antiviral agents developed in recent years that have activity against respiratory infections in experimental animals and man. Consequently, there are little comparative data on which to judge the predictability of the animal model infections. The experimental animal infections that have been most widely used, or developed for use in evaluating antiviral agents are listed in table 11. A wide range of animal species including mice, cotton rats, ferrets, hamsters, guinea pigs, dogs, pigs and non-human primates have been used, however, those most commonly used for evaluation of antiviral agents have been mice and ferrets. Mice are not very susceptible to infection with influenza virus unless the virus is adapted by passage through mouse lung. Ferrets are highly susceptible to human influenza viruses without adaptation and develop symptoms similar to those seen in humans. One main disadvantage to the use of ferrets as an animal model for antiviral studies is their size and cost.

Amantadine, which is approved for both prophylaxis and treatment of influenza A virus infection has reasonably good activity in mice depending on virus strain, inoculum size, route of challenge, dosage, time, and route of drug administration.[50,51] In contrast to the results

Table 11. Animal Models for Respiratory Virus Infections

Virus	Animal	Disease	Measure of efficacy
Influenza A or B	mice	respiratory	mortality virus titers in lung
	cotton rat	respiratory	mortality virus titers
	ferrets	respiratory	morbidity virus titers
RSV	cotton rat hamster	respiratory	virus titers histopathology

in most other animals or humans, amantadine treatment is not effective in ferrets,[52] which suggest that this may not be the most predictable model. In experimental infections of mice, Rimantidine is as active if not more so than Amantadine, but in human disease, Amantadine appears to be the more active of the two drugs. Ribavirin also has been shown to have activity against influenza virus infections of mice, cotton rats, ferrets, hamsters and squirrel monkeys,[53] and in human studies aerosolized ribavirin has been reported to alter clinical symptoms and reduce shedding of influenza virus from the respiratory tract.[54]

Respiratory syncytial virus (RSV) is the causative agent of severe bronchiolitis in infants and young children. Mice, hamsters, cotton rats, ferrets and non-human primates infected with RSV have been utilized as models for this disease. The cotton rat appears to be the animal of choice for antiviral studies directed against RSV. After intranasal inoculation RSV replicates to high titer in the nose and lungs with lower titers found in the trachea.[55] In this model ribavirin delivered by aerosol decreased RSV titers in lung 10-50 fold.[56] In a trial of RSV infection of infants with lower respiratory tract disease, aerosolized ribavirin significantly increased improvement of clinical severity scores, and at the end of therapy treated infants had lower virus titers in nasal wash samples than placebo recipients.[57]

Rhinoviruses are the most widespread cause of the common cold syndrome in man. Although there have been reports of mouse models for rhinovirus infections using highly adapted virus, there is currently no accepted animal model for these infections.

Table 12. Animal Models for Retrovirus Infections

Virus	Animal	Disease	Measure of efficacy
HIV, HTLV-III, LAV	chimpanzee	viremia lymphadenopathy no immunodeficiency	virus titers antibody pathogenesis
Murine leukemia	mice	lymphomas immunodeficiency	unknown
Feline leukemia	cats	lymphoma leukemia immunosuppression	unknown
FTLV	cats	T-lymphotropic chronic immunosuppression	unknown
STLV-III SIV	monkeys	T-lymphotropic immunodeficiency	unknown

ANIMAL MODELS FOR RETROVIRUS INFECTIONS

The emergence of acquired immunodeficiency disease syndrome (AIDS) and the subsequent isolation of human immunodeficiency virus (HIV, HTLV-III, LAV) as the etiologic agent of this disease has intensified the search for animal models of human retrovirus infections. The experimental animal systems that are currently being investigated are summarized in table 12. The chimpanzee is the only animal that has been successfully infected with HIV. Viremia can be readily and consistently demonstrated and seroconversion occurs. Although many of the animals have a persistent infection, an AIDS-like disease has not developed.[58,59] A transient lymphadenopathy syndrome or alterations in T4 or T8 levels have been reported but are a rare occurrence. Due to the shortage of chimpanzees it is unlikely that this animal model will be available for use in testing antiviral agents, but will be utilized for evaluation of HIV vaccines. Therefore, other experimental retrovirus systems will have to be utilized for initial determination of antiviral efficacy in vivo. Animal model infections using murine,[60] feline,[61,62] and simian[63,64] retroviruses all have potential for use in antiviral research. There are obvious advantages and disadvantages of each of the model systems and variables such as sensitivity of the virus to an antiviral in vitro, relatedness of the virus to HIV, similarity of the animal disease to AIDS in humans, interaction of the virus with components of the immune system, and the availability of animals and appropriate biological containment facilities must all be considered when selecting an animal model for HIV

Table 13. Animal Models for Togavirus and Arenavirus Infections

Virus	Animal	Disease	Measure of efficacy
Alphaviruses	mice	encephalitis	mortality
Flaviviruses	mice	encephalitis	mortality
Lassa Fever	old world monkeys guinea pigs	lassa fever	mortality virus isolation pathogenesis
Machupo	old world monkeys guinea pigs	hemorrhagic fever	mortality virus isolation pathogenesis
Junin	old world monkeys guinea pigs	hemorrhagic fever	mortality virus isolation pathogenesis

infections. Since there currently are no comparative data regarding the efficacy of an antiviral in an animal model and humans, no predictions can be made as to which animal systems may be the most appropriate.

ANIMAL MODELS FOR TOGAVIRUSES AND ARENAVIRUSES

The model infections usually used for these arthropod transmitted diseases are listed in table 13. Of the approximately 400 known togaviruses, about 10% of these, mainly alphaviruses and flaviviruses cause severe disease in man. The clinical syndromes include encephalitis and hemorrhagic fevers which are often fatal in the accidently infected human host. The suckling mouse is the animal of choice for these viral infections. Since these diseases are not routinely and commonly found in the U.S. or Europe, there has been little progress in the development of antivirals for these infections. The arenaviruses are the cause of severe, often fatal hemorrhagic fever in man and are limited geographically to Africa and South America. Although mice can be infected with many of these viruses, old world monkeys and guinea pigs appear to be the best model infections for studying these diseases. Ribavirin has been the most widely tested antiviral for these diseases and these studies are reviewed elsewhere in this volume (Peter G. Canonico, Exotic RNA viruses).

Table 14. Animal Models for Other Important Virus Infections

Virus	Animal	Disease	Measure of efficacy
Enteroviruses	mice	encephalomyelitis cardiac diabetes	death histopathology
Hepatitis B	chimpanzee	antigenemia chronic carriers	viremia
Woodchuck hepatitis	woodchucks	chronic hepatitis hepatomas	viremia
Human papilloma	?	?	?
Bovine papilloma	cows	warts	size and severity
Rotavirus	mice cows	diarrhea	severity virus titers

ANIMAL MODELS FOR OTHER IMPORTANT VIRAL INFECTIONS

There are a number of other viral infections of man that warrant development of antiviral therapy (table 14). Investigations concerning the pathogenesis and treatment of these diseases has been hampered by the inability to infect tissue culture cells or experimental animals with many of these viruses. The enteroviruses include the polioviruses, coxsackieviruses and echoviruses. While poliovirus infections have essentially been controlled through immunization, the coxsackieviruses and echoviruses are the cause of a number of serious infections including aseptic meningitis and cardiomyopathies. Mice can be infected experimentally with many of these viruses, and are used for evaluation of antiviral agents. The one animal model that has been shown to be susceptible to infection with hepatitis B virus is the chimpanzee. These animals regularly develop an antigenemia and some go on to be chronic carriers of surface antigen. The animals do not, however, generally develop cirrhosis, and do not develop hepatomas. The primary disadvantage in using the chimpanzee as an animal model is its cost and lack of availability. In recent years a number of animal hepatitis viruses have been identified that share many of the features of human hepatitis B infection. Woodchuck hepatitis, ground squirrel hepatitis, and duck hepatitis are potential models that could be utilized for evaluating antiviral agents that have potential use against hepatitis B. Of these species, the woodchuck appears to be the animal of choice as it most closely mimics the spectrum of disease caused by the human virus and is available commercially. There is currently no animal model available for human papillomaviruses, although there are experimental systems under development. For viral chemotherapy studies, the bovine and rabbit papilloma systems has been the most widely used. The rotaviruses are another example of the failure of the human virus to readily and reproducibly infect the non-human host, necessitating the use of animal virus infections such as murine, porcine, or bovine systems for studying the prevention or treatment of these diseases. There is little data concerning antiviral efficacy in man and experimental models for any of these important viral infections, so predictability can not be assessed for any of the experimental models.

SUMMARY

Animal model systems have been developed for a large number of viral infections of humans. With the exception of the herpesviruses, there have not been enough antiviral agents evaluated in man to establish the predictability of most experimental animal systems. The current problem is that most viral infections in man that have the greatest need for an effective antiviral agent such as those caused by HIV, CMV, hepatitis B, and papillomaviruses exhibit species specificity and do not infect experimental animals. For these viral infections, investigations will need to concentrate on developing natural animal infections that most closely represent the corresponding human disease. For animal models where predictability has been established, i.e., HSV and certain respiratory infections, these models should be utilized to their greatest potential in developing more effective viral chemotherapeutic agents.

ACKNOWLEDGEMENTS

The author wishes to express his appreciation to Peggy Vogt for technical assistance and to Sharon Seely for typing the manuscript. This is publication No. 94 from the Cooperative Antiviral Testing Group, Development and Applications Branch, National Institute of Allergy and Infectious Diseases (NIAID), National Institutes of Health (NIH), Bethesda, MD. This work was supported in part by Public Health Service Contract No. NO1-AI-62518 from the Antiviral Substances Program, Development and Applications Branch, NIAID, NIH.

REFERENCES

1. R.J. Whitley, S.-J. Soong, R. Dolin, G.J. Galasso, L.T. Ch'ien, C.A. Alford, and the Collaborative Study Group. Adenine arabinoside therapy of biopsy-proved herpes simplex encephalitis. N. Engl. J. Med. 297:289 (1977).

2. R.J. Whitley, S.-J. Soong, M.S. Hirsch, A.W. Karchmer, R. Dolin, G. Galasso, J.K. Dunnick, C.A. Alford, and the NIAID Collaborative Antiviral Study Group. Herpes simplex encephalitis. Vidarabine therapy and diagnostic problems. N. Engl. J. Med. 304:313 (1981).

3. E.R. Kern, J.T. Richards, L.A. Glasgow, J.C. Overall, Jr., and P. De Miranda. Optimal treatment of herpes simplex virus encephalitis in mice with oral acyclovir. Symposium on acyclovir. Am. J. Med. 73:125 (1982).

4. R.J. Whitley, C.A. Alford, M.S. Hirsch, R.T. Schooley, J.P. Luby, F.Y. Aoki, D. Hanley, A.J. Nahmias, S.-J. Soong, and the NIAID Collaborative Antiviral Study Group. N. Engl. J. Med. 314:144 (1986).

5. E.R. Kern, J.T. Richards, J.C. Overall, Jr., and L.A. Glasgow. Alteration of mortality and pathogenesis of three experimental Herpesvirus hominis infections of mice with adenine arabinoside 5'- monophosphate, adenine arabinoside, and phosphonoacetic acid. Antimicrob. Agents Chemother. 13:53 (1978).

6. E.R. Kern, J.C. Overall, Jr., and L.A. Glasgow. Herpesvirus hominis infection in newborn mice: comparison of the therapeutic efficacy of 1-β-D-Arabinofuranosylcytosine and 9-β-D-Arabinofuranosyladenine. Antimicrob. Agents Chemother. 7:587 (1975).

7. R.J. Whitley, A.J. Nahmias, S.-J. Soong, G.T. Galasso, C.L. Fleming, and C.A. Alford. Vidarabine therapy of neonatal herpes simplex virus infections. Pediatrics 66:495 (1980).

8. R.J. Whitley, A. Yeager, P. Kartus, Y. Bryson, J.D. Connor, A. Nahmias, and S.-J. Soong. Neonatal herpes simplex virus infection. Follow-up evaluation of vidarabine therapy. Pediatrics 72:778 (1983).

9. W.R. Hubler, T.D. Felber, D. Troll, and M. Jarratt. Guinea pig model for cutaneous herpes simplex virus infection. J. Invest. Dermatol 62:92 (1974).

10. T.W. Schaefer, M. Lieberman, J. Everitt, and P. Came. Cutaneous herpes simplex virus infection as a model for antiviral chemotherapy. Ann. N.Y. Acad. Sci. 284:624 (1977).

11. S.L. Spruance, D.J. Freeman, and N.V. Sheth. Comparison of topical foscarnet, acyclovir (ACV) and ACV ointment in the treatment of experimental cutaneous herpes simplex virus (HSV) infection. Antimicrob. Agents Chemother. 30:196 (1986).

12. S.L. Spruance, M.B. McKeough, and J.R. Cardinal. Penetration of guinea pig skin by acyclovir in different vehicles and correlation with the efficacy of topical therapy of experimental cutaneous herpes simplex virus infection. Antimicrob. Agents Chemother. 25:10 (1984).

13. E.R. Kern, L.A. Glasgow, J.C. Overall, Jr., J.M. Reno, and J.A. Boezi. Treatment of experimental herpesvirus infections with Phosphonoformate and some comparison with Phosphonoacetate. Antimicrob. Agents Chemother. 14:817 (1978).

14. E.R. Kern. Acyclovir treatment of experimental genital herpes simplex virus infections. Symposium on acyclovir. Am. J. Med. 73:100 (1982).

15. A.D. Pronovost, H.L. Lucia, P.R. Dann, and G.D. Hsiung. Effect of acyclovir on genital herpes in guinea pigs. J. Infect. Dis. 145:904 (1982).

16. L.R. Stanberry, E.R. Kern, J.T. Richards, T.A. Abbott, and J.C. Overall, Jr. Genital herpes in guinea pigs: Pathogenesis of the

primary infection and description of recurrent disease. J. Infect. Dis. 146:397 (1982).

17. G.D. Hsiung, D.R. Mayo, H.L. Lucia, and M.L. Landry. Genital herpes: Pathogenesis and chemotherapy in the guinea pig model. Rev. Infect. Dis. 6:33 (1984).

18. E.B. Fraser-Smith, D.F. Smee, and T.R. Matthews. Efficacy of the acyclic nucleoside 9-(1,3-dihydroxy-2-propoxymethyl) guanine against primary and recrudescent genital herpes simplex virus type 2 infections in guinea pigs. Antimicrob. Agents Chemother. 24:883 (1983).

19. E.R. Kern. Treatment of genital herpes simplex virus infections in guinea pigs, In: "Herpesvirus", F. Rapp, ed., Alan R. Liss, Inc., N.Y. (1984).

20. L. Corey, H.G. Adams, Z.A. Brown, and K.K. Holmes. Genital herpes simplex virus infections: Clinical manifestations, course, and complications. Ann Intern. Med. 98:958 (1983).

21. L. Corey, A.J. Nahmias, M.E. Guinan, J.K. Benedetti, C.W. Critchlow, and K.K. Holmes. A trial of topical acyclovir in genital herpes simplex virus infections. N. Engl. J. Med. 306:1313 (1982).

22. R.C. Reichman, G.J. Badger, M.E. Guinan, A.J. Nahmias, R.E. Keeney, L.G. Davis, T. Ashikaga and R. Dolin. Topically administered acyclovir in the treatment of recurrent herpes simplex genitalis: A controlled trial. J. Infect. Dis. 147:336 (1983).

23. Y.J. Bryson, M. Dillon, M. Lovett, G. Acuna, S. Taylor, J.D. Cherry, B.L. Johnson, E. Wiesmeier, W. Growdon, T. Creagh-Kirk, and R. Keeney. Treatment of first episodes of genital herpes simplex virus infection with oral acyclovir. N. Engl. J. Med. 308:916 (1983).

24. L. Corey, K.H. Fife, J.K. Benedetti, C.A. Winter, A. Fahnlander, J.D. Connor, M.A. Hintz, and K.K. Holmes. Intravenous acyclovir for the treatment of primary genital herpes. Ann. Intern. Med. 98:914 (1983).

25. Y.J. Bryson, M. Dillon, M. Lovett, D. Bernstein, E. Garratty, and J. Sayre. Treatment of first episode genital HSV with oral acyclovir: Long term follow-up of recurrences. A preliminary report. Scand. J. Infect. Dis. (Suppl.) 47:70 (1985).

26. G.J. Mertz, C.W. Critchlow, J. Benedetti, R.C. Reichman, R. Dolin, J. Connor, D.C. Redfield, M.C. Savoia, D.D. Richman, D.L. Tyrrell, L Miedzinski, J. Portnoy, R.E. Keeney, and L. Corey. Double-blind placebo controlled trial of oral acyclovir for first episode genital herpes. J. Am. Med. Assoc. 252:1147 (1984).

27. R.C. Reichman, G.J. Badger, G.J. Mertz, L. Corey, D.D. Richman, J.D. Connor, D. Redfield, M.C. Savoia, M.N. Oxman, Y. Bryson, D.L. Tyrrell, J. Portnoy, T. Creigh-Kirk, R. Keeney, T. Ashikaga, and R. Dolin. Treatment of recurrent genital herpes simplex infections with oral acyclovir. A controlled trial. J. Am. Med. Assoc. 251:2103 (1984).

28. J.M. Douglas, C. Critchlow, J. Benedetti, G.J. Mertz, J.D. Connor, M.A. Hintz, A. Fahnlander, M. Remington, C. Winter, and L. Corey. A double-blind study of oral acyclovir for suppression of recurrences of genital herpes simplex virus infection. N. Engl. J. Med. 310:1551 (1984).

29. L.A. Glasgow, J.T. Richards, and E.R. Kern. Effect of acyclovir treatment on acute and chronic murine cytomegalovirus infection. Symposium on acyclovir. Am. J. Med. 73:132 (1982).

30. F.J. Bia, B.P. Griffith, C.K.Y. Fong and G.D. Hsiung. Cytomegalovirus infections in the guinea pig: Experimental models for human disease. Rev. Infect. Dis. 5:177 (1983).

31. C.K.Y. Fong, S.D. Cohen, S. McCormick, and G.D. Hsiung. Antiviral effect of 9-(1,3-dihydroxy-2-propoxymethyl) guanine against cyto-

megalovirus infection in a guinea pig model. Antiviral Res. 7:11(1987).

32. J. Mills. 9-(1,3-dihydroxy-2-propoxymethyl) guanine (DHPG) for treatment of cytomegalovirus infections, In: "Antiviral Chemotherapy, New Directions for Clinical Application and Research", J. Mills and L. Corey, eds., Elsevier, New York (1986).

33. Collaborative DHPG Treatment Study Group. Treatment of serious cytomegalovirus infections with 9-(1,3-dihydroxy-2-propoxymethyl) guanine in patients with AIDS and other immunodeficiencies. N. Engl. J. Med. 314:801 (1986).

34. D.H. Shepp, P.S. Dandliker, P. de Miranda, T.C. Burnette, D.M. Cederberg, L.E. Kirk, and J.D. Meyers. Activity of 9-[2-hydroxy-1-(hydroxymethyl)ethoxymethyl] guanine in the treatment of cytomegalovirus pneumonia. Ann. Intern. Med. 103:368 (1985).

35. M.G. Myers, H.L. Duer, and C.K. Hausler. Experimental infection of guinea pigs with varicella-zoster virus. J. Infect. Dis. 142:414 (1980).

36. A.D. Felsenfeld, and N.J. Schmidt. Antigenic relationships among several simian varicella-like viruses and varicella-zoster virus. Infect. Immun. 15:807 (1977).

37. K.F. Soike, A.D. Felsenfeld, S. Gibson, and P.J. Gerone. Ineffectiveness of adenine arabinoside and adenine arabinoside 5'-monophosphate in simian varicella infection. Antimicrob. Agents Chemother. 18:142 (1980).

38. K. F. Soike, A.D. Felsenfeld, and P.J. Gerone. Acyclovir treatment of experimental simian varicella infection of monkeys. Antimicrob. Agents Chemother. 20:291 (1981).

39. K.F. Soike, and P.J. Gerone. Acyclovir in the treatment of simian varicella virus infection of the African Green monkey. Symposium on Acyclovir. Am. J. Med. 73:112 (1982).

40. K.F. Soike, M.J. Kramer, and P.J. Gerone. In vivo antiviral activity of recombinant type α interferon A in monkeys with infections due to simian varicella virus. J. Infect. Dis. 147:933 (1983).

41. K.F. Soike, S. Gibson, and P.J. Gerone. Inhibition of simian varicella virus infection of African Green monkeys by (E)-5-(2-bromovinyl)-2'-deoxyuridine (BVDU). Antiviral Res. 1:325 (1981).

42. R.J. Whitley, M. Hilty, R. Haines, Y. Bryson, J.D. Connor, S.-J. Soong, C.A. Alford, Jr., and NIAID Collaborative Antiviral Study Group. Vidarabine therapy of varicella in immunocompromised patients. J. Pediatrics. 101:125 (1982).

43. R.J. Whitley, S.-J. Soong, R. Dolin, R. Betts, C. Linnemann, Jr., C.A. Alford, Jr., and the NIAID Collaborative Antiviral Study Group. Early vidarabine therapy to control the complications of herpes zoster in immunocompromised patients. N. Engl. J. Med. 307:971 (1982).

44. C.G. Prober, L.E. Kirk, and R.E. Keeney. Acyclovir therapy of chickenpox in immunosuppressed children - A collaborative study. J. Pediatrics. 101:622 (1982).

45. H.H. Balfour, Jr., B. Bean, O.L. Laskin, R.F. Ambinder, J.D. Meyers, J.C. Wade, J.A. Zaia, D. Aeppli, L.E. Kirk, A.C. Segreti, and R.E. Keeney. Acyclovir halts progression of herpes zoster in immunocompromised patients. N. Engl. J. Med. 308:1448 (1983).

46. D.H. Shepp, P.S. Dandliker, and J.D. Meyers. Treatment of varicella- zoster virus infection in severely immunocompromised patients. A randomized comparison of acyclovir and vidarabine. N. Engl. J. Med. 314:208 (1986).

47. A.M. Arvin, J.H. Kushner, S. Feldman, R.L. Baehner, D. Hammond, and T.C. Merigan. Human leukocyte interferon for the treatment of varicella in children with cancer. N. Engl. J. Med. 306:761 (1982).

48. Y. Benoit, G. Laureys, M.-J. Delbeke and E. De Clercq. Oral BVDU treatment of varicella in children with cancer. Eur. J. Pediatr. 143:198 (1985).

49. Z. Nagy, T.A. Jennings, T.G. Brady, H.L. Lucia, J.A. Armstrong, and G.D. Hsiung. Effect of cyclosporin A immunosuppression on primary lymphotropic herpes virus infection in the guinea pig. Intervirol. In press.

50. R.R. Grunert, J.W. McGahen, and W.L. Davies. The in vivo antiviral activity of 1-adamantanamine (amantadine) 1. prophylactic and therapeutic activity against influenza viruses. Virology 26:262 (1965).

51. J.S. Walker, E.L. Stephen, and R.O. Spertzel. Small particle aerosols of antiviral compounds in treatment of type A influenza pneumonia in mice. J. Infect. Dis. 133:A140 (1976).

52. C.W. Potter and J.S. Oxford. Animal models of influenza virus infection as applied to the investigation of antiviral compounds. In: "Chemoprophylaxis and Virus Infections of the Respiratory Tract". J.S. Oxford, ed., CRC Press, Cleveland (1977).

53. F.G. Hayden. Animal models of influenza virus infection for evaluation of antiviral agents. In: "Experimental Models in Antimicrobial Chemotherapy", vol. 3, O. Zak and M.A. Sande eds., Academic Press, London (1986).

54. V. Knight and B.E. Gilbert. Ribavirin aerosol treatment of Influenza. In: "Infectious Disease Clinics of North America", V. Knight and B.E. Gilbert, eds. W.B. Saunders, Philadelphia (1987).

55. G.A. Prince, A.B. Jenson, R.L. Horswood, E. Camargo, and R.M. Chanock. The pathogenesis of respiratory syncytial virus infection in cotton rats. Am. J. Pathol. 93:771 (1978).

56. P.R. Wyde, S.Z. Wilson, R. Petrella, and B.E. Gilbert. Efficacy of high dose - short duration ribavirin aerosol in the treatment of respiratory syncytial virus infected cotton rats and influenza B virus infected mice. Antiviral Res. 7:211 (1987).

57. C.B. Hall, J.T. McBride, C.L. Gala, S.W. Hildreth, and K.C. Schnabel. Ribavirin treatment of respiratory syncytial viral infection in infants with underlying cardiopulmonary disease. J. Am. Med. Assoc. 254:3047 (1985).

58. H.J. Alter, J.W. Eichberg, H. Masur, W.C. Saxinger, R. Gallo, D.M. Macher, H.C. Lane, and A.S. Fauci. Transmission of HTLV-III infection from human plasma to chimpanzees: An animal model for AIDS. Science 226:549 (1984).

59. P.N. Fultz, H.M. Mclure, R.B. Swenson, C.R. McGrath, A. Brodie, J.P. Getchell, F.C. Jensen, D.C. Anderson, J.R. Broderson, and D.P. Francis. Persistent infection of chimpanzees with human T-lymphotropic virus type III/lymphadenopathy associated virus: a potential model for acquired immunodeficiency syndrome. J. Virol. 58:116 (1986).

60. D.E. Mosier, R.A. Yetter, and H.C. Morse, III. Retroviral induction of acute lymphoproliferative disease and profound immunosuppression in adult C57BL/6 mice. J. Exp. Med. 161:766 (1985).

61. W.D. Hardy. Feline acquired immune deficiency syndrome: A feline retrovirus-induced syndrome of pet cats. In: "Animal Models of Retrovirus Infection and Their Relationship to AIDS". L.A. Salzman ed., Academic Press, London. (1986).

62. N.C. Pedersen, E.W. Ho, M.L. Brown, and J.K. Yamamoto. Isolation of a T-lymphotropic virus from domestic cats with an immunodeficiency-like syndrome. Science 235:790 (1987).

63. P.J. Kanki, and M. Essex. Animal models of HTLV-III/LAV infection and AIDS. In: "AIDS, Modern Concepts and Therapeutic Challenges", S. Broder, ed. Marcel Dekker Inc. (1987).

64. R.C. Desrosiers and N.L. Letvin. Animal models for acquired immunodeficiency syndrome. Rev. Infect. Dis. 9:438 (1987).

EXPERIMENTAL ASPECTS OF ANTIVIRAL PHARMACOLOGY

William H. Prusoff and Tai-Shun Lin

Department of Pharmacology
Yale University School of Medicine
333 Cedar Street
New Haven, Connecticut 06510

Experimental pharmacology of antiviral agents include:
1. Design and synthesis of antiviral agents
2. Determination of antiviral activity in cell cultures, experimental animals and humans
3. Study of absorption, distribution, metabolism and excretion
4. Elucidation of the molecular basis for efficacy or toxicity.

Effective clinical use requires proper diagnosis of the infection, proper selection of the drug, and a knowledge of the pharmacokinetics of the drug. The drug must be absorbed into the body, transported to the infected cell, and maintain an effective concentration at the target site for the time required to exert its antiviral activity. Ideally this is to be achieved without the production of toxicity to the uninfected cells or organ function, or the production of a resistant population of virus.

1. HOW ARE DRUGS DEVELOPED?

Most of the antiviral agents that are at present clinically useful antiviral drugs evolved from a program of random screening or serendipity. Thus for example amantadine was discovered during a screening program for antiviral activity; idoxuridine (5-iodo-2'-deoxyuridine), trifluridine (5-trifluoromethyl-2'-deoxuridine), and vidarabine (9-β-D-arabinofuranosyladenine) were originally developed as anticancer agents; acyclovir [9-(2-hydroxyethoxymethyl)-guanine] evolved from a program designed to find inhibitors of the enzyme, adenosine deaminase.

Although serendipity and random screening still play a major role in producing effective antiviral agents, the hope is that the rapid progress being made in understanding of the molecular biology of viral replication and of virus-host interrelationships will uncover unique targets for antiviral chemotherapy. Rapid progress is being made in how we may determine the structural and spatial requirements for a compound to interact uniquely with the viral target.

Rational design of antiviral agents based on our understanding of gene structure and function is a very promising area. Some success has already been achieved in the synthesis of oligonucleotides complementary to critical regions of viral m-RNA which result in specific hybridization affecting their function (Smith et al., 1986; Toulme et al., 1986; Zamecnik et al., 1986; Lemaitre et al., 1987; and earlier references cited in Prusoff et al., 1986). The necessity for chiral purity to ensure the required complementarity was discussed by Goldanskii et al. (1986).

Computer-assisted molecular design for the synthesis of drugs is becoming of increasing importance, and has been successful in the design of analogs of trimethoprim by Kuyper and coworkers (1982) following determination of the three-dimensional molecular structure of the enzyme, dihydrofolate reductase, by X-ray crystallography.

The complete three dimensional structure of the human rhinovirus-14 was determined by Rossman and coworkers (1985) by use of the Cyber 205 supercomputer into which they loaded X-ray pictures of the crystallized virus. Such studies have provided an understanding of the spatial geometry of the atoms required for drug-virus interactions which result in prevention of uncoating of the virus. Nuclear magnetic resonance in addition may define the nature of the drug when it is bound to a macromolecule, such as an enzyme. This has been well discussed by Markley and coworkers (1984). Thus ideally we would like to be able to elucidate by X-ray crystallography and NMR studies how an antiviral agent interacts with the viral target - whether it be a surface receptor or a macromolecule such as DNA, RNA or protein. In the meantime structural modifications of lead antiviral agents will be used to provide a rational approach to serendipitous discoveries.

2. HOW IS ANTIVIRAL ACTIVITY DETERMINED?

Assay of antiviral activity has been well discussed in this volume by Dr. H. Eggers, Dr. E.R. Kern and others.

3. HOW ARE DRUGS ABSORBED, DISTRIBUTED, METABOLIZED AND EXCRETED?

A. Absorption

In order for a compound to reach an intracellular target site it must have some solubility in an aqueous solution. Evaluation of relatively insoluble compounds for antiviral activity in cell culture may be accomplished for example by first dissolving the compound in dimethyl sulfoxide (DMSO) followed by aqueous dilution to non-toxic levels of the solvent. It is essential to determine the effect of the aqueous DMSO or other solvent, in order to eliminate the possibility that the antiviral activity produced is merely a primary effect of the solvent on the host cell. Thus there are several questions that one may wish to ask:

(1) What effect does the solvent have on the morphology of the cell? This may be examined by both light- and electron microscopy.
(2) What effect does the solvent have on the generation time of the host cell?
(3) What effect does the solvent have on the viability of the cell as determined by cloning in soft agar?
(4) What effect does the solvent have on the metabolic activity of the cell? Even though a treated cell may appear microscopically to be unchanged, metabolic pathways required for viral replication may have been disturbed to the extent that virus yield is decreased.

There are three major routes that have been used to administer antiviral drugs: topical, oral and systemic.

(1)Topical. The skin is an effective barrier to drug penetration, and this has been circumvented by the use of various ointments and solvents to facilitate transport of topically administered antiviral drugs such as idox-uridine or acyclovir. Before the drug can reach the infected cells in the lower layers of the epidermis, it must penetrate the stratum corneum. No ideal approach has as yet been found, and what is ideal may vary with the physico-chemical properties of the ointment or cream formulation, the anti-viral drug in question, the area and thickness of the applied preparation and the permeability of the specific area of the skin applied. We do need more research on the molecular factors which affect trans-dermal penetration.

Therapy of herpetic keratitis is effectively treated by the topical administration of idoxuridine, trifluoridine, or vidarabine in solution or as an ointment. In this situation there is direct contact between the infected cell and the administered drugs which are readily transported into the infected epithelial cell of the eye.

Solvents such as DMSO and various ointments have been used for the topical administration of idoxuridine and other drugs. Thus Juel-Jensen and his colleagues (1970) used a 40% solution of idoxuridine for therapy of herpes zoster in patients and found beneficial results. DMSO not only af-fords a high concentration of idoxuridine (40%), but also has the property of facilitating transport of the drug into the cell. Spruance et al. (1984) found acyclovir in DMSO to be more effective than acyclovir in polyethylene glycol for treatment of cutaneous herpes in the guinea-pig.

The use of DMSO as a vehicle for vidarabine (araA) was mildly inhibito-ry to HSV-1 cutaneous infections in experimental animals (Sidwell et al., 1987, and references therein). Caution should be taken in evaluation of a new vehicle for an antiviral agent, since for example, Azone was found by Leonard et al. (1987) to be antiviral by itself when applied topically to HSV-1 cutaneous infections in guinea pigs.

(2) Oral. The oral route of administration has unique pharmacokinetic problems. Absorption of a drug will depend on the solution process if given as a solid, its stability to pH of the gastric and intestinal secretions, its resistance to degradative enzymes, the rate of passage through the gastro-intestinal tract as well as whether it is absorbed into the body via simple diffusion, active transport, facilitated transport or pinocytosis. Thus, nucleoside drugs such as idoxuridine, even if it were non-toxic to uninfected cells, cannot be given orally because of rapid degradation either by the intestinal phosphorylases, or by similar hepatic enzymes after pas-sage from the g.i. tract into the liver via the portal vein (Welch and Prusoff, 1960).

(E)-5-(2-Bromovinyl)-2'-deoxyuridine (BVdUrd), although readily ab-sorbed by the oral route, is rapidly catabolized to the free pyrimidine (E)-5-(2-bromovinyl)-uracil (BVUra) presumably by phosphorolysis of the deoxyri-bose moiety. The formed BVUra is very slowly eliminated from the blood stream and its presence can be detected for as long as 24 hours in the plasma. Of interest is the ability to partially reconstitute the nucleo-side, BVdUrd, by the addition of a nucleoside as a deoxyribose donor (Des-granges et al., 1984).

Acyclovir being an acyclic nucleoside is stable to enzymic cleavage as well as to pH variation in the g.i. tract. However it is poorly absorbed since only about 15-30% is bioavailable. An approach to overcome this prob-lem is the development of the 6-deoxy analog of acyclovir, a prodrug of acyclovir which not only is about 18-fold more soluble, but also is well absorbed when given orally. It is converted by xanthine oxidase to acyclovir, and produced plasma levels of acyclovir which were considerably higher than that produced when acyclovir itself was given even at more than 3-fold the dose of the prodrug (Krenitsky et al., 1984).

Amantadine and rimantadine are well absorbed when given orally to treat respiratory infections due to the influenza A virus. Ribavirin administered via aerosol has recently been approved by the FDA for therapy of severe respiratory syncytial virus infections in children (The Medical Letter, 1986; Hall et al., 1983; Gilbert and Knight, 1986). The concentration in lungs is about 100 fold that in the blood stream. The topical intranasal administration of interferon has also been evaluated for the prophylactic treatment of rhinovirus respiratory infection, and although reasonably effective, there was a high rate of side effects such as bloody mucus (Farr et al., 1984). The intranasal administration of enviroxime or dichlorflavin, although effective _in vitro_ against the rhinovirus in cell culture, was not effective in man. Whether the mucus presented a formidable barrier or whether other factors prevented efficacy in this situation is not clear.

Peptide antiviral agents are potentially very important antiviral drugs not only to interact with viral or cellular receptors to prevent their attachment to the cell, but also to prevent viral uncoating, to inhibit viral proteases and to prevent the function of viral regulatory proteins. Saffran and coworkers (1986) have reported that peptide drugs, which when given orally are inactivated by enzymes in the stomach and small intestine, may be given orally when coated with polymers cross-linked with azoaromatic groups. Although protected from digestion in the stomach and small intestine, the azo bonds are reduced by the microflora normally present in the large intestine, the cross-links are broken and the drug released and absorbed: (R-C_6H_4-N=N-C_6H_4-R $\rightarrow$ 2R-C_6H_4-NH_2). These investigators reported that the polymer coated vasopressin and insulin administered orally to rats did indeed produce the expected biological responses of antidiuresis and hypoglycemia respectively.

(3) Systemic. Systemic administered drugs may present problems which are also applicable to orally administered drugs once absorbed into the body. Drugs which have poor solubility and are administered intravenously, such as vidarabine, are given in a large fluid volume and this has created problems in patients with impaired renal function or cerebral edema because of the risks of fluid overload. The use of the monophosphate of vidarabine markedly increases its solubility about 50-fold, however, increased toxicity has also been reported (Lok et al., 1984). Whether other soluble derivatives of vidarabine can be developed with retention of antiviral activity but without concomitant increase in toxicity is an obvious objective.

Another approach to decrease the volume of fluid required to solubilize the drug is to increase the biological half-life of the antiviral agent. One approach is to modify the drug so that it is no longer susceptible to metabolic inactivation. An example is the synthesis of cyclaridine, the carbocylic analog of vidarabine, which is resistant to several enzymes that degrade vidarabine such as adenosine deaminase, nucleoside phosphorylase and nucleoside hydrolase (Vince and Daluge, 1977). Cyclaridine is active not only _in vitro_ but also _in vivo_ against lethal herpesvirus infection in mice. The increased stability to enzyme degradation affords a longer biological half-life _in vivo_.

The biological half-life of an active drug can also be increased by prevention of its metabolic degradation by inhibition of the catabolic enzymes involved. Thus deamination of vidarabine can be decreased by inhibition of adenosine deaminase by 2'-deoxycoformycin, an irreversible inhibitor, or erythro-9-(2-hydroxy-3-nonyl)adenine (EHNA), a reversible inhibitor of this enzyme. However this approach has not been established to be of clinical significance.

Acyclovir could also present a problem if administered i.v. too rapidly in high concentration because of nephrotoxicity due to crystallization of the drug in the kidneys.

B. Distribution

The ideal distribution of the antiviral drug after absorption is to be transported specifically to the viral infected cell. Unfortunately this problem has not been solved, although preferential retention of certain antiviral drugs in viral infected cells has been attained. Although the basis for this will be expanded upon in a later section, the process concerns the presence of viral encoded enzymes which preferentially interact with the drug.

There are a number of approaches which have been or are being explored to improve transport of a drug to the target tissue.

(1) Liposomes. Liposomes are aqueous compartments that are enclosed by a lipid bilayer. Drugs can be trapped in the aqueous compartment by forming the vesicles in the presence of the drugs. The liposome can fuse with the plasma membrane of a cell, and thereby release the contents of the vesicle into the cell, or the liposome may be taken up into the cell via endocytosis. This procedure not only protects the drug from destruction in the bloodstream, but also affords the opportunity to introduce into the cell highly ionized or other impermeable substances. Another desirable feature, afforded by this technique, is the increase in the time that the drug is in circulation, thereby functioning as an intravascular sustained release preparation of the drug (Gregoriadis, 1973; Juliano, 1981).

Idoxuridine has been incorporated into liposomes and its therapeutic efficacy against experimental herpes keratitis in rabbits was compared with a similar amount of idoxuridine given in solution. The use of liposomes as carriers of idoxuridine was reported to enhance its efficacy significantly relative to the use of the drug alone. Both were given topically (Smolin et al., 1981).

Ribavirin encapsulated in liposomes was more effective than ribavirin alone in protecting mice against an intranasal challenge with ten LD_{50} of influenza virus (Gangemi et al., 1987). Similarly Kende et al. (1985) found increased efficacy when ribavirin was administered encapsulated in liposomes to mice infected with Rift Valley fever virus.

(2) Liposomes plus hyperthermia. The use of temperature-sensitive liposomes containing the anticancer drugs methotrexate (MTX) or cis-dichlorodiamineplatinum (II) have been used successfully to effect release of the drug preferentially to the target tissues (Weinstein et al., 1980; Yatvin et al., 1978, 1981). Mice bearing tumors on both hind legs were injected with liposomes containing [^{3}H]MTX and one leg was heated to 42°C in a water bath. Within 4 h the heated tumor contained about 14 times as much [^{3}H]MTX as did the unheated tumor. Since leucovorin blocked the uptake of MTX, the increase was suggested to be a consequence of release from the liposomes and uptake into the cell by a carrier-mediated transport. This approach could be of value in the systemic therapy of genital herpes, for example, by local application of heat to the external genitalia.

(3) Antibody targeted liposomes. The specific targeting of a liposome based on immunological recognition involves the preparation of drug containing liposomes that have the appropriate antibody associated with the lipid complex of the liposome. Thus a liposome associated with a specific viral antibody conceivably could preferentially interact with the viral antigen on

the surface of the infected cell (Gregoriadis and Neerunjun, 1975; Jansons and Mallett, 1981). The liposome containing an antiviral drug and a monoclonal antibody would constitute a two pronged attack on the viral infected cell.

(4) Monoclonal Antibodies. Most viruses have antigens that evoke a strong immune response. Herpes simplex viruses produce HSV antigens which are inserted into the surface membrane of cells that are acutely infected or transformed. Five glycoproteins have been identified in HSV virion envelopes and in the surface plasma membrane of HSV-infected cells: gC, gB, gA, gE and gD. Specific antisera can immunoprecipitate such infected cells, and antisera specific for either gC, gA and gB or gD are cytolytic for HSV-infected cells. Thus these glycoproteins can be targets for immunocytolysis.

Large quantities of a specific antibody can now be made which react with the target antigen. Monoclonal antibody production requires the formation of hybridomas, which are a fusion of cells that produce the antibody with mouse myeloma cells. Clones are grown from the individual hybrids and these are screened for their ability to produce the desired antibody which will react with cells containing the target antigen but do not react with normal cells.

Recent studies by Dix et al. (1981) show that mouse monoclonal antibody against glycoprotein gC or gD reduced the dissemination of HSV as well as the severity of acute neurological disease in virus-infected mice. HSV was inoculated into the footpads of mice and the virus spread has been shown to be neural and not viraemic. The mice developed ipsilateral monoplegia on day 5 post-infection, and then for the next 2 days flaccid paraplegia with ascending myelitis with death within 7-10 days. The monoclonal antibody afforded protection from virus-induced illness when administered either 2 h before infection or at 24 h post-inoculation with HSV.

Scheinberg and Strand (1982) prepared a monoclonal antibody against the Rauscher virus envelope glycoprotein, gp 70, and found that it reacted with the cell surface of Rauscher or Friend leukemia cells or RLV-infected fibroblasts, but not with normal cells in vitro. More important, however, was their study of the effect of this monoclonal antibody given to mice 72 h after inoculation i.v. with the Rauscher leukemia virus. Complete cures of the RLV-induced erythroleukemia were obtained.

(5) Magnetically targeted albumin microspheres. Albumin microspheres containing a drug and ferrosoferric oxide (Fe_3O_4) have been shown in experimental animals to be retained at the target organ by use of an external magnetic field. Albumin microspheres containing the drug doxorubicin and ferrosoferric oxide were injected i.v. into rats bearing the Yoshida sarcoma tumor in their tail. Those animals in which the tumor was exposed externally to a magnetic field had marked tumor regressions and all survived, whereas the tumors in the control animals all increased in size and most animals died. Between 50% and 80% of the microspheres could be retained in the target area by this technique. By retaining the drug at the tumor site, lower concentrations of drug can be used which also serves to decrease toxicity (Widder et al., 1981). Magnetic targeting of drugs thus affords another procedure for selective or at least preferential drug delivery to the target site (Senyei and Widder, 1981).

(6) Retrograde Axonal Flow. Both herpes simplex and varicella viruses are neurotropic and can establish a latent state in ganglia from which reactivation of the virus can occur by a number of stimuli. To prevent recurrences attempts have been made to deliver the antiviral drug by retrograde

axonal transport to the cell body of the ganglion.

Horseradish peroxidase is readily taken up by pinocytosis by axonal terminals at sensory nerve endings, synapses and neuromuscular end-plates, and subsequently transported into the nerve cell body by retrograde axonal flow. Horseradish peroxidase, a low molecular weight (40,000 daltons) protein, has been shown to be transported to the trigeminal ganglion of mice when applied topically to the eye or nose of mice (Fox and White, 1980; Kristensson, 1978).

If the viral genome in the host cell DNA is periodically induced to replicate by any one of a variety of stimuli with subsequent formation of the virion, then introducing the antiviral agent into the neuronal cell body would have little or no value. However, if it is given immediately following a prodromal signal, then viable virions may be prevented from reaching the surface epithelium where damage is created.

If the latency were to involve a slow but continuous formation of virions, then replication of such virions could be prevented if this technique of drug delivery were successful.

The carbohydrate portion of horseradish peroxidase was partially oxidized with sodium metaperiodate to produce aldehyde moieties which could form a Schiff base with the 5'-amino group of 5-iodo-5'-amino-2',5'-dideoxyuridine (AIdUrd). The ratio of AIdUrd to protein ranged from 2 to 9. Retrograde transport of the complex was established by autoradiography of the trigeminal ganglion neuron 24 hours after corneal injection of the [^{125}I]AIdUrd-protein Schiff base (Haschke et al., 1980).

The obvious questions include: 1. Does the radioactivity present in the ganglion represent [^{125}I]AIdUrd? 2. How readily does the complex dissociate to form free AIdUrd? 3. Can an antiviral concentration be established by this technique? If answers to these questions are favorable, then obviously one would attempt to complex horseradish peroxidase with more potent compounds such as FIAC, BVdUrd or acyclovir.

Iwasaki et al. (1986) investigated the efficacy of retrograde axoplasmic flow in rats by injection of doxorubicin into the nerve previously injected with HSV-1 and demonstrated to be present in the trigeminal ganglia. A marked decrease in the isolation of virus from the ganglia was found from 84% in the non-drug treated rats to 8% in the rats injected with doxorubicin. Unfortunately this drug is very cytotoxic and although retrograde axonal flow was established, a specific antiviral drug should be used.

(7) Iontophoresis. Iontophoresis is a procedure whereby ionized drugs may be transported by an electric field created by a direct current electrode placed on the surface of the skin or mucosal membrane. A ground pad is attached to the patient at the shoulder, neck, mastoid area or near the virus-infected target tissue, and a thin probe with cotton saturated with a solution of the drug is applied to the lesion. The electrical energy is responsible for the transport of the ionized drug across the membrane or skin barrier. The drug, depending on its charge, can be delivered from either a negative or positive terminal.

Idoxuridine (0.1 %) has been applied by this procedure to herpes simplex virus mucocutaneous infections using 0.5 to 0.8 mA for 10 min. Complete resolution was reported by Lekas (1979) to occur within 36 h, whereas the normal course is 10-14 days. Gangarosa et al. (1982) have treated patients with herpes orolabialis by iontophoresis with idoxuridine (0.1%), and reported a 63% reduction in healing time. Ara-AMP was also effective when given by cathodal iontophoresis for treatment of herpesvirus infection of the skin of mice (Park et al., 1978), and efficacy was also found in patients treated for herpes orolabialis (Gangarosa et al., 1986).

The advantage of this procedure is that systemic toxicity is minimized since only small amounts of drug are transported into the target area.

(8) Continuous infusion devices. A major advance in drug delivery is the development of devices for the control of the rate at which a drug can enter the systemic circulation. This approach can eliminate the problem of fluctuation of drug levels which peak at a toxic concentration and have a nadir at a sub-therapeutic concentration. A number of these devices have been developed for ocular, intrauterine, transdermal and oral delivery of drugs in humans.

Recently the use of a portable infusion pump worn by the patient affords the opportunity to provide continuous chemotherapy to ambulatory outpatients. Sommadossi and his colleagues (1987) at the University of Alabama have developed an interesting approach for long term therapy with potentially toxic drugs. 9-(2,3-Dihydroxy-2-propoxymethyl)-guanine (DHPG, 2'NDG; BW 759U; BIOLF 62) has as a major side effect neutropenia. Since these investigators reasoned that this toxicity resulted from the dose and/or the regimen, they investigated the use of continuous drug infusion by use of a portable pump associated with an individualized regimen. These investigators were able to demonstrate inhibition of cytomegalovirus infection in a patient with T-cell immunodeficiency syndrome and CMV colitis, hepatitis, viremia and retinitis. This was accomplished with a steady-state plasma level that was less than the ID_{50} for CMV and equally as important with development of no adverse toxicity. Sommadossi and coworkers believe this approach has great value for long term therapy with potentially toxic drugs.

(9) Polymeric carriers. The complex formation of antiviral drugs with a protein has been investigated in experimental animals (Fiume et al., 1983). Trifluridine was converted to the glutarate and coupled via its hydroxysuccinimide ester to the epsilon amino of lysine residues of asialofetuin at a molar ratio of drug to protein of 8 to 1 and injected into mice infected with Ectromelia. It was reported that viral DNA synthesis was inhibited in the liver cells, however DNA synthesis in bone marrow was also affected.

Because of the limitations of transport into cells of oligonucleotides, Lemaitre (1987) coupled their antisense synthetic oligonucleotide to poly(L-lysine) via the epsilon-amino moiety. The oligonucleotide sequence, complementary to the initiation region of the vesicular stomatitis virus-mRNA for its N-protein, specifically inhibited viral protein synthesis and viral replication.

Vidarabine as the glutarate or 5'-monophosphate, the latter forming a phosphamide bond between the epsilon amino of lysine of the protein and the 5'-phosphate of ara-AMP. Since asialofetuin is immunogenic, conjugates of ara-A and ara-AMP with lactosaminated serum albumin were prepared. This neoglycoprotein is not immunogenic in mice and enters hepatocytes specifically. Fiume et al. (1986) hope this approach may be of value for therapy of chronic hepatitis B. The potential importance of these studies is that an antiviral agent may be selectively delivered into an infected cell without affecting cellular DNA synthesis in intestine or bone marrow, but whether applicable to man, remains to be evaluated.

C. Metabolism

A major problem is how to measure the concentration of an antiviral agent at the site where it exerts its activity. Thus measurement of the drug concentration in the blood stream may or may not reflect the amount of drug present within the cell, or more importantly the concentration at the active site. For example, even though an apparent adequate intracellular concentration is achieved because of good equilibrium between the drug in the blood stream and the cellular cytoplasm, the drug may not be effective because it does not achieve an adequate concentration in the nucleus or organelle where the target site may be located. What is meant, by the drug concentration at the target site, is the active form of the drug which may require metabolic conversion of a nucleoside to the nucleoside triphosphate in the case of antiviral agents such as 5-iodo-2'-deoxyuridine (idoxuridine) or acyclovir, or the drug may not require metabolic conversion as with phosphonoformate (Foscarnet).

The antiviral effect of the drug requires achievement and maintenance of the drug concentration for an adequate period of time, and this is related to how it interacts with the target site. For example, there is a direct correlation between the concentration of idoxuridine in the media, its uptake into the DNA of herpes simplex virus and the resultant antiviral activity. Thus once an adequate amount of idoxuridine has been incorporated into the viral DNA, there is no longer need to maintain an antiviral concentration of the drug within the cell. However in those situations where a covalent interaction does not occur with the target, as in the case of Foscarnet, one must maintain the concentration for a prolonged period of time. Thus, a knowledge of the pharmacokinetics as well as pharmacodynamics of an antiviral drug is very important for effective use. What are some of the problems affecting the attainment and maintenance of an effective antiviral concentration at the target site?

Studies of the metabolic fate of an antiviral drug are critical to understanding of:

(a) How the drug exerts its effect-- What is the active form of the drug?
(b) How the drug should be administered-- orally, systemically, topically?
(c) How frequently the drug should be administered. What is its half-life at the active site?

The synthetic antiviral drugs may be divided into two categories-- those that exert their antiviral activity without undergoing metabolic alteration, and those which must be converted metabolically into the active molecular form.

(1) Antiviral agents not metabolically altered. (a) Phosphonoformate (Foscarnet). Oberg (1983) and Boezi (1979) have written excellent reviews on that antiviral agents phosphonoformate and phosphonoacetate. Phosphonoformate, like phosphonoacetate, exerts antiviral activity by interaction directly with nucleic acid polymerases. Although synthesized in 1924 by Nylen (1924), it was found over fifty years later to inhibit influenza virus RNA polymerase by Helgstrand et al. (1978). Subsequently it was found to inhibit a variety of RNA- and DNA polymerases, including reverse transcriptase (Sundquist and Oberg, 1979). Oberg (1983) has tabulated the various enzymes inhibited by phosphonoformate. In addition to RNA polymerases, DNA polymerases and reverse transcriptase, Derse and Cheng (1981) found the compound inhibited the HSV-1 3',5'-exonuclease.

Extensive studies of the metabolism and excretion of [3H] or [14C] phosphonoacetate indicated no metabolic alteration of the compound. Orally administered PAA is poorly absorbed in the rat (14%), rabbit (2%), monkey (8%) and dog(60%). It is rapidly excreted in urine and feces, but none in bile or expired air. However it is deposited in the bone. PAA is excreted with a half-life of 5.8 days in the rat, 8.7 days in the rabbit and 13.9 days in the monkey and dog. In the rabbit radiolabelled PAA was retained in the bone for over 200 days.

The pharmacokinetics of phosphonoformate were reported by Helgstrand et al. (1980), and in the mouse about 30% is deposited in bone and cartilage. The half-life in blood varies from 0.7 to 3.5 hours depending on the animal species, and no metabolic derivatives were found (Helgstrand et al., 1980).

(b) Amantadine and rimantadine. Hoffman (1980) and Oxford and Gailbraith (1980) have reviewed these compounds. Amantadine (1-aminoadamantane, 1-adamantanamine HCl) and rimantadine (α-methyl-1-adamantanamine HCl) have been used for therapy and prophylaxis of respiratory infections caused by the influenza A virus in the United States and the U.S.S.R. respectively. However, rimantadine is now in clinical trial in the United States.

There is no evidence for metabolic alteration of amantadine, and when incubated with cells in vitro, it is rapidly transported into the cell and concentrated in the lysosomal and cytosol fractions (Richman et al., 1982). Even when the compound is removed from the incubation media, a significant portion of amantadine remains in these cellular fractions. However this cell-associated amantadine does not appear to be critical for the antiviral activity, since removal of the drug before or at the time of infection decreases its antiviral activity. The compound must be present during the first two hours postinfection in order for it to exert its antiviral activity. Removal of the drug after 2 hours of incubation does not decrease its antiviral activity.

After oral administration amantadine is well absorbed, is not metabolized and about 56% is secreted within 24 hours in the urine unchanged by glomerula filtration (Bleidner et al., 1965). By 4 days about 86% of a single oral dose is recovered in the urine indicating almost complete bioavailability (Bleidner et al., 1965). After absorption the upper respiratory tract accumulates a significant deposit.

Amantadine has a mean half-life (t 1/2) of about 15 hours, hence a long time is required to achieve in humans a steady state blood level. However there is considerable individual variation of the t 1/2 which may range from 10 to 31 hours.

Rimantadine, like amantadine, is well absorbed after oral administration; however, in contrast to amantadine, it is largely metabolized and has an average half-life of about 30 hours which is twice that of amantadine. Pharmacokinetic studies by Hayden et al. (1985) of rimantadine showed that the proportion of the dose recovered in the urine within 24 hours varied from 5.5 to 49%. Although no metabolites of rimantadine were detected in the plasma, the urine contained in addition to the parent drug, ortho-, meta- and parahydroxy metabolites. Hence in contrast to amantadine, rimantadine is subject to metabolic alteration. The peak plasma levels of rimantadine are about one third that of amantadine.

(2) Antiviral agents requiring metabolic alteration

(a) 5-Iodo-2'-deoxyuridine. 5-Iodo-2'-deoxyuridine (Idoxuridine; IdUrd; IUdR; IDU) was synthesized in 1959 (Prusoff, 1959). It is an analog of thymidine in which the methyl group in the 5-position of the pyrimidine moiety is substituted by an iodine atom.

IdUrd utilizes the same anabolic and catabolic pathways as thymidine, and this is depicted below.

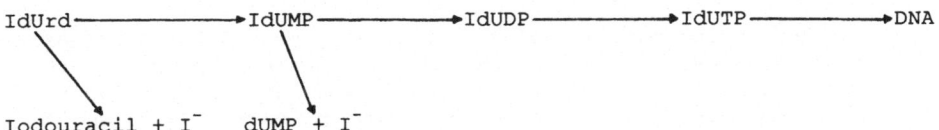

The catabolic enzyme responsible for the cleavage of IdUrd to IUra is thymidine phosphorylase and this activity did not vary much in cells post virus infection. In order for IdUrd to be active, it is required to be first phosphorylated to the monophosphate derivative, IdUMP. Herpes simplex virus type 1, herpes simplex virus type 2, varicella zoster virus and Epstein-Barr virus-infected cells, induce a new species of virus specified thymidine kinase which, in addition to the cellular thymidine kinase, is capable of phosphorylating IdUrd and other nucleosides (Littler et al., 1986; Turenne-Tessier, 1986; also see references cited by Larder and Darby, 1984). The binding affinity of IdUrd to the virus-induced thymidine kinase is better than that of host thymidine kinase. Thus, IdUrd should phosphorylate more efficiently in virus-infected cells than noninfected host cells. Once it is converted to IdUMP, thymidylate synthetase could dehalogenate this nucleotide analog to form the normal metabolite dUMP. However IdUMP is primarily phosphorylated by TMP kinase to the diphosphate derivative, IdUDP, and then further to the triphosphate derivative, IdUTP, by nucleoside diphosphate kinase. As the triphosphate, IdUTP, it is incorporated into both cellular and viral DNA.

(b) 5-Trifluoromethyl-2'-deoxyuridine. 5-Trifluoromethyl-2'-deoxyuridine (Trifluoridine; F_3dThd, F_3TdR, TFT) was synthesized by Heidelberg et al. (1964) and is an analog of thymidine. The methyl group of thymidine with Van der Waals' radii of 2.00 A° was replaced with a trifluoromethyl moiety which has radii of 2.44 A°.

5-Trifluoromethyl-2'-deoxyuridine is metabolized in viral infected and uninfected cells as depicted below.

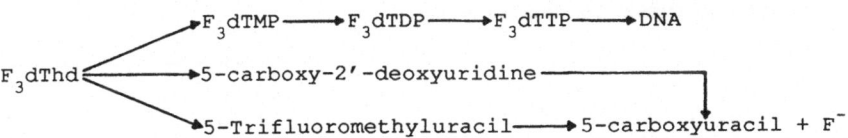

The conversion of the trifluoromethyl moiety to the carboxyl group is a chemical process without involvement of known enzymes. 5-Carboxy-2'-deoxyuridine does not have antiviral activity but is cytotoxic to HEp 2 cells by production of a metabolic block to the de novo pyrimidine biosynthesis pathway at either orotate phosphoribosyl transferase and/or orotidine-5'-phosphate decarboxylase.

Like IdUrd, F_3dThd is phosphorylated to F_3dTMP by thymidine kinase. The herpes simplex virus- or varicella zoster virus-infected phosphorylate F_3dThd more efficiently than uninfected cells not only because of the higher activity of thymidine kinase in the infected cells, but also because of the better affinity of F_3dThd toward the virus-specified thymidine kinase present only in the infected cells. F_3TMP is a potent inhibitor or thymidylate synthetase and hence decreases the formation of de novo dTMP. The decreased pool of dTMP leads to less competition to the phosphorylation F_3dTMP and its subsequent incorporation into DNA. Virus DNA polymerase utilizes F_3dTTP more efficiently as a substrate than the host enzymes. Additional details of F$_3$dThd pharmacology were presented by Heidelberger and King (1979) and Carmine et al. (1982).

(c) 9-β-D-Arabinofuranosyl-9H-purine-6-amine. 9-β-D-Arabinofuranosyl-9-H-purine-6-amine (Vidarabine; adenine arabinoside; spongoadenosine; araA) was first synthesized by Lee et al. (1960) and later identified as an anti-biotic produced by Streptomyces antibioticus by Parke-Davis and Co. (cited by North and Cohen, 1979). North and Cohen (1979) have written an excellent review of araA and of several other ara-nucleosides.

The metabolic reactions to which araA is subjected are depicted below.

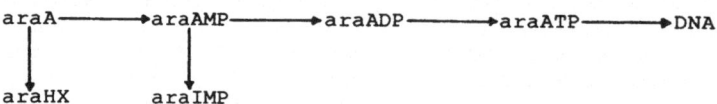

The deamination of araA by adenosine deaminase results in the formation of the arabinoside of hypoxanthine (araHX) which is less active as an anti-viral agent. None of the enzymes concerned with the metabolism of araA are viral encoded, hence it is not surprising that the formation of araATP in HSV-infected cells is essentially the same as that in the uninfected cells. The binding affinity of araATP is significantly less for mammalian α or β DNA polymerases than for the viral encoded HSV-1 DNA polymerases, and the ratio of the Ki for araATP relative to the Km for dATP is similar for both the human and viral encoded DNA polymerase (Ostrander and Cheng, 1980; Reinke et al., 1978). Ostrander and Cheng (1981) found araATP to be a competitive inhibitor of dATP with HSV-1 and HSV-2 encoded DNA polymerase, and to have a Ki of 0.33 uM and 0.42 uM respectively; whereas the Km of dATP for these two viral enzymes was 0.11 uM and 0.14 respectively. Thus the Ki value of araATP is similar to the Km value of dATP, its competing substrate.

The 3'-exonuclease associated with HSV-1 encoded DNA polymerase can remove the araAMP incorporated at the terminal of the DNA chain. Hence araATP could act as a pseudo chain terminator by continuous incorporation and removal of ara AMP from the terminal position. Thus the rate of elonga-tion of the DNA polymer is decreased.

AraA is not administered orally because it is very poorly absorbed, hence it is given intravenously but in large fluid volumes because of poor solubility. Intravenously administered araA has a half-life in the blood stream of about 3 to 5 hours. About 50 to 60% of the drug is excreted in the urine within 24 hours, partly as unchanged araA (1 to 3%), but primarily as hypoxanthine arabinoside (araHX) (about 50%). The maximum plasma levels of araHX are about 10-fold greater than that of araA. These levels are about 5% of an inhibitory dose for HSV and presumably a higher concentration is found intracellularly to account for its antiviral efficacy. The drug can pass through the blood brain barriers and is present in the cerebrospi-nal fluid at about 50 percent of the concentration in plasma.

(d) Acyclovir. 9-[(2-Hydroxyethoxy)methyl]guanine (Acyclovir, ACV, ACG) is a structural analog of guanosine in which the 2' and 3' carbons and their associated hydroxyl groups have been deleted.

Cells infected with HSV-1, HSV-2 and varicella zoster induce the viral encoded enzyme thymidine kinase which phosphorylates acyclovir to the mono-phosphate derivative (ACV-MP) (Elion et al., 1977; Fyfe et al., 1978). This virus-specified enzyme is induced in cells postinfection. Virus mutants which lack the ability to induce such an enzyme, or induce an altered thymi-dine kinase which does not accept ACV as a substrate, lose their sensitivity to the inhibitory effects of ACV.

Uninfected cells may also phosphorylate ACV, but to a very limited extent. This is due to ability of certain cellular enzymes to phosphorylate ACV albeit poorly. Colby et al. (1981) found the formation of ACV-MP in both EBV genome-negative and -positive B-lymphoblastoid cells. Chen et al. (1978) have reported that EBV induces thymidine kinase, and Littler et al. (1986) and de Turenne-Tessier et al. (1986) have evidence that EBV does indeed encode this enzyme. Burns et al. (1981) found that although mouse cytomegalovirus does not induce a thymidine kinase, the virus is sensitive to the antiviral activity of ACV with an ED_{50} of 0.23 uM.

The monophosphate of acyclovir (ACV-MP) once formed is then phosphorylated to the diphosphate derivative, ACV-DP, by the cellular enzyme, GMP kinase (Miller and Miller, 1980). Miller and Miller (1982) subsequently showed that acyclovir diphosphate (ACV-DP) is capable of being phosphorylated by each of seven cellular enzymes to the triphosphate derivative (ACV-TP). None of these enzymes had higher activity in the viral infected cell. They found that the relative rate of phosphorylation of ACV-DP by these seven enzymes in Vero cells was phosphoglycerate kinase >>> pyruvate kinase > phosphoenolpyruvate carboxykinase > nucleoside diphosphate kinase > succinyl-CoA synthetase > creatine kinase > adenylosuccinate synthetase. Even though the Km of ACV-DP (2.4 mM) for phosphoglycerate kinase was 10-fold greater than that for phosphoenolpyruvate carboxykinase (Km, 0.23 mM), the Vm for phosphoglycerate kinase was 80-fold greater and may be the cellular enzyme primarily involved.

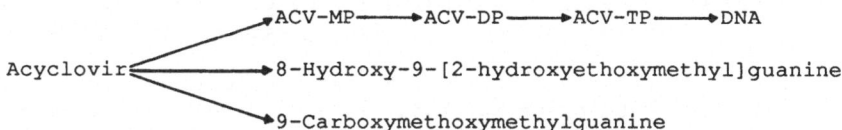

Metabolism of acyclovir

The pharmacokinetics of acyclovir has been extensively studied in man and animals (de Miranda et al., 1981; Furman et al., 1981; Biron et al., 1982; Laskins, 1983; de Miranda and Blum, 1983; Laskins, 1984; Barry et al., 1986). Oral administration of acyclovir results in only partial absorption which is dose limited. The bioavailability being between 15 and 30%. Intravenous infusion produced peak levels within 1 hour with plasma half-lives that ranged from 2 to 4 hours. The compound appears to be excreted in the urine by both tubular excretion and glomerular filtration. The total urinary recovery ranged from 71 to 99%, and was essentially unmetabolized acyclovir. A minor metabolite 9-(carboxymethoxymethyl)guanine represented 8.5 to 14%, and 8-hydroxy-9-(2-hydroxymethyl) guanine less than 2%.

Whereas 9-(carboxymethoxymethyl) guanine represented about 8.5 to 14 % of the urinary excretion in man, it amounted to about 30 % in the rabbit and guinea pig, but less than 2% in the mouse, rat and dog. The 8-hydroxy derivative of ACV although a minor metabolite in humans was present in the urine of the rhesus monkey to the extent of 25 to 30%.

Acyclovir is not subjected to phosphorolysis by purine nucleoside phosphorylase, however the diphosphate of acyclovir (ACV-DP) is a very potent inhibitor of this enzyme having a Ki of 0.0087 uM, whereas acyclovir has a Ki of 100 uM (Tuttle and Krenitsky, 1984), ACV-MP 6.6 uM and ACV-TP 0.31 uM.

Because of the relatively poor absorption of acyclovir following oral administration, Krenitsky et al. (1984) synthesized and evaluated an analog of ACV which lacked the 6-hydroxy moiety [6-deoxyacyclovir; 2-amino-9-(2-hydroxyethoxymethyl)-9H-purine]. This compound was well absorbed following oral administration and was rapidly oxidized by the enzyme xanthine oxidase to ACV. This prodrug also has the advantage of being about 18 times more soluble in water than ACV. Additional pharmacokinetic studies in humans were reported by Rees et al. (1986).

(e) <u>Ribavirin</u>. Ribavirin (1-β-D-ribofuranosyl-1,2,3-triazole-3-carbox-amide; virazole) is a triazole nucleoside, and X-ray crystallography revealed striking similarities to guanosine and inosine. Ribavirin inhibits a wide variety of DNA and RNA viruses, however it is approved in the U.S.A. only for therapy of severe respiratory infections in infants and children caused by the respiratory syncytial virus (Sidwell et al., 1979; Smith and Kirkpatrick, 1980; Hall et al., 1983; Robins, 1986). Several reviews have been presented (Smith et al., 1980; Smith et al., 1984; Robins, 1986; Canonico et al., 1984; Gilbert and Knight, 1986; Sidwell et al., 1979).

Ribavirin is metabolized to the mono-, di- and triphosphates by cellular enzymes (Streeter et al., 1973; Streeter et al., 1974; Miller et al., 1977; Sidwell et al., 1979). Smee and Matthews (1986) studied the metabolism of ribavirin in respiratory syncytial virus-infected and uninfected cells and found the extent of phosphorylation related to the concentration of ribavirin in the media, with the major species being the triphosphate derivative. The triphosphate of ribavirin was present up to 2.6-fold greater than in the uninfected cells. The initial phosphorylation was originally attributed to deoxyadenosine kinase (Streeter et al., 1974), however later studies by Willis et al. (1978) identified adenosine kinase to be the major if not only enzyme involved. The binding affinity for ribavirin monophosphate for IMP dehydrogenase is about 70 times greater than for IMP.

Ribavirin is soluble in water and is absorbed following oral administration, however for therapy of RSV infections in infants it is given as a continuous aerosol for 3 to 6 days (Hall et al., 1983). When ribavirin was administered by Jahrling et al. (1980) in high doses systematically to monkeys it protected them from the lethal effects of Lassa Fever virus, and clinical trials in patients infected with Lassa Fever virus indicate significant efficacy in reducing mortality (McCormick et al., 1986; Robins, 1986).

Ribavirin is almost completely absorbed when given orally to rats, and has an excretion half-life of 7 to 8 hours. By 24 hours about 80% of the drug was in urine, and by 72 hours 93%. About 3% appeared as carbon dioxide.

Metabolic derivatives found after i.v. administration include 1,2,4-triazole-3-carboxamide, 1,2,4-triazole-3-carboxylic acid, 1-β-D-ribofuranosyl-1,2,4-triazole-3-carboxylic acid, in addition to the mono-, di- and triphosphate derivatives of ribavirin.

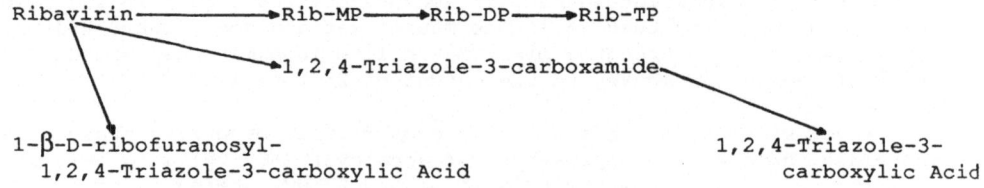

Metabolism of Ribavirin

In man the percentage composition of the urinary metabolites of ribavirin in urine collected in a 22- to 24-hr time interval was 17% as ribavirin, 50% as 1,2,4-triazole-3-carboxamide and 22% as 1,2,4 triazole carboxylic acid. A 72-hr collection accounted for about 40% of the total labeled drug. It is postulated that 20 to 30% of the total dose is retained in the body beyond 72 hr, but precisely where is uncertain. Patients receiving 43.2 mg/kg of ribavirin had serum levels of 5.3 ug/ml 24 hr later. The red blood cells appear to accumulate about 3% of a total dose of ribavirin which is in excess of plasma level, and has a half-life of about 40 days. The intracellular form of ribavirin is primarily as the triphosphate derivative.

Although ribavirin when administered by aerosol can be detected in the systemic circulation, the amount present in respiratory secretion is about 100 times higher (The Medical Letter, 1986).

(f) 3'-Azido-3'-deoxythymidine. 3'-Azido-3'-deoxythymidine (AZT) was first synthesized by Horwitz et al. (1964) and later by a slight modification by Lin and Prusoff (1978). Dyatkina et al. (1986) reported the synthesis of the α and β anomers of AZT in a ratio of 1.5 to 1. Activity of AZT against a retrovirus, Friend virus, was first reported by Ostertag et al. (1974 a,b) and by Dube et al. (1974, 1975). Ostertag et al. (1974 b) suggested that AZT could be phosphorylated to the triphosphate derivative and can only be added terminally to DNA because of the presence of the 3'-azido moiety.

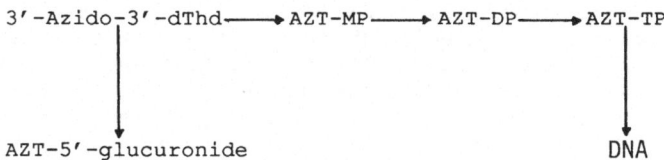

Metabolism of AZT

The metabolism of AZT was investigated by Furman and coworkers (1986). They found AZT to be phosphorylated to the 5'-mono-, di- and triphosphate derivatives and to the same extent in uninfected and HIV-infected cells. Phosphorylation of AZT to the three phosphorylated derivatives is performed by cellular enzymes, since HIV does not encode for a viral specific thymidine kinase, thymidylate kinase or a nucleoside diphospho kinase. Furman et al. (1986) found the K_m values for phosphorylation of dThd and of AZT by thymidine kinase was 2.9 uM and 3.0 uM respectively and the Vm of AZT to be 60% of that of thymidine.

The striking finding by Furman et al. (1986) was although the Km for AZT-monophosphate (8.6 uM) was twice as great as that for dTMP, the Vm was only 0.3% of that for dTMP. This supports their conclusion that AZT-monophosphate is an alternate-substrate inhibitor of dTMP kinase and explains the reduced intracellular pool size of dTTP. Thus the rate limiting step in the metabolism of AZT is the phosphorylation of the monophosphate of AZT by thymidylate kinase to the diphosphate derivative of AZT.

More recently AZT has been found by Mitsuya et al. (1985) to inhibit various strains of human T-lymphotropic virus (HTLV-III), also termed lymphadenopathy-associated virus (LAV), AIDS-associated retrovirus (ARV), and at present identified as human immunodeficiency virus (HIV). The therapeutic index of AZT in vitro in H9 cells or peripheral blood leukocytes for inhibition of HIV replication is about 10^4 which indicates very high selectivity for viral inhibition (Furman et al., 1986).

Elwell and coworkers (1987) reported AZT to have potent bacteriocidal activity against a variety of bacteria: E. coli, Salmonella typhimurium Shigella flexneri, Enterobacter aerogenes, etc. Combination therapy in many instances has been found to be synergistic, particularly if the two compounds have different sites of inhibition.

Hartshorn et al. (1987) combined AZT with recombinant alpha A Interferon (rIFN-α_A) and observed a synergistic inhibition of HIV replication in peripheral blood mononuclear cells. Since the concentration of each drug is stated to be easily obtainable in patients, a clinical trial with this combination was a logical conclusion. Similarly Mitsuya and Broder (1987) found AZT and acyclovir to be synergistic in the inhibition of HTLV-III when cultured in ATH8 cells. Although ribavirin and AZT each inhibit the replication of HIV, when combined an antagonistic effect was observed which is attributed to the inhibition of the phosphorylation of AZT by ribavirin (Vogt et al., 1987). Since this virus is not genetically stable, combination therapy might also prevent the emergence of viral resistance to these drugs.

AZT in a preliminary 6-week clinical trial in patients with AIDS or AIDS related complex was reported by Yarchoan et al. (1986) to be well absorbed from the g.i. tract, to have a half-life of about 1 hr, to cross the blood brain barrier and to produce in some patients positive clinical improvement such as partial restoration of the immune functions, weight gain, and reduced mortality. Therapeutic levels were maintained with 5 mg given intravenously or 10 mg given orally every 4 hours. AZT is metabolized to the 5'-glucuronide in the liver and excreted in the kidneys, probenecid slows both metabolism and excretion (Medical Letter, 1986). Although AZT is not a cure, its ability to prolong life of some AIDS patients, in spite of bone marrow suppression observed in many patients, has encouraged the evaluation of other nucleosides for anti-HIV activity (Mitsuya et al., 1985; Mitsuya and Broder, 1986; Lin et al., 1987; Balzarini et al., 1986). Recently Yarchoan et al. (1987) obtained evidence in two patients that AZT could reverse certain neurological dysfunctions-- specifically dementia and peripheral neuropathy.

4. WHAT IS THE MOLECULAR BASIS FOR ANTIVIRAL ACTIVITY?

We know a number of targets where antiviral agents exert their effect and these have been well reviewed (Prusoff et al., 1986; De Clercq, 1985; Cheng et al., 1985). However the ultimate molecular event responsible for the inhibitory effect is more difficult to elucidate.

Amantadine

The molecular basis for the action of amantadine is yet to be elucidated. Several proposals have been made:

(1) Interaction with the external plasma membrane of the cell, thereby altering the surface charge of the membrane and hence inhibition of endocytosis or uncoating of the virus (Kato and Eggers, 1969; Koff and Knight, 1979; Schlegel et al., 1982).

(2) Inhibition of the dissociation of the influenza virus M-protein from the nucleocapsid or ribonucleoprotein transcription complex present in subviral particles. Only ribonucleoproteins free of the M protein may enter the nuclei where the viral RNA synthesis occurs (Bukrinskaya et al., 1982).

(3) The virus enters the cell by endocytosis, into vesicles called endosomes which then fuse with a lysosome. Fusion of the viral membrane with the membrane of the vesicle requires the pH within the vesicle to be decreased to about pH 5. Amantadine or rimantadine, being primary amines, increase the pH, thereby preventing the acid catalyzed conformational rearrangement of the capsid protein required for the fusion process (Skehel et al., 1977; Skehel et al., 1982; Beyer et al., 1986; Stegman et al., 1986).

Idoxuridine

5-Iodo-2'-deoxyuridine is phosphorylated by cellular as well as viral encoded thymidine kinase to the monophosphate derivative. There is very little difference sterically between thymidine and IdUrd since the size of the iodine atom as defined by its van der Waals' radius is 215 pm whereas that of the methyl group is 200 pm. However the electron configuration of the pyrimidine moiety is altered because the halogen may produce two different effects. The first of these results from the high electronegativity of the halogen atom, so that the electrons, by the inductive effect, are pulled away from the carbon atom to which it is attached. The second effect is a consequence of the halogen atom having an unshared pair of electrons which can initiate a resonance effect in the pyrimidine ring by increasing the electron density of the ring. The inductive effect of the halogen is probably the dominant effect responsible for the labilization of the proton on the N3 position of IdUrd. This inductive effect of the halogen is responsible for the lower pKa for IdUrd (8.2) relative to that for thymidine (9.8). The physicochemical and biological properties of IdUrd were reviewed by Langen (1975), Prusoff and Goz (1975), Goz (1978), and Prusoff et al. (1979).

Three major areas have been found where IdUrd or the appropriate phosphorylated derivative exert marked effects: (1) competitive inhibition of several enzymes concerned with the biosynthesis of DNA-thymine (thymidine kinase, thymidylate kinase and DNA polymerase); (2) allosteric or feedback inhibition by the triphosphate of IdUrd of the regulatory enzymes thymidine kinase, deoxycytidylate deaminase and ribonucleoside diphosphate reductase; (3) alteration of gene expression subsequent to the incorporation of IdUrd into DNA. However the herpesvirus encoded ribonucleoside diphosphate reductase, in contrast to the cellular ribonucleoside diphosphate reductase, in general is poorly susceptible to feedback inhibition by the triphosphates of nucleoside analogs (Nakayama et al., 1982). The two types of enzyme inhibition are competitive and readily reversible, whereas incorporation into DNA is not.

Incorporation of IdUrd into DNA produces a variety of physical and biological effects, and the consequences of such incorporation have been reviewed by Prusoff et al. (1979), Prusoff et al. (1984), as well as in references cited in these reviews.

Incorporation into DNA is believed to be responsible for toxicity as well as the antiviral effect of IdUrd. Incorporation of halogenated nucleosides (BrdUrd or IdUrd) into viral DNA has been found in vaccinia virus (Prusoff et al., 1963; Easterbrook and Davern, 1963), in pseudorabies (Kaplan et al., 1965), in herpes simplex virus (Schiek and Schiek, 1969; Matsumura, 1973; Chen et al., 1976; Fischer et al., 1980), in SV40 virus (Buettner and Werchau, 1973; Calothy et al., 1973), in polyoma virus (Hirt, 1966), and in adenovirus (Wigand and Klein, 1974). Buettner and Werchau (1973) isolated infectious DNA from SV40 grown in the presence of IdUrd and showed a relationship between loss of infectivity and the extent of substitution of SV40 DNA-thymidine by IdUrd. A direct correlation between uptake of IdUrd into the DNA of herpes simplex virus type 1 (HSV-1) and loss of infectivity was found by Fisher et al. (1980).

The incorporation of IdUMP into the DNA chain does not prevent further elongation, but the substituted DNA does not serve as a good template for either DNA replication or RNA transcription.

Incorporation of IdUrd into HSV-1 DNA produced no apparent effect on the physical integrity of the substituted DNA molecules since Fisher et al. (1980) found no evidence of single-strand or double-strand breaks when subjected to alkaline and neutral sucrose density gradient centrifugation.

Incorporation of IdUrd into the herpesvirus DNA produced pronounced effects on viral gene expression of those processes which rely on newly synthesized viral DNA, that is late RNA and protein synthesis.

Thus treatment of herpes simplex virus type 1 (HSV) infected Vero cells with IdUrd causes a 65% reduction in the amount of total viral RNA present at late times of infection. This decrease is apparent as early as 6 hours post infection, affecting the RNA viral cytoplasmic polyadenylated (poly A^+) species to a greater extent than non-polyadenylated (poly A^-) RNA. Cytoplasmic viral poly A^+ RNA is present in treated cultures at a decreased amount relative to the control poly A^+ levels, yet there is no evidence of a direct inhibition of RNA polyadenylation (Otto et al., 1984).

The RNA effects are reflected in the protein synthesized by HSV-1. Otto et al. (1982) found a reduction in the synthesis of late (β and γ) but not early (α) proteins. There was a wide range of inhibition of β and γ proteins ranging from 46% to 100% of control levels. However two proteins (ICP 35 and 39) were not affected, and one (ICP 36) which migrated as thymidine kinase was actually increased. Also, the progeny virions had altered protein patterns (Prusoff et al., 1986; Prusoff et al., 1979; Prusoff and Otto, 1983; Otto et al., 1984; Cheng and Prusoff, 1986).

Trifluoridine

5-Trifluoromethyl-2'-deoxyuridine (CF_3dUrd) is phosphorylated by cellular as well as viral encoded thymidine kinase. Like 5-iodo-2'-deoxyuridine, CF_3dUrd utilizes the metabolic pathways concerned with thymidine metabolism. The initial phosphorylated product is the monophosphate, CF_3dUMP, which like IdUMP interacts with thymidylate synthetase, the enzyme responsible for the de novo synthesis of thymidine monophosphate (dTMP) from dUMP. However CF_3dUMP inhibits thymidylate synthesis, whereas the monophosphate of IdUrd not only does not inhibit this enzyme, but upon interaction is dehalogenated to dUMP and then is methylated to thymidylate (dTMP).

At the monophosphate level CF_3dUMP is phosphorylated to the di- and triphosphate (CF_3dUTP) and is incorporated into the viral DNA in substitution of thymidylate.

CF_3dUrd is a competitive inhibitor with respect to thymidine for thymidine kinase, and the various phosphorylated derivatives (CF_3dUMP; CF_3dUDP; CF_3dUTP) are competitive inhibitors of the corresponding thymidine nucleotides (dTMP, dTDP, dTTP). CF_3dUTP competes with dTTP for incorporation into viral DNA. The consequences of incorporation of CF_3dUrd into vaccinia virus DNA include the formation of morphologically defective viruses which are non-infectious and an inability of these viruses to properly transcribe late messenger RNA. If fragmentation of DNA, or chemical conversion of F_3dTMP in DNA to 5-carboxy dUMP with subsequent distortion of the DNA structure were to occur, then these events could also contribute to the antiviral activity (Heidelberger and King, 1979; Shannon, 1984; Carmine et al., 1982).

Vidarabine

Vidarabine (araA) does not require a viral encoded enzyme for conversion to the active form of the drug. Cellular enzymes convert it to the mono-, di- and triphosphate derivatives. The formation of araATP in herpesvirus-infected cells is about the same as that in uninfected cells.

There are multiple sites of inhibition, but the inhibition causally related to the antiviral activity has not yet been established:

190

(1)   AraATP is a potent inhibitor of ribonucleoside diphosphate reductase from both infected and uninfected cells, but that from HSV-infected cells is more sensitive to inhibition.

(2)   AraATP prevents post-transcriptional addition of polyA to viral mRNA.

(3)   AraATP inhibits RNA-dependent RNA polymerase of vesicular stomatitis virus.

(4)   AraATP is a competitive inhibitor of dATP for both mammalian as well as viral DNA-polymerases; however, HSV DNA-polymerase is about 10-fold more sensitive with a K_i of 0.3 uM versus 5.0 uM for DNA polymerase .

(5)   AraATP is also a substrate for DNA polymerase and is incorporated in internucleotide linkage in both viral and cellular DNA.

(6)   Slowing of DNA elongation has been reported and this could be a consequence of 3'-exonuclease associated with HSV DNA polymerase. Thus by continuous incorporation and removal of araAMP from the terminal position, araATP could act as a pseudo chain terminator.

(7)   AraATP inhibits terminal deoxynucleotidyl transferase.

(8)   AraA, without being phosphorylated, inhibits S-adenosylhomocysteine hydrolase, and this results in an accumulation of S-adenosylhomocysteine. The consequence of this is an inhibition of S-adenosylmethionine-dependent methylation reactions. However this site of inhibition is believed to play a more major role in producing cytotoxicity rather than antiviral activity (North and Cohen, 1979; Drach, 1983; Prusoff and Otto, 1983; Shannon, 1984; Cheng and Prusoff, 1986).

Acyclovir (ACV; acyclo-G)

Elion (1982) reviewed the molecular basis for the antiviral activity of ACV and appropriate references are cited therein. Acyclovir is phosphorylated to the monophosphate (acyclo-GMP) by a viral encoded thymidine kinase, and subsequently by cellular enzymes to the di- and triphosphate derivatives. Acyclovir is an extremely poor substrate for cellular thymidine kinase which explains its high selectivity for the viral infected cell.

Acyclovir triphosphate (ACV-TP) is a very potent and selective inhibitor of HSV-induced DNA polymerases (Furman et al., 1979; Derse et al., 1981). ACV-TP is a competitive inhibitor of the utilization of GTP for the synthesis of DNA (Allaudeen et al., 1982; Derse et al., 1981; Furman et al., 1979), and also is incorporated into the viral DNA (Elion et al., 1977; Furman et al., 1979). Once incorporated into the DNA it functions as a chain terminator since there is no 3'-carbon hydroxide for another dNTP to enter. In addition acyclo-GMP in the terminal position of DNA not only can not be removed by the 3',5'-exonuclease associated with HSV-1 DNA polymerase, but also Furman et al. (1984) showed acyclo-GTP during the process of catalysis by HSV DNA polymerase formed an irreversible complex (Enzyme + ACV-MP + DNA template). However, Larsson et al. (1986) found that inhibition of the HSV-2 DNA polymerase by acyclo-GTP could be reversed by an excess of dGTP, whereas Furman et al. (1984) found reversal of the inhibited HSV-1 polymerase reaction required an excess of dGTP and fresh enzyme. Whether the difference is due to properties of HSV-1 and HSV-2 enzymes remains to be clarified. Thus acyclovir as the triphosphate is a suicide inhibitor of HSV DNA polymerase.

Further evidence that acyclovir is a DNA chain terminator was provided by McGuirt et al. (1984) by demonstration of small DNA fragments when HSV-1 was grown in the presence of ACV. Acyclovir may also be phosphorylated, albeit very inefficiently by cytoplasmic 5'-nucleotidase in uninfected cells, and was competitively inhibited by inosine (Keller et al., 1985).

In contrast to the effect of acyclo-GTP on HSV-1 DNA polymerase (K_i, 0.03 uM), Allaudeen et al. (1982) found the K_i for EBV DNA polymerase to be 9.8 uM. Hence they proposed that the primary mechanism for the inhibition of EBV replication may not involve inhibition of the EBV DNA polymerase. Acyclo-GTP, however, is a very potent inhibitor of cytomegalovirus DNA polymerase (Wahren et al, 1987).

A consequence of viral growth in the presence of acyclovir is a marked effect on viral protein synthesis (Furman and McGuirt, 1983). They found a reduction in the synthesis by HSV-1 of γ proteins but not of α and β proteins. The synthesis of glycoproteins was also inhibited.

Ribavirin

As indicated above ribavirin is metabolized to the mono-, di- and tri-phosphate by cellular enzymes. The monophosphate of ribavirin interacts with IMP dehydrogenase thereby inhibiting the conversion of inosine-5'-monophosphate (IMP) to xanthosine monophosphate (XMP), an action which could decrease the availability of GTP and dGTP for nucleic acid synthesis. The binding affinity of ribavirin monophosphate for IMP dehydrogenase is about 70 times greater than for IMP. However this is believed not to be involved in viral inhibition by ribavirin, because guanosine can restore the pool of GTP but yet not affect the inhibition of viral replication. Also ribavirin inhibits viral but not cellular polypeptide synthesis.

Ribavirin triphosphate can inhibit the influenza virus-associated RNA polymerase as well as the HSV-encoded DNA polymerase by preventing the utilization of GTP and dGTP respectively (Eriksson et al., 1977; Robins, 1986).

Presumably a major action of ribavirin triphosphate is prevention of the capping of viral mRNA by inhibition of mRNA guanyltransferase and N^7-methyl transferase (Goswami et al., 1979; Robins, R.K., 1986) which are enzymes required for the addition of guanosine-5'-phosphate from GTP to the 5'-end of viral mRNA and subsequent methylation of the N-7 position.

Toltzis and Huang (1986) in a study of the mechanism of inhibition by ribavirin of the vesicular stomatis virus (VSV) found primary transcription was not affected, but viral RNA synthesis decreased by 40 to 60%. However the viral RNA so produced is translated poorly in an in vitro translation system, and all proteins normally made were decreased by about 95%. The viral mRNA was full length, hence chain termination was not involved. Whether the inhibition is a consequence of interference with the 5'-cap formation is a strong possibility but remains to be established.

3'-Azido-3'-deoxythymidine (AZT; 3'-N$_3$dThd)

Furman et al (1986) found AZT to be phosphorylated to the mono-, di-and triphosphate derivatives. At the triphosphate level AZT is a very potent inhibitor of the HIV-encoded reverse transcriptase being a 100-fold (Furman et al., 1986) or 1000-fold (Vrang et al., 1987) more potent than the inhibition of DNA polymerase α. Cheng et al. (1987) found 3'-N$_3$dTTP to be 100-fold more inhibitory to HIV reverse transcriptase than to cellular DNA polymerase γ, the IC_{50} being 0.04 uM and 39.0 uM respectively. It is not clear whether AZT is incorporated into the growing DNA, but if it were then chain termination should ensue since AZT has no 3'-hydroxyl (Furman et al., 1986; Ostertag et al., 1974 b).

An X-ray analysis of AZT was performed by Birnbaum et al. (1987) to determine its three-dimensional structure. The crystals were found to belong to the monoclinic space group $\underline{P2}_1$, and two crystallographic independent molecules were found in the asymmetric unit. One (molecule A) is a common conformation among nucleosides: C2' endo/C3' exo pucker of the furanose ring with a glycosidic torsion angle of X_{CN}= 53.4°. The other, molecule B, is unusual in that the pentose pucker is C3' exo/C4' endo and the glycosidic torsion angle is 2.3°. This high-energy conformation may be critically involved biologically, and if true, will be of value in the design of other inhibitors of HIV.

Phosphonoformate (Foscarnet)

The mechanism of action of phosphonoacetate (PAA) was carefully investigated by Mao and Robishaw (1975) and Leinbach et al. (1976). Studies of enzyme kinetics by Mao and Robishaw (1975) found PAA to be a noncompetitive inhibitor of the utilization of deoxyribonucleotide triphosphates with a Ki value of about 0.45 uM, and uncompetitive inhibition with respect to the template nucleic acid. Leinbach et al. (1976) found PAA inhibited the herpesvirus DNA polymerase by interacting with the pyrophosphate binding site thereby being a competitive inhibitor of the dNTP-pyrophosphate exchange reaction.

Phosphonoformate has a similar mechanism of action with HSV-1 DNA polymerase showing noncompetitive inhibition, with respect to dNTPs and uncompetitive with respect to the nucleic acid template (Eriksson et al., 1980; Ostrander and Cheng, 1980). However Reno et al. (1978) reported turkey-herpesvirus DNA polymerase exhibits noncompetitive kinetics with respect to dNTP and the DNA template.

Although RNA- and DNA-polymerases have pyrophosphate binding sites, the differences in inhibition of these enzymes by PFA or PAA indicate that specific variation at the active site in the atomic or steric parameters must be involved.

In order to exert maximum antiviral activity, phosphonoformate is required to be present within 3 hours postinfection and continuous exposure throughout the whole cycle of virus replication. Because covalent interaction is not involved at the target site, its antiviral activity is dependent on the duration of exposure.

ACKNOWLEDGEMENTS

The work done in this laboratory was supported by the United States Public Health Service Grant, CA-05262, from the National Cancer Institute. The authors thank Ms. Betsy Mahaffey for the excellent secretarial help.

REFERENCES

Allaudeen, H.S., Descamps, J. and Sehgal, R.K., 1982, Mode of Action of Acyclovir Triphosphate on Herpesviral and Cellular DNA Polymerase, Antiviral Res. 2:123.

Balzarini, J., Pauwels, R., Herdewijn, P., De Clercq, E., Cooney, D.A., Kang, G.-J., Dalal, M., Johns, D.G., and Broder, S., 1986, Potent and Selective Anti-HTLV-III/LAV Activity of the 2',3'-Unsaturated Derivative of 2',3'-Dideoxycytidine, Biochem. Biophys. Res. Commun., 140:735.

Barry, D.W., Nusinoff-Lehrman, S. and Nixon Ellis, M., 1986, Clinical and Laboratory Experience With Acyclovir-Resistant Herpes Virus, J. Antimicrob. Chemother. 18:Suppl. B, 75.

Beyer, W.E.P., Ruigrok, R.W.H., van Driel, H. and Masurel, N., 1986, Influenza Virus Strains With a Fusion Threshold of pH 5.5 or Lower Are Inhibited by Amantadine, Arch. Virol 90:173.

Birnbaum, G.I., Giziewicz, J., Gabe, E.J., Lin, T.-S. and Prusoff, W.H., 1987, Structure and Conformation of 3'-Azido-3'-deoxythymidine (AZT), An Inhibitor of the HIV (AIDS) Virus, Canad. J. Chem. In Press.

Biron, K.K., Noblin, J.E., de Miranda, P. and Elion, G.B., 1982, Uptake, Distribution and Anabolism of Acyclovir in Herpes Simplex Virus-Infected Mice, Antimicrob. Agents Chemother. 21:44.

Bleidner, W.E., Harmon, J.B., Hewes, W.E., Lynes, T.E. and Hermann, E.C.; 1965, Absorption, Distribution and Excretion of Amantadine Hydrochloride, J. Pharmacol. Exper. Ther. 150:484.

Boezi, J.A., 1979, The Antiherpes Action of Phosphonoacetate, Pharmac. Ther. 4:231.

Buettner, W. and Werchau, H.; 1973, Incorporation of 5-Iodo-2'-deoxyuridine (IUdR) into SV40 DNA. Virology 52:553.

Bukrinskaya, A.G., Vorkunova, N.K., Kornilayeva, G.V., Narmanbetova, R.A. and Vorkunova, G.K., 1982, Influenza Virus Uncoating in Infected Cells and Effects of Rimantadine, J. Gen. Virol. 60:49.

Burns, W.H., Wingaid, J.R., Bender, W.J. and Saral, R., 1981, Thymidine Kinase Not Required for Antiviral Activity of Acyclovir Against Mouse Cytomegalovirus, J. Virol. 39:889.

Calothy, C., Hirai, K., and Defendi, V., 1973, 5-Bromodeoxyuridine Incorporation Into Simian Virus 40 Deoxyribonucleic Acid. Effects on Simian Virus 40 Replication in Monkey Cells. Virology 55:329.

Canonico, P.G., Kende, M., Luscri, B.J. and Huggins, J.W., 1984, In-Vivo Activity of Antivirals Against Exotic RNA Viral Infections, J. Antimicrob. Chemother. 14, Suppl. A:27.

Carmine, A.A., Brogden, R.N., Heel, R.C., Speight, J.M. and Avery, G.S., 1982. Trifluridine, a Review of its Antiviral Activity and Therapeutic Use in the Topical Treatment of Viral Eye Infections. Drugs 23:329.

Chen, M.S., Summers, W.P., Walker, J., Summers, W.C. and Prusoff, W.H., 1979. Characterization of Pyrimidine Deoxyribonucleoside Kinase (Thymidine Kinase) and Thymidylate Kinase as a Multifunctional Enzyme in Cells Transformed by Herpes Simplex Virus Type 1 and in Cells Infected with Mutant Strains of Herpes Simplex Virus. J. Virol. 30:942.

Chen, M.S., Ward, D.C. and Prusoff, W.H., 1976, Specific Herpes Simplex Virus-induced Incorporation of 5-Iodo-5'-amino-2',5'-dideoxyuridine into Deoxyribonucleic Acid. J. Biol. Chem. 251:4833.

Chen, S.-T., Estes, J.E., Huang, E.S. and Pagano, J.S., 1978, Epstein-Barr Virus Associated Thymidine Kinase, J. Virol. 26:203.

Cheng, Y.-C., Bastow, K., Frank, K., Nutter, L., Chiou, J.-F. and Grill, S., 1984. Enzymes as Antiviral Targets in: Proceedings Vol. 1, IUPHAR 9th International Congress of Pharmacology. eds. W. Paton, J. Mitchell, P. Turner, Macmillan Press Ltd., London.

Cheng, Y.-C., Dutschman, G.E., Bastow, K.F., Sarngadharan, M.G. and Ting, R.Y.C., 1987, Human Immunodeficiency Virus Reverse Transcriptase: General Properties and its Interactions With Nucleoside Triphosphate Analogs, J. Biol. Chem. 262:2187.

Cheng, Y.-C. and Prusoff, W.H., 1986, Antiviral Chemotherapy, in "CRC Handbook of Chemotherapeutic Agents, Vol. II." ed. M. Verderame, CRC Press, Boca Raton, Florida.

Colby, B.M., Furman, P.A., Shaw, J.E., Elion, G.D., and Pagano, J.S., 1981, Phosphorylation of Acyclovir [9-(2-Hydroxyethoxymethyl)guanine] in Epstein-Barr Virus-Infected Lymphoblastoid Cell Lines, J. Virol. 38:606.

De Clercq, E., 1985, Targets for the Antiviral and Antitumor Activities of Nucleoside, Nucleotide and Oligonucleotide Analogues, Nucleosides & Nucleotides 4:3.

de Miranda, P. and Blum, M.R., 1983, Pharmacokinetics of Acyclovir After

Intravenous and Oral Administration, J. Antimicrob. Chemother. 12 Suppl. B:29.

de Miranda, P., Good, S.S., Laskin, O.L., Krasny, H.C., Connor, J.D. and Lietman, P.S., 1981, Disposition of Intravenous Radioactive Acyclovir, Clin. Pharmacol. Therapy 30:662.

Derse, D. and Cheng, Y.-C., 1981, Herpes Simplex Virus Type 1 DNA Polymerase, J. Biol. Chem. 256:8525.

Derse, D., Cheng, Y.-C., Furman, P.A., St. Clair, M.H. and Elion, G.B., 1981, Inhibition of Purified Human and Herpes Simplex Virus-Induced DNA Polymerases by 9-(2-Hydroxyethoxymethyl)guanine Triphosphate, J. Biol. Chem.256:11447.

Desgranges, C., Razako, G., Drouillet, F., Bricaud, H., Herdewijn, P. and De Clercq, E., 1984, Regeneration of the Antiviral Drug (E)-5-(2-Bromovinyl)-2-deoxyuridine, Nucleic Acids Res. 12:20811.

de Turenne-Tessier, M., Ooka, T., de The, G. and Daillie, J., 1986, Characterization of an Epstein-Barr Virus-Induced Thymidine Kinase, J. Virol. 57:1105.

Dix, R.D., Pereira, L. and Baringer, J.R., 1981, Use of Monoclonal Antibody Directed Against Herpes Simplex Virus Glycoproteins to Protect Mice Against Acute Virus-Induced Neurological Disease, Infection and Immunity 34:192.

Drach, J.C., 1983, Purine Nucleoside Analogs as Antiviral Agents, in "Targets for the Design of Antiviral Agents," eds. R.T. Walker and E. De Clercq, Plenum Press, London.

Dube, S.K., Gaedicke, G., Kluge, N., Weimann, B.J., Melderis, H., Steinheider, G., Crozier, T., Beckmann, H., Ostertag, W., 1974, in Proceeding of the 4th International Symposium of the Princess Takamatsu Cancer Research Fund, Tokyo, 1973, Differentiation and Control of Malignancy of Tumor Cells, Nakahara, W., Ono, T., Sugimura, T., Sugano, H., Ed.; University of Tokyo Press: Tokyo, p. 99.

Dube, S.K., Pragnell, I.B., Kluge, N., Gaedicke, G., Steinheider, G., Ostertag, W., 1975, Proc. Natl. Acad. Sci. U.S.A. 72:1863.

Dyatkina, N.B., Krayevsky, A.A. and Azhayev, A.V., 1986, Aminonucleosides and Their Derivatives, XIV. A General Method for Synthesis of 3'-Azido-2',3'-dideoxynucleosides, Bioorg. Khim. 12:1048.

Elion, G.B., 1982, Mechanism of Action and Selectivity of Acyclovir, Amer. J. Med., 73, No. 1a:7.

Elion, G.B., Furman, P.A., Fyfe, J.A., de Miranda, P., Beauchamp, L. and Schaeffer, H.J., 1977, Selectivity of Action of an Antiherpetic Agent, 9-(2-Hydroxyethoxymethyl)guanine, Proc. Natl. Acad. Sci. U.S.A. 74:5716.

Elwell, L.P., Ferone, R., Freeman, G.A., Fyfe, J.A., Hill, J.A., Ray, P.H., Richards, C.A., Singer, S.C., Knick, V.B., Rideout, J.L. and Zimmerman, T.P., 1987, Antibacterial Activity and Mechanism of Action of 3'-Azido-3'-Deoxythymidine (BW A509U), Antimicrob. Agents Chemother. 31:274.

Ericksson, B., Helgstrand, E., Johansson, N.G., Larsson, A., Misiorny, A., Noren, J.O., Philipson, L., Stenberg, K., Stening, G., Stridh, S. and Oberg, B., 1977, Inhibition of Influenza Virus Ribonucleic Acid Polymerase by Ribavirin Triphosphate, Antimicrob. Agents Chemother., 11:946.

Eriksson, B., Larsson, A., Helgstrand, E., Johansson, N.-G., and Oberg, B., 1980, Pyrophosphate Analogues as Inhibitors of Herpes Simplex Virus Type 1 DNA Polymerase, Biochem. Biophys. Acta 607:53.

Farr, B.M., Gwaltney, J.M. Jr., Adams, K.F. and Hayden, F.G., 1984, Intranasal Interferon-Alpha 2 for Prevention of Natural Rhinovirus Colds, Antimicrob. Agents Chemother. 26:31.

Fischer, P.H., Chen, M.S. and Prusoff, W.H., 1980, The Incorporation of 5-Iodo-5'-amino-2',5'-dideoxyuridine and 5-Iodo-2'-deoxyuridine Into Herpes Simplex DNA, Relationship Between Antiviral Activity and Effects on DNA Structure, Biochem. Biophys. Acta 606:236.

Fiume, L., Bassi, B., Busi, C., Mattioli, D., and Spinosa, G., 1986, Drug Targeting In Antiviral Chemotherapy. A Chemically Stable Conjugate of 9-β-D-Arabinofuranosyl-Adenine-5'-Monophosphate with Lactosaminated Albumin Accomplishes A Selective Delivery of the Drug to Liver Cells, Biochem. Pharmacol. 35:967.

Fiume, L., Bassi, B., Busi, C., Mattioli, D. and Wieland, T., 1985, A Study on the Pharmacokinetics of Adenine-9-β-arabinofuranoside 5-Monophosphate Conjugated with Lactosaminated Albumin, Experientia, 41:1326.

Fiume, L., Busi, C. and Mattioli, A., 1983, Targeting of Antiviral Drugs by Coupling With Protein Carriers, FEBS Lett. 153:6.

Fox, J.S. and White, D.O., 1980, Delivery of Antiviral Chemotherapeutic Agents to Neurons by Retrograde Axonal Transport, Medical Hypothesis 6:773.

Furman, P.A., de Miranda, P., St. Clair, M.H. and Elion, G.B., 1981, Metabolism of Acyclovir in Virus-Infected and Uninfected Cells, Antimicrob. Agents Chemother. 20:518.

Furman, P.A., Fyfe, J.A., St. Clair, M.H., Weinhold, K., Rideout, J.L., Freeman, G.A., Nusinoff Lehrman, S., Bolognesi, D.P., Broder, S., Mitsuya, H., and Barry, D.W., 1986, Phosphorylation of 3'-Azido-3'-deoxythymidine and Selective Interaction of the 5'-Triphosphate with Human Immunodeficiency Virus Reverse Transcriptase, Proc. Natl. Acad. Sci. U.S.A. 83:8333.

Furman, P.A., St. Clair, M.H., Fyfe, J.A., Rideout, P.M., Keller, P.M. and Elion, G.B., 1979, Inhibition of Herpes Simplex Virus-Induced DNA Polymerase Activity and Viral DNA Replication by 9-(2-Hydroxyethoxymethyl)guanine and its Triphosphate, J. Virol. 32:72.

Furman, P.A., St. Clair, M.H. and Spector, T., 1984, Acyclovir Triphosphate Is a Suicide Inactivator of the Herpes Simplex Virus DNA Polymerase, J. Biol Chem. 259:9575.

Fyfe, J.A., Keller, P.M., Furman, P.A., Miller, R.L. and Elion, G.B., 1978, Thymidine Kinase from Herpes Simplex Virus Phosphorylates The New Antiviral Compound, 9-(2-Hydroxyethoxymethyl)guanine, J. Biol. Chem. 253:8721.

Gangarosa, L.P., Hill, J.M., Thompson, B.L., Leggett, C. and Rissing, J.P., 1986, Iontophoresis of Vidarabine Monophosphate for Herpes Orolabialis, J. Infect. Dis., 154:930.

Gangarosa, L.P., Park, N.H., Kwon, B.S. and Hill, J.M., 1982, Iontophoretic Application of Antiviral Drugs. In "Herpesvirus: Clinical, Pharmacological and Basic Aspects," Eds. H. Shiota, Y.-C. Cheng and W.H. Prusoff, Excerpta Medica International Congress Series 571:201.

Gangemi, J.D., Nachtigal, M., Barnhart, D., Krech, L. and Jani, P., 1987, Therapeutic Efficacy of Liposome-Encapsulated Ribavirin and Muramye Tripeptide In Experimental Infection with Influenza or Herpes Simplex Virus, J. Infect. Dis., 155:510.

Gilbert, B.E. and Knight, V., 1986, Biochemistry and Clinical Applications of Ribavirin, Antimicrob. Agents Chemother. 30:201.

Goldanskii, V.I., Avetisov, V.A. and Kuz'min, V.V., 1986, Chiral Purity of Nucleosides as a Necessary Condition of Complementarity, FEBS Lett., 207:181.

Goswami, B.B., Borek, E., Sharma, O.K., Fujitaki, and Smith, R.A., 1979, The Broad Spectrum Antiviral Agent Ribavirin Inhibits Capping of mRNA, Biochem. Biophys. Res. Commun. 89:830.

Goz, B., 1978, The Effects of Incorporation of 5-Halogenated Deoxyuridines Into the DNA of Eukaryote Cells. Pharmac. Rev. 29:249.

Gregoriadis, G., 1973, Drug Entrapment in Liposomes, FEBS Letters 36:292.

Gregoriadis, G. and Neerunjun, E.D., 1975, Homing of Liposomes to Target Cells, Biochem. Biophys. Res. Comm. 65:537.

Hall, C.B., McBride, J.T., Walsh, E.E., Bell, D.M., Gala, C.L., Hildreth, S., Ten Eck, L.G. and Hall, W.J., 1983, Aerosolized Ribavirin Treatment of Infants With Respiratory Syncytial Viral Infections, New England J. Med. 308:1443.

Hall, C.B., Walsh, E.E., Hruska, J.F., Betts, R.F. and Hall, W.J., 1983, Ribavirin Treatment of Experimental Respiratory Viral Infection. A Controlled Double-Blind Study In Young Adults, J. Amer. Med. Assoc. 249:2666.

Hartshorn, K.L., Vogt, M.W., Chou, T.-C., Blumberg, R.S., Byington, R., Schooley, R.T. and Hirsch, M.S., 1987, Synergistic Inhibition of Human Immunodeficiency Virus In Vitro by Azidothymidine and Recombinant Alpha A Interferon, Antimicrob. Agents Chemother. 31:168.

Haschke, R.H., Ordronneau, J.M. and Bunt, A.H., 1980, Preparation and Retrograde Axonal Transport of an Antiviral Drug/Horseradish Peroxidase Conjugate, J. Neurochem. 35:1431.

Hayden, F.G., Minocha, A., Spyker, D.A. and Hoffman, H.E., 1985, Comparative Single-Dose Pharmacokinetics of Amantadine Hydrochloride and Rimantadine Hydrochloride in Young and Elderly Adults. Antimicrob. Agents and Chemother. 28:216.

Heidelberger, C. and King, D.H., 1979, Trifluorothymidine. Pharmac. Ther. 6:427.

Heidelberger, C., Parsons, D.G. and Remy, D.C., 1964, Synthesis of 5-Trifluoromethyluracil and 5-Trifluoromethyl-2'-deoxyuridine. J. Med. Chem. 7:1.

Helgstrand, E., Eriksson, B., Johansson, N.G., Lannero, B., Larsson, A., Misiorny, A., Noren, J.O., Sjoberg, B., Stenberg, K., Stening, G., Stridh, S., Oberg, B., Alenius, S. and Philipson, L., 1978, Trisodium Phosphonoformate, a New Antiviral Compound, Science 201:819.

Helgstrand, E., Flodh, H., Lernestedt, J.-O., Lundstrom, J. and Oberg, B., 1980, Trisodium Phosphonoformate: Antiviral Activities, Safety Evaluation and Preliminary Clinical Results. In "Developments in Antiviral Therapy," eds. L.H. Collier and J. Oxford, Academic Press, London.

Henderson, N.L., 1983, Recent Advances in Drug Delivery System Technology, Ann. Rev. Med. Chem. 18:275.

Hirt, B., 1966, Evidence for Semiconservative Replication of Circular Polyoma DNA, Proc. Natl. Acad. Sci. U.S.A. 55:997.

Hoffman, C.E., 1980, Structure, Activity and Mode of Action of Amantadine HCl and Related Compounds, Antibiot. Chemother. 27:233.

Horwitz, J.P., Chua, J. and Noel, M., 1964, Nucleosides V. The Monomysylates of 1-(2'-deoxy-β-D-lyxofuranosyl)thymine, J. Org. Chem. 29:2076.

Iwasaki, Y., Yamamoto, T., Konno, H., Iizuka, H. and Kudo, H., 1986, Eradication of Herpes Simplex Virus Persistence in Rat Trigeminal Ganglia by Retrograde Axoplasmic Transport, J. Virol. 59:242.

Jahrling, P.B., Hesse, R.A., Eddy, G.A., Johnson, K.M., Callis, R.F., Stephen, E.L., 1980, Lassa Fever Virus Infection of Rhesus Monkeys: Pathogenesis and Treatment With Ribavirin, J. Infect. Dis. 141:580.

Jansons, V.K. and Mallett, P.L., 1981, Targeted Liposomes: A Method for Preparation and Analysis, Analytical Biochem. 111:54.

Juel-Jensen, B.E., MacCallum, F.O. and MacKenzie, A.M.R., 1970, Treatment of Zoster With Idoxuridine In Dimethyl Sulphoxide: Results of Two Double Blind Controlled Trials, Brit. Med. J. 4:776.

Juliano, R.L., 1981, Liposomes as a Drug Delivery System, Trends in Pharm. Sci. 2:39.

Kato, M. and Eggers, H.J., 1969, Inhibition of Uncoating of Fowl Plague Virus by 1-Adamantanamine Hydrochloride, Virology 37:632.

Keller, P.M., McKee, and Fyfe, J.A., 1985, Cytoplasmic 5'-Nucleotidase Catalyzes Acyclovir Phosphorylation, J. Biol. Chem. 260:8664.

Kende, M., Alving, C.R., Rill, W.L., Swartz, G.M. Jr. and Canonico, P.G., 1985, Enhanced Efficacy of Liposome-Encapsulated Ribavirin Against Rift Valley Fever Virus Infection in Mice, Antimicrob. Agents Chemother., 27:903.

Koff, W.C. and Knight, V., 1979, Inhibition of Influenza Virus Uncoating by Rimantadine Hydrochloride, J. Virology 31:261.

Krenitsky, T.A., Hall, W.W., de Miranda, P., Beauchamp, L.M., Schaeffer, H.J., and Whiteman, P.D., 1984, 6-Deoxyacyclovir: A Xanthine Oxidase-

Activated Prodrug of Acyclovir, Proc. Natl. Acad. Sci. U.S.A. 81:3209.

Kristensson, K., 1978, Retrograde Transport of Macromolecules in Axons, Annual Rev. Pharm. and Tox. 18:97.

Langen, P., 1975, "Antimetabolites of Nucleic Acid Metabolism." Gordon and Breach, New York.

Larder, B.A. and Darby, G., 1984, Virus Drug-Resistance: Mechanisms and Consequences, Antiviral Res. 4:1.

Larsson, A., Sundqvist, A. and Parnerud, A.-M., 1986, Inhibition of Herpes Simplex Virus-Induced DNA Polymerases and Cellular DNA Polymerase α by Triphosphates of Acyclic Guanosine Analogs, Molec. Pharmacol. 29:614.

Laskin, O.L., 1984, Acyclovir, Rational Drug Therapy 18:Number 5.

Laskin, O.L., 1983, Clinical Pharmacokinetics of Acyclovir, Clin. Pharmacokinetics 8:187.

Lee, W.W., Benitez, A., Goodman, L., and Baker, B.R., 1960, Potential Anti-cancer Agents XI. Synthesis of the β-anomer of 9-(D-arabinofuranosyl)adenine, J. Am. Chem. Soc. 82:2648.

Lemaitre, M., Bayard, B. and Lebleu, B., 1987, Specific Antiviral Activity of a Poly(L-lysine)-conjugated Oligodeoxyribonucleotide Sequence Complementary to Vesicular Stomatitis Virus N-Protein mRNA Intitiation Site, Proc. Natl. Acad. Sci. U.S.A., 84:648.

Leonard, M.F., Kumar, A., Murray, D.L., Beaman, D.C., 1987, Inhibitory Effect of Azone[R] (1-Dodecylazacycloheptan-2-one) on Herpes Simplex Viruses, In Vivo and In Vitro Studies, Chemother., 33:151.

Lin, T.-S., Chen, M.S., Mclaren, C., Gao, Y.S., Ghazzouli, I. and Prusoff, W.H., 1987, Synthesis and Antiviral Activity of Various 3'-Azido, 3'-Amino, 2',3'-Unsaturated and 2',3'-Dideoxy Analogues of Pyrimidine Deoxyribonucleosides Against Retroviruses, J. Med. Chem., 30:440.

Lin. T.-S. and Prusoff, W.H., 1978, Synthesis and Biological Activity of Several Amino Analogs of Thymidine, J. Med. Chem. 21:109.

Lin, T.-S., Schinazi, R.F., Chen, M.S., Kinney-Thomas, E. and Prusoff, W.H., 1987, Antiviral Activity of 2',3'-Dideoxycytidin-2'-ene (2',3'-Dideoxy-2',3'-didehydrocytidine) Against Human Immunodeficiency Virus In Vitro, Biochem. Pharmacol. 36:311.

Littler, E., Zeuthen, J., McBride, A.A., Trost-Sorenson, E., Powell, K.L., Walsh-Arrand, J.E. and Arrand, J.R., 1986, Identification of an Epstein-Barr Virus-coded Thymidine Kinase, The EMBO J. 5:1959.

Lok, A.S.F., Wilson, L.A. and Thomas, H.C., 1984, Neurotoxicity Associated with Adenine Arabinoside Monophosphate In The Treatment of Chronic Hepatitis B Virus Infection, J. Antimicrob. Chemother. 14:93.

Markley, J.L., Westler, W.M., Chan, T.-M., Kojiro, C.L. and Ulrich, E.L., 1984, Two-Dimensional NMR Approaches to the Study of Protein Structure and Function, Fed. Proc. 43:2648.

Matsumura, K., Fujimoto, M., and Mitsui, Y., 1973, Micro-autographic Studies on Incorporation of 5-Iodo-2'-deoxyuridine Into Herpes Simplex Virus, Jap. J. Ophthal. 17:125.

McCormick, J.B., King, I.J., Webb, P.A., Scribner, C.L., Craven, R.B., Johnson, K.M., Elliott, L.H. and Belmont-Williams, R., 1986, Lassa Fever, Effective Therapy with Ribavirin, New Engl. J. Med. 314:20.

McGuirt, P.V., Shaw, J.E., Elion, G.B. and Furman, P.A., 1984, Identification of Small DNA Fragments Synthesized in Herpes Simplex Virus-Infected Cells In The Presence of Acyclovir, Antimicrob. Agents Chemother. 25:507.

Medical Letter, 1986, Azidothymidine for AIDS, 28:107.

Medical Letter, 1986, Ribavirin (Virazole), 28:46.

Miller, J.P., Kigwana, L.J., Streeter, D.G., Robins, R.K., Simon, L.N. and Roboz, J., 1977, The Relationship Between the Metabolism of Ribavirin and its Proposed Mechanism of Action, Ann. N.Y. Acad. Sci. 284:211.

Miller, W.H. and Miller, R.L., 1980, Phosphorylation of Acyclovir (Acycloguanosine)monophosphate by GMP Kinase, J. Biol. Chem. 255:7204.

Mitsuya, H. and Broder, S., 1986, Inhibition of the In Vitro Infectivity and Cytopathic Effect of Human T-lymphotropic Virus Type III/Lymphadenopathy-Associated Virus (HTLV-III/LAV) by 2',3'-dideoxynucleosides, Proc. Natl. Acad. Sci. U.S.A. 83:1911.

Mitsuya, H. and Broder, S., 1987, Strategies for Antiviral Therapy in AIDS, Nature 325:773.

Mitsuya, H., Weinhold, K.J., Furman, P.A., St. Clair, M.H., Nusinoff Lehrman, S., Gallo, R.C., Bolognesi, D., Barry, D.W., and Broder, S., 1985, 3'-Azido-3'-deoxythymidine (BW A509U): An Antiviral Agent That Inhibits the Infectivity and Cytopathic Effect of Human T-lymphotropic Virus Type III/Lymphadenopathy-Associated Virus In Vitro, Proc. Natl. Acad. Sci. U.S.A. 82:7096.

Nakayama, K., Ruth, J.L., and Cheng, Y.-C., 1982, Differential Effect of Nucleoside Analog Triphosphates on Ribonucleotide Reductase From Uninfected and Herpes Simplex Virus-Infected HeLa Cells, J. Virol. 43:325.

North, T.W. and Cohen, S., 1979, Aranucleosides and Aranucleotides in Viral Chemotherapy, Pharmac. Ther. 4:81.

Nylen, P., 1924, Beitrag zur Kenntnis der Organischen Phosphor-Verbindungen, Chem. Ber. 57:1023.

Oberg, B., 1983, Antiviral Effects of Phosphonoformate (PFA, Foscarnet Sodium), Pharmac. Ther. 19:387.

Ostertag, W., Cole, T., Crozier, T., Gaedicke, G., Kind, J., Kluge, N., Krieg, J.C., Roselser, G., Steinheider, G., Weimann, B.J., Dube, S.K. in Proceeding of the 4th International Symposium of the Princess Takamatsu Cancer Research Fund, Tokyo 1973, Differentiation and Control of Malignancy of Tumor Cells, Nakahara, W., Ono, T., Sugimura, T., Sugano, H., Ed.; University of Tokyo Press: Tokyo, 1974, p. 485.

Ostertag, W., Roesler, G., Kreig, C.J., Cole, T., Crozier, T., Gaedicke, G., Steinheider, G., Kluge, N., Dube, S.K., 1974, Proc. Natl. Acad. Sci. U.S.A. 71:4980.

Ostrander, M. and Cheng, Y.-C., 1980, Properties of Herpes Simplex Virus Type 1 and Type 2 DNA Polymerase, Biochem. Biophys. Acta 609:232.

Otto, M.J., Goz, B. and Prusoff, W.H., 1984, Antiviral Activity of Iodinated Pyrimidine Deoxyribonucleosides. In "Antiviral Drugs and Interferon: The Molecular Basis of Their Activity." ed. Y.C. Becker, Martinus Nijhof, Hingham, MA.

Otto, M.J., Lee, J.J. and Prusoff, W.H., 1982, Effects of Nucleoside Analogues on the Expression of Herpes Simplex Type 1 Induced Protein, Antiviral Res. 2:267.

Oxford, J.S. and Gailbraith, A., 1980, Antiviral Activity of Amantadine: A Review of Laboratory and Clinical Data, Pharmac. Ther. 11:181.

Park, N.H., Gangarosa, L.P., Kwon, B.S., Hill, J.M., 1978, Iontophoretic Application of Adenine Arabinoside Monophosphate to Herpes Simplex Virus Type-I-infected Hairless Mouse Skin, Antimicrob. Agents Chemother. 14:604.

Prusoff, W.H., Bakhle, Y.S. and McCrea, J.F., 1963, Incorporation of 5-Iodo-2'-deoxyuridine Into the Deoxyribonucleic Acid of Vaccinia Virus, Nature, Lond. 199:1310.

Prusoff, W.H., Chen, M.S., Fischer, P.H., Lin, T.-S., Mancini, W.R., Otto, M.J., Shiau, G.T., Schinazi, R.F. and Walker, J., 1984, Antiviral Iodinated Pyrimidine Deoxyribonucleosides: 5-Iodo-2'-Deoxyuridine; 5-Iodo-2'-Deoxycytidine; 5-Iodo-5'-Amino-2',5'-Dideoxyuridine. In "Internat. Encyclopedia of Pharmacol. and Therap., Section III, Viral Chemotherapy", Ed. D. Shugar. Pergamon Press, N.Y.

Prusoff, W.H., Chen, M.S., Fischer, P.H., Lin, T.-S., Shiau, G.T., Schinazi, R.F. and Walker, J., 1979, Pharmac. Ther. 7:1.

Prusoff, W.H. and Goz, B., 1975, Halogenated Pyrimidine Deoxyribonucleosides, in "Antineoplastic and Immunosuppressive Agents." Vol. 2. eds. A.C. Sartorelli and D.G. Johns. Springer, Berlin.

Prusoff, W.H., Lin, T.-S. and Zucker, M., 1986, Potential Targets for Antiviral Chemotherapy, Antiviral Res. 6:311.

Prusoff, W.H. and Otto, M.J., 1983, Problems in the Pharmacology and Pharmacokinetics of Antivirals, in "Problems of Antiviral Therapy," eds. C.H.

Stuart-Harris and J. Oxford. Academic Press, New York.

Rees, P.J., Selby, P., Prentice, H.G., Whiteman, P.D. and Grant, D.M., 1986, A5I5U: A Prodrug of Acyclovir With Increased Oral Bioavailability, J. Antimicrob. Chemother. 18:Suppl. B, 215.

Reinke, C.M., Drach, J.C., Shipman, C.Jr. and Weissbach, A., 1978, in "Oncogenesis and Herpesviruses III" (Part 2). eds. G. De The, W. Henle, and F. Rapp. IARC, Lyon, France.

Reno, J.M., Lee, L.F. and Boezi, J.A., 1978, Inhibition of Herpesvirus Replication and Herpesvirus-Induced Deoxyribonucleic Acid Polymerase by Phosphonoformate, Antimicrob. Agents Chemother. 13:188.

Richman, D.D., Yazaki, P. and Hostetler, K.Y., 1982, The Intracellular Distribution and Antiviral Activity of Amantadine, Virology, 112:81.

Robins, R.K., 1986, Chem. Engineer. News Jan. 27:28.

Rossman, M.G., 1985, Cited in Chem. Engineer. News Sept 16:81.

Saffran, M., Kumar, G.S., Savariar, C., Burnheim, J.C., Williams, F. and Neckers, D.C., 1986, A New Approach to the Oral Administration of Insulin and Other Peptide Drugs, Science 233:1081.

Scheinberg, D.A. and Strand, M., 1982, Leukemic Cell Targeting and Therapy by Monoclonal Antibody in a Mouse Model System, Cancer Res. 42:44.

Schiek, W. and Schiek, E., 1969, Untersuchung Uber Infektioses Bromodesoxy-uridinhaltiges Herpes Virus Hominis. Bistimmung der Dichte und der Sedimentationkonstanten is $CsCl-H_2O$ Dichtergradienten, Arch. ges. Virusforsch 28:229.

Schlegel, R., Dickson, R.B., Willingham, M.C. and Pastan, I.H., 1982, Amantadine and Dansylcadaverine Inhibit Vesicular Stomatitis Virus Uptake and Receptor-Mediated Endocytosis of α_2-Macroglobulin, Proc. Natl. Acad. Sci. U.S.A. 79:2291.

Senyei, A.E. and Widder, K.J., 1981, Drug Targeting: Magnetically Responsive Albumin Microspheres-- A Review of the System to Date, Gyn. Onc. 12:1.

Shannon, W.M., 1984, Mechanisms of Action and Pharmacology: Chemical Agents, in "Antiviral Agents and Viral Diseases of Man," Eds. G.J. Galasso, T.C. Merigan and R.A. Buchanon, Raven Press, New York.

Shaw, J.E., 1980, Drug Delivery Systems, Ann. Rept. Med. Chem. 15:302.

Sidwell, R.W., Huffman, J.H., Call, E., Alaghamandan, H., Dixon, G.J., 1987, Effect of Vidarabine in Dimethyl Sulfoxide Vehicle on Type 1 Herpes-virus-Induced Cutaneous Lesions in Laboratory Animals, Chemother., 33:141.

Sidwell, R.W., Robins, R.K. and Hillyard, I.W., 1979, Ribavirin: An Antiviral Agent, Pharmacol. Ther. 6:123.

Skehel, J.J., Baley, P.M., Brown E.B., Martin, S.R., Waterfield, M.D., White, J.R., Wilson, I.A. and Wiley, D.C., 1982, Changes in the Conformation of Influenza Virus Hemagglutinin and the pH Optimum of the Virus-Mediated Membrane Fusion, Proc. Natl. Acad. Sci. U.S.A. 79:968.

Skehel, J.J., Hay, A.J. and Armstrong, J.A., 1977, On the Mechanism of Inhibition of Influenza Virus Replication by Amantadine Hydrochloride, J. Gen. Virol. 38:97.

Smee, D.F. and Mathews, T.R., 1986, Metabolism of Ribavirin In Respiratory Syncytial Virus-Infected and Uninfected Cells, Antimicrob. Agents Chemother. 30:117.

Smith, C.C., Aurelian, L., Reddy, M.P., Miller, P.S. and Ts'o, P.O.P., 1986, Antiviral Effect of an Oligo(Nucleoside Methylphosphonate) Complementary to the Splice Junction of Herpes Simplex Virus Type 1 Immediate Early Pre-mRNAs 4 and 5, Proc. Natl. Acad. Sci. U.S.A., 83:2787.

Smith, R.A. and Kirkpatrick, W., eds., 1980, "Ribavirin, A Broad Spectrum Antiviral Agent," Academic Press, New York.

Smith, R.A., Knight, V., Smith, J.A.D., eds., 1984, "Clinical Applications of Ribavirin", Academic Press, New York, New York.

Smolin, G., Okumoto, M., Feiler, S. and Condon, D., 1981, Idoxuridine-liposome Therapy for Herpes Simplex Keratitis, Am. J. Opthal. 91:220.

Sommadossi, J.P., Barnes, D.W., Miller, L.R., Markiewicz, M.A. and Whitley,

R.J., 1986, Novel Pharmacologic Strategies in the Treatment of Life-Threatening Infections: Clinical Experience with 9-(1,3-Dihydroxy-2-Propoxy-Methyl)-Guanine (DHPG), Abstract: <u>Amer. Fed. Clin. Res.</u>, In Press.

Spruance, S.L., McKeough, B. and Cardinal, J.R., 1984, Penetration of Guinea Pig Skin by Acyclovir in Different Vehicles and Correlation With the Efficacy of Topical Therapy of Experimental Cutaneous Herpes Simplex Virus Infection, <u>Antimicrob. Agents Chemother.</u> 25:10.

Stegman, T., Hoekstra, D., Scherphof, G. and Wilschut, J., 1986, Fusion Activity of Influenza Virus. A Comparison Between Biological and Artificial Target Membrane Vesicles, <u>J. Biol. Chem.</u> 261:10,966.

Streeter, D.G., Simon, L.N., Robins, R.K. and Miller, J.P., 1974, The Phosphorylation of Ribavirin by Deoxyadenosine Kinase from Rat Liver. Differentiation Between Adenosine and Deoxyadenosine Kinase, <u>Biochemistry</u> 13:4543.

Streeter, D.G., Witkowski, J.T., Khare, G.P., Sidwell, R.W., Bauer, R.J., Robins, R.K. and Simon, L.N., 1973, Mechanism of Action of 1-β-D-ribofuranosyl-1,2,4-triazole-3-carboxamide (Virazole), a New Broad-Spectrum Antiviral Agent, <u>Proc. Natl. Acad. Sci. U.S.A.</u> 70:1174.

Sundquist, B. and Oberg, B., 1979, Phosphonoformate Inhibits Reverse Transcriptase, <u>J. Gen. Virol.</u> 45:273.

Toltzis, P. and Huang, A.S., 1986, Effect of Ribavirin on Macromolecular Synthesis in Vesicular Stomatitis Virus-Infected Cells, <u>Antimicrob. Agents Chemother.</u> 29:1010.

Toulme, J.J., Krisch, H.M., Loreau, N., Thuong, N.T. and Helene, C., 1986, Specific Inhibition of mRNA Translation by Complementary Oligonucleotides Covalently Linked to Intercalating Agents, <u>Proc. Natl. Acad. Sci. U.S.A.</u>, 83:1227.

Turenne-Tessier, M., Ooka, G., Daillie, J., 1986, Characterization of an Epstein-Barr Virus-Induced Thymidine Kinase, <u>J. Virol.</u>, 57:1105.

Tuttle, J.V. and Krenitsky, T.A., 1984, Effects of Acyclovir and Its Metabolites on Purine Nucleoside Phosphorylase, <u>J. Biol. Chem.</u> 259:4065.

Vince, R. and Daluge, S., 1977, Carbocyclic Arabinosyladenine, an Adenosine Deaminase-Resistant Antiviral Agent, <u>J. Med. Chem.</u> 20:612.

Vogt, M.W., Hartshorn, K.L., Furman, P.A., Chou, T.-C., Fyfe, J.A., Coleman, L.A., Crumpacker, C. Schooley, R.T. and Hirsch, M.S., 1987, Ribavirin Antagonizes the Effect of Azidothymidine on HIV Replication, <u>Science</u> 235:1376.

Vrang, L., Bazin, H., Remaud, G., Chattopadhyaya and Oberg, B., 1987, Inhibition of the Reverse Transcriptase from HIV by 3'-Azido-3'-deoxythymidine Triphosphate and Its Threo Analogue, <u>Antiviral Res.</u> 7:139.

Wahren, B., Larsson, A., Ruden, V., Sundquist, A., and Solver, E., 1987, Acyclic Guanosine Analogs as Inhibitors of Human Cytomegalovirus, <u>Antimicrob. Agents Chemother.</u> 31:317.

Weinstein, J.N., Magin, R.L., Cysyk, R.L. and Zaharko, D.S., 1980, Treatment of Solid L1210 Murine Tumours With Local Hyperthermia and Temperature Sensitive Liposomes Containing Methotrexate, <u>Cancer Res.</u> 40:1388.

Welch, A.D. and Prusoff, W.H., 1960, A Synopsis of Recent Investigations of 5-Iodo-2'-Deoxyuridine, <u>Cancer. Chemother. Rept.</u> 6:29.

Widder, K.J., Morris, R.M., Poore, G., Howard, Jr., D.P. and Senyei, A.E., 1981, Tumor Remission in Yoshida Sarcome-bearing Rats by Selective Targeting of Magnetic Albumin Microspheres Containing Doxorubicin, <u>Proc. Natl. Acad. Sci. U.S.A.</u> 78:579.

Wigand, R. and Klein, W., 1974, Properties of Adenovirus Substituted with Iododeoxyuridine. <u>Arch. ges. Virusforsch.</u> 45:298.

Yarchoan, R., Brouwers, P., Spitzer, A.R., Grafman, J., Safai, B., Perno, C.F., Larson, S.M., Berg, G., Fischl, M.A., Wichman, A., Thomas, R.V., Brunetti, A., Schmidt, P.J., Myers, C.E., Broder, S., 1987, Response of Human-Immunodeficiency-Virus-Associated Neurological Disease to 3'-Azido-3'-deoxythymidine, <u>Lancet</u> i:132.

Yarchoan, R., Weinhold, K.J., Lyerly, H.K. Lyerly, Gelmann, E., Blum, R.M.,

Shearer, G.M., Mitsuya, H., Collins, J.M., Myers, C.E., Klecker, R.W., Markham, P.D., Durack, D.T., Nusinoff Lehrman, S., Barry, D.W., Fischl, M.A., Gallo, R.C., Bolognesi, D.P., Broder, S., 1986, Administration of 3'-Azido-3'-deoxythymidine, an Inhibitor of HTLV-III/LAV Replication, to Patients With AIDS or AIDS-related Complex, The Lancet i:575.

Yatvin, M.B., Muhlensiepen, H., Porschen, W., Weinstein, J.N. and Feinendegen, L.G., 1981, Selective Delivery of Liposome-associated Cis-dichlorodiamineplatinum (II) by Heat and its Influence on Tumor Drug Uptake and Growth, Cancer Res. 41:1602.

Yatvin, M.B., Weinstein, J.N. and Blummenthal, R., 1978, Design of Liposomes for Enhanced Local Release of Drugs by Hyperthermia, Science 202:1290.

Zamecnik, P.C., Goodchild, J., Taguchi, Y. and Sarin, P.S., 1986, Inhibition of Replication and Expression of Human T-Cell Lymphotropic Virus Type III in Cultured Cells by Exogenous Synthetic Oligonucleotides Complementary to Viral RNA, Proc. Natl. Acad. Sci. U.S.A., 83:4143.

VIRUS DRUG RESISTANCE

Hugh J. Field and Lindsey J. Owen

Cambridge University
Department of Clinical Veterinary Medicine
Madingley Road, Cambridge, UK

INTRODUCTION

A Definition of Resistance

Drug resistance in viruses is generally considered to be an acquired heritable change which is characterised by relief from inhibition by a particular drug. This is always the result of one or more mutations in the virus genome, causing changes in targets for the drug or its metabolites which are reflected in a measurable difference in sensitivity between the mutant and parental strain. Such resistance has now been observed in widely divergent virus families to a variety of specific inhibitors; some important examples are summarised in Table 1.

The isolation of drug-resistant variants has long been considered to indicate that the drug has a virus-specific element in its mode of action and that it is not simply acting as a cytotoxic agent (Herrmann & Herrmann, 1977). However, it is important to note that the isolation of resistant mutants does not preclude host-mediated factors which are essential to the virus inhibition. For example interferon is known to act by a two stage process in which interaction between interferon molecules and a cell receptor results in the activation of cellular genes and the induction of cell products which interfere with virus replication. Even so, interferon-resistant and sensitive variants have been described for several virus families and imply changes in particular targets (for example sensitivity to a nuclease (Simon et al, 1976)) causing the interferon to become ineffective. That interferon-resistant variants are not generally observed may relate to the multiple mechanisms involved in its mode of action.

Usually, changes in viruses in response to inhibitors are manifested by the acquisition of resistance but in some cases the alteration in virus can be reflected in hypersensitivity. This may mean that resistance to one compound increases susceptibility to a different drug. An example of this is the observation of herpes simplex virus (HSV) mutants which were

selected for resistance to phosphonoacetic acid (PAA) and found
to be hypersensitive to aphidicolin (Coen *et al*, 1984). There
are also instances where drug-dependent mutants have been
selected, for example guanidine-dependent mutants of polio
(Loddo *et al*, 1962), see below.

Table 1. Examples of viruses in which drug resistance
has been demonstrated.

Virus Family		Compounds[*]	Animals[†]	Man[‡]
Pox	vaccinia	thiosemicarbazones nucleoside analogues pyrophosphate analogues	yes	
Herpes	HSV VZ CMV	numerous nucleoside and pyrophosphate analogues, aphidicolin [ribavirin]	yes	yes yes
Picorna	rhinovirus foot & mouth polio	guanidine HBB, arildone dichloroflavan rhodanine, RMI chalcones [flavone] [enviroxime]	yes	
Orthomyxo	influenza A fowl plague	amantadine rimantadine norakin [ribavirin]	yes	yes
Retro	murine retroviruses	PFA coumermycin		

* : Resistance obtained by passage in tissue culture in the
presence of drug.
† : Resistance obtained from experimentally infected, drug-
treated animals.
‡ : Resistance observed among clinical isolates.
[]: Resistance sought but not observed to date.

The Importance of Failure to Isolate Resistant Mutants

The successful isolation of drug-resistant mutants is very
important evidence for a true virus-specific mechanism for the
drug, however, the converse is not necessarily true. There are
several possible reasons for the inability to isolate mutants.
One is that the change required for relief from inhibition
invariably leads to a lethal mutation in the virus and this may
be because the virus target or its genetic locus has a multiple
function. Alternatively, there may be several targets for drug
action such that simultaneous changes in several genetic loci
would be required for resistance. However, in cases where
prolonged attempts to detect resistance have failed, for example
the absence of HSV- resistance to ribavirin (Allen & Fingal,
1977) or bovine herpesvirus-1 to a thiosemicarbazone (Field &

Reading, 1987), this may be included among data which suggest a lack of true selective toxicity in the mode of action.

The Occurrence of Clinical Resistance

The detection of resistant clinical isolates depends on a knowledge of the spectrum of sensitivity among natural isolates and the ability to detect a significant increase above the normal range. The term carries the additional implication that the reduced sensitivity of the virus strain will result in reduced efficiency of chemotherapy. The existence of resistant strains may represent natural variants which have reduced sensitivity or they may again be mutants selected by exposure to the drug (or a different drug with the same target for inhibition). This may best be seen when sequential isolates from the same patient (Crumpacker et al,1982; Wade et al,1982; McLaren et al, 1983; Christophers & Sutton, 1987) or from a single epidemic (Webster et al, 1986) are found to have acquired resistance. Resistance in vivo may involve more factors than are readily modelled in a tissue culture drug assay and this is a problem in the design of screening tests for large numbers of clinical isolates during drug trials. Furthermore, the development of "treatment resistance" may involve other factors unrelated to the mode of action of the drug such as the metabolism of a drug to an inactive form or the development of a hypersensitivity reaction which exacerbates the virus lesion following topical therapy. These are among the ideas which have been put forward to explain the phenomenon of "treatment resistance" in relation to herpes keratitis therapy which could not be accounted for by "biochemical resistance" in the virus isolates (Coleman et al, 1968; Jawetz et al, 1970). However, it should be remembered that the mode of action of the drugs in question were less well understood at the time when these studies were carried out and there remains a further possibility that true virus resistance was overlooked.

The Pathogenicity of Drug-Resistant Virus Strains

In relation to the clinical impact of resistance it is very important to consider the question of how the resistant strains interact with the multicellular host. This is especially important when the mutants have been genetically engineered or selected for resistance in tissue culture systems and may not thrive in vivo. In turn this can give important clues about the role of particular virus genes in pathogenesis. The idea that resistant strains have a selective disadvantage in the absence of the inhibitor is one which has been considered for several other types of organisms, including bacteria, protozoa and fungi but is an area which has not yet been well researched. Viruses, having relatively simple genetic constitution can provide useful models for the investigation of the phenomenon. This aspect will be given particular attention in relation to the herpes and poxviruses below.

Comparison With Drug-Resistance in Other Micro-Organisms

Drug resistance has been observed among all the groups of micro-organisms including mycoplasma, bacteria, fungi, protozoa and neoplastic cells. There appear to be at least 5 general mechanisms by which resistance can develop: i. A modification

occurs in a target for the drug or there is a loss of an enzyme involved in drug activation. ii. There is decreased permeability of the microbe for the drug. iii. An enzyme is induced which modifies (or inactivates) the drug. iv. An alternative metabolic pathway is introduced by the micro-organism. v. There is an increase in a metabolite antagonizing the drug.
Within each general class there are a number of variations possible for particular microbes and the genetic basis for the same mechanism can differ. Bacteria are among the most notorious organisms for the development of resistance. A notable feature is their capacity to acquire simultaneously multiple resistance to several different (often unrelated) drugs (O'Brien, 1986). This was explained by the discovery of insertion sequences which can mobilize genes giving rise to transposable genetic elements (Grinsted, 1986). In turn these can be carried in extra-chromosomal elements of DNA such as plasmids or bacteriophage. The resistance encountered in eukaryotic cells may have closer parallels with virus resistance. For example the development of resistance to nucleoside analogues such as 5-fluorocytosine in fungi (Kerridge & Nicholas, 1986) is analogous to the development of resistance to nucleoside analogues in DNA viruses.

The chief resistance mechanisms in neoplastic cells (which is a major factor in failure of cancer chemotherapy to cytostatic and cytotoxic drugs) appear to be: altered drug transport, drug-induced metabolic alterations, cell cycle kinetic insensitivity, drug-induced modification of tumour cell antigenicity and modulation of the immune response and finally drug inaccessibility (reviewed by Hill, 1986). Commonly there is an overproduction of a metabolic product either through change in regulation of a gene or amplification of the gene copy number. Many of these mechanisms are peculiar to eukaryotic cells, however the stage in the cell cycle affects the susceptibility of neoplastic cells to drugs and this parallels the situation with latent viruses which can become inert within the host cells and thus become refractory to inhibitors (Field, 1985). This is a problem already encountered with herpesviruses and likely to be a factor in AIDS chemotherapy in the future.

Because of their relatively simple genetic constitution, viruses are restricted in the range of resistance mechanisms available. There is no direct way in which pre-formed genes can be readily acquired by viruses similar to those associated with transposons in bacteria nor can the complex reorganisation of genes and their expression occur as in eukaryotic cells. (Of course there is scope for viruses to undergo recombination, for example, in the picornaviridae and retroviridae or reassortment as in the negative strand, segmented-genome viruses). Generally viruses acquire resistance by single mutations in the genome. The alteration of drug targets by point mutation also accounts for some resistant strains encountered in other microbes, these appear to have much less medical importance by comparison and other mechanisms dominate the literature.

Thus the precise mechanisms of drug resistance in viruses appear to be much more limited, but some other aspects relating to drug-resistance may be similar to other microbes. One example is the change in pathogenicity sometimes encountered in drug resistant strains. This is an area which is suited to study in relatively simple virus systems and which could have wider

206

significance. In general terms however, the wealth of data
available from the study of resistance in other organisms,
especially bacteria has little direct relevance to viruses; this
is a new field now rapidly developing from first principles.

The Reasons for Studying Virus Resistance

It is clear that drug-resistant mutants have themselves
served as important tools during the development of virology.
The most important areas may be summarised as follows:

Elucidation of Drug Mode of Action. Drug-resistant mutants
can be used to identify gene products (and thus genetic loci)
which interact with the inhibitor or its metabolites. This
enables virus target molecules to be identified and drugs to be
compared for similarities in their modes of action. Several
examples of this will be discussed below.

Genetic Analysis. By selection for drug resistance it is
often possible to obtain a series of readily identifiable
genetic markers. Co-transfer of such markers and techniques of
"marker-rescue" can then provide convenient methods for the
selection of "silent" genetic events. When the drug mode of
action is well understood the study of gene products from the
resistant loci may elucidate gene function in the virus. For
example the evidence for vaccinia-induced (Moss & Cooper, 1982)
and herpesvirus-induced DNA-pol (Purifoy et al, 1977) were first
obtained with PAA and the resistance of adenovirus to PAA has
focused attention of the role of the 72K DNA- binding protein
(Foster et al, 1982). By defining precise changes in mutant
virus genomes by sequence analysis the functional domains of
virus enzymes and interactions with their substrates can be
approached. Recently, this strategy has been extremely fruitful
in relation to the herpesvirus- induced TK (Darby et al, 1986)
and DNA polymerase (Larder et al, 1987) and this will be
discussed below; a similar approach is now being pursued in
relation to the reverse transcriptase of the retroviridae.

Study of Pathogenesis. By selecting resistant strains
using drugs with well-defined modes of action virus mutants may
be obtained with lesions in one or more specific gene products.
An important advantage of this strategy is that (unless the
agent is itself mutagenic) the selection of specific mutants is
possible with minimal chance of adventitious mutations in
unrelated genes. Such drug-selected mutants may then be used to
investigate the role of those genes in pathogenesis. By far the
most extensive work of this kind has been done on herpes and pox
viruses with lesions in TK and DNA polymerase (see below). The
use of this kind of mutant does suffer from the disadvantages of
i. difficulty of completely removing low level contamination
with wild-type virus, ii. unwanted reversion, iii. additional
specific mutations in unpredicted sites of drug action.
However, such preliminary experiments can identify interesting
mutants which paves the way for the painstaking development of
specific deletion mutants or site-directed mutants which do not
suffer from these disadvantages.

Clinical Importance. The laboratory selection of drug-
resistant strains indicates the likely range of resistant
viruses that may be encountered among clinical isolates. Such

prototype strains then form important reference strains when developing tests for screening clinical specimens. Well-characterised laboratory mutants are also helpful in devising strategies to counteract resistance, for example by using alternative drugs with different modes of action (Field et al,1981; Field & Neden, 1982; Gauri, 1979; Larder & Darby, 1986) or the simultaneous administration of drug combinations (Schinazi, 1986) reviewed by Hall, (1986). Finally experiments can be carried out in model systems to examine the circumstances which may lead to clinical resistance and several important examples of this will be discussed below in the sections on influenza and herpes.

These then are the ways in which the phenomenon of drug resistance has actively approached for a number of viruses. The remainder of this review will comprise a discussion of several specific examples concentrating on those viruses for which there are available inhibitors and the compounds which are currently in use or likely to be used clinically in the foreseeable future.

LABORATORY METHODS FOR THE SELECTION OF DRUG-RESISTANT VIRUS

Tissue Culture Methods

The traditional method for the selection of drug- resistant strains of virus involves infecting cells in the presence of a subinhibitory concentration of the inhibitor. The yield is harvested and the process repeated in gradually increasing concentrations of drug until a population of virus is established with the markedly increased ability to replicate in the presence of drug.

This method suffers from a major disadvantage; the mutation frequency for herpes and many other viruses is high, being of the order $1/10^4$ (Coen, 1986). As a consequence an accumulation of mutations may be selected during several rounds of virus replication in a selective medium. For example, herpesvirus mutants defective in thymidine kinase (TK) may readily be selected with a single passage in the presence of a TK-mediated inhibitor. Thus further virus replication cycles (which may occur in the first passage if low multiplicity of infection is used) are likely to result in the selection of mutations in further sites leading to double or multiple lesions in the virus DNA. This means that passage in increasing levels of compound makes further mutations more likely for example in both HSV TK and DNA polymerase (Field et al, 1980). A different example of this is to be found with the two levels of amantadine resistance in influenza (Hay et al, 1986) to be discussed below. For many purposes it is therefore best to first try a single step selection by culturing the virus at a concentration of drug slightly above the inhibitory level (although for many virus/inhibitor combinations this will not be successful) and if the virus forms plaques, infected cells from a single focus can be aspirated at this stage followed by further rounds of plaque purification (Klein, 1975; Parris & Harrington, 1982). A note of caution which should be mentioned is that in practice it is very difficult to obtain homogeneous populations of HSV (Field, 1982) or influenza (Patterson & Oxford, 1986) even after several successive single plaque isolations. Therefore the possible presence of small proportions of wild-type virions must always be considered. These virions may derive from the original virus stock or may result from reversion. Such contamination may be of the order 0.1% or less and difficult to detect. However, this level of wild type virus can be very significant in an animal experiment where inocula are commonly of the order 10^6 infectious units or more. This problem of mixtures is one which affects much work which is carried out with drug-resistant virus mutants.

When several different loci are involved in the development of resistance to a particular compound (see below) it may be possible to design specific strategies to isolate particular types of resistant mutant. Some examples of strategies which have proved successful in this laboratory for the selection of HSV mutants resistant to acyclovir are summarised in Table 2 and several methods are elaborated on below.

In some cases cellular pathways involved in the activation of virus inhibitors can mask the potential resistance in the virus, for example, the inhibition of HSV by trifluorothymidine (TFT) which is activated by HSV TK but is also readily phosphorylated by cellular deoxypyrimidine kinases. Mutant viruses which fail to induce TK remain sensitive to TFT in normal cells because the nucleoside analogue is converted to its respective nucleotides by cellular enzymes. However, if the cells are themselves first made resistant to TFT (or by use of TK-defective cells) they can be used to demonstrate the virus resistance (Field et al, 1981). This could be of more than academic interest since such a test system may be more closely similar to the resting, differentiated cells encountered by the virus in the infected tissue, for example the human eye, than the rapidly dividing cells usually employed for tissue culture. In general terms when the usual methods for selection of resistant virus are unsuccessful, it may be worth attempting to adapt cells to high concentrations of the drug before carrying out the virus selection procedure. It is interesting to note that this strategy was employed recently in relation to HSV resistance to aphidicolin. In this case, resistant cells were obtained in which the inhibition of virus by the drug could no longer be demonstrated (Rapazzo et al 1986).

Animal Systems for the Selection of Resistant Virus *In Vivo*

Another way to obtain resistant virus is the use of chemotherapy in infected animals which are susceptible to the virus in question. This has the advantage of selecting viruses (or mixtures of mutant and wild-type viruses) which are pathogenic and may be particularly useful when investigating the likely patterns of resistance development in man.

TABLE 2. Strategies which have proved useful for the isolation of several different kinds of acyclovir-resistant mutants of HSV.

Strategy	Kinds of Mutant Isolated
Single passage in low concentration of ACV	Usually TK$^-$ Occasionally TK^r or DNApol^r
Repeated passage, or Passage of TK^r mutant in high concentration of ACV	Double mutants: TK$^-$+DNApol^r
Recombination of double mutant with wild-type	DNApol^r
Passage in low concentration of ACV in serum-starved cells	TK^r
Passage in low concentration of ACV in TK-transformed cells	TK^r and DNApol^r
Passage in increasing concentration of PAA	DNApol^r
Recombinant DNA techniques	TK$^-$/TK^r and DNApol^r

ACV: acyclovir (low concentration is <0.1μg/ml; high concentration is >1.0μg/ml).

PAA: phosphonoacetic acid.
TK$^-$: thymidine kinase defective (usually <3% wild-type TK).
TK^r : thymidine kinase with altered substrate specificity (usually >30% wild-type TK)

One approach is to passage the virus from animal to animal in the presence of an effective (or subeffective) dose of chemotherapy in a fashion analogous to the tissue culture system described above. This was an early method used for the selection of poxvirus resistant to a thiosemicarbazone (Appleyard & Way, 1966) and was later used to obtain amantadine-resistant influenza (Oxford *et al*, 1970). A similar technique proved satisfactory for the isolation of acyclovir-resistant mutants of HSV (Field, 1982). A more natural kind of experiment involving the spread of influenza from bird to bird in a flock of amantadine-treated chickens has also produced positive results (Webster *et al*, 1986).

An adaptation of this technique which is also highly relevant to the development of clinical resistance in man is to immunosuppress individual animals by means of X- irradiation, chronic cyclosporin A therapy (Field & Efstathiou, unpublished observations) or by using athymic "nude" mice (Ellis *et al*, 1986). In all these systems inoculation with HSV results in persistent virus replication in the skin for several weeks. When acyclovir chemotherapy is administered to such mice resistant virus can be recovered from the appropriate biopsies or necropsies without the need to passage the infection from animal to animal. One potential problem that must be ruled out in all these methods is the possibility of carry-over of significant concentrations of the inhibitor from the infected tissue into the *in vitro* virus isolation system.

Laboratory Methods for Detecting Resistant Strains of Virus

Determination of ED$_{50}$. The most common method for testing the antiviral activity of drugs is to determine the drug concentration to give a 50% reduction in virus yield, cytopathic effect, or plaque formation. Focal assays such as plaque enumeration are attractive since the method is quantitative and employs low multiplicity of infection; the shape of the response curve can give information on the homogeneity of the virus population under test. The major disadvantage of the test is that it is somewhat subjective at the plaque counting stage. This is particularly important if the plaques gradually decrease in size with increasing drug concentrations. Yield reduction is a more objective test but if high multiplicity is used the presence of a mixed population of virions may be obscured.

Both plaque and yield methods are labour-intensive and less appropriate for automation. Several large surveys of HSV have employed the "dye-uptake" technique which is well suited to automatic colorimetric analysis (McLaren *et al*, 1983). This test relies on the ability of viable cells to exclude a vital stain such as phenol red and effectively measures the protection from cytotoxicity afforded by the virus inhibitor. The method gives ED$_{50}$ concentrations approximately 10-fold higher than those obtained for plaque or yield reduction (Barry *et al*, 1986). Again the use of high multiplicity is a disadvantage of the method and the possibility that virus strains differ in their cytotoxic properties has yet to be considered. Some drugs have been found to have a marked multiplicity dependence and this may have to be taken into account (Harmenberg *et al*, 1985c). Clearly the usefulness of the method for large scale screens is dependent on the inclusion of an adequate number of carefully chosen, well- characterised prototype strains of both sensitive and resistant viruses by way of controls. In addition to this screening assay a number of other types of test to detect resistance have been described including the use of hybridization to measure inhibition of DNA synthesis and "ELISA" methods for the detection of virus antigens. Several different methods of this kind were compared recently by Harmenberg *et al*, (1986) and further developments along these lines are likely in the near future.

The Use of Plaque Autoradiography

In the particular case of HSV strains which are resistant because of loss or change in the virus-induced TK a particularly appropriate strategy for screening isolates involves the use of autoradiography to identify the proportion of plaque-forming virus which is able to readily phosphorylate thymidine or thymidine analogues such as iododeoxycytidine (Van-Dyke & Connor, 1985) or acyclovir (Martin *et al*, 1985). The method involves inclusion of labelled nucleosides in the infection medium and later exposing the fixed monolayer to X- ray film.

As described by Martin *et al*, (1985) it is informative to use alternately both the natural substrate (dThd) and the particular substrate under consideration, (acyclovir). This is because mutants exist which have selectively lost the ability to phosphorylate acyclovir while retaining the ability to phosphorylate dThd (Darby *et al*, 1981); this emphasises the point that the actual analogue substrate must be used in preference to a convenient substitute. Analysis of specimens of HSV by this means led to the observation that clinical isolates contained mixtures of virions some of which were either TK-defective, or induced TK with altered substrate specificity (Martin *et al*, 1985). These observations may have great significance for the emergence of clinical resistance to acyclovir and similar nucleosides in man but of course the method gives no information about possible resistance in loci other than TK.

The Clinical Relevance of *In Vitro* Tests for Resistance

It has been learnt from the study of bacteria that it is of paramount importance to design drug sensitivity tests which accurately mimic the situation *in vivo* (Philips, 1986). For example the minimum inhibitory concentration for an antibiotic can be significantly higher when measured in urine or serum compared with the test carried out in defined synthetic media. Similarly a tissue culture method designed to test the action of a virus inhibitor on sensitive strains may not properly reflect the situation of virus resistance *in vivo*. The cell type needs to be considered and at least one factor is the multiplication state of the cells and the presence of natural metabolites such as dThd. For example it was observed that a mutant strain of HSV was highly resistant to BVDU when tested in a plaque assay using pre-formed monolayers of BHK cells. However, the same mutant strain became completely sensitive in an assay which was identical except that the cells were infected in suspension prior to their forming a monolayer (Field & Neden, 1982).

The presence of competing nucleosides such as dThd and other natural nucleosides occurring in host tissues (especially inflamed tissue) should also be considered (Larsson *et al*, 1983; Harmenberg *et al*, 1985a,1985b). There is still relative ignorance about how a certain degree of resistance (and the particular mechanism involved may be important here) as measured in any *in vitro* test will be reflected *in vivo* in terms of clinical response. In fact reasonably good correlation between resistance measured in a tissue culture plaque reduction assay and response to chemotherapy in a murine skin infection model was obtained here with a series of different kinds of acyclovir-resistant **mutants** derived from several parental strains of HSV and showing various degrees of resistance to acyclovir (Field & Darby, 1980). However, it must be stressed that these data were obtained in an animal model and it is possible that quite different factors will be operating in, for example the human cornea during herpes keratitis. This subject of the correlation between virus inhibition at enzyme and tissue culture level in relation to pharmacology and clinical response *in vivo* has been addressed by Datema *et al*, (1987) but clearly more progress is still desirable in the design of tests which are informative yet suitable for mechanization and large-scale screening of clinical isolates. This approach has been relatively straightforward for the herpesviruses where there are effective tissue culture

methods and a variety of useful animal models. Similarly with
influenza but the situation is much more difficult with other
human pathogens and in particular the human retroviruses. This
is an area which will develop rapidly in the near future.

RESISTANCE IN HERPES AND POXVIRIDAE

The herpes and poxviridae are families of large DNA
viruses whose complexity provides,in theory at least, numerous
targets for selective toxicity. Individual members of both
these families contain DNA genomes which encode in excess of 100
polypeptides, several of which form proteins with enzyme
activity in the infected cells. It is not surprising then that
these viruses have been found more or less susceptible to a very
wide range of different chemical inhibitors. Of the compounds
which have been of most practical benefit almost all interfere
either directly or indirectly with virus DNA, DNA-synthesis, or
DNA precursor pathways (De Clercq, 1982). To date most
attention has focused on the use of nucleoside analogues which
inhibit the replication of members of the herpesvirus family.

Herpesviruses and Nucleoside Analogues

All herpesviruses encode a DNA-polymerase whose function is
essential for DNA replication and the expression of late virus-
induced proteins (Allaudeen, 1985). Of the four nucleoside
triphosphates it is the production of TTP which appears to be of
particular importance (Harmenberg et al, 1985a) and the
herpesviruses (especially the neurotropic ones) appear to have
evolved strategies for maintaining high thymidine pools in the
infected cells. This is partly achieved by means of the virus-
induced TK. We may not yet fully appreciate the full
significance of this versatile enzyme which in the case of HSV
is capable of phosphorylating dThd, dCyd, and a wide variety of
nucleoside analogues to their respective monophosphates. It also
has thymidylate kinase activity and can convert TMP to TDP. The
biochemical assay for TK is relatively simple and perhaps this
has led us to overlook the possibility that the enzyme has other
biological activities in vivo which are not immediately
apparent? It is now well-known that HSV TK is the key to the
selectivity of acyclovir and many other "second generation"
nucleoside analogues which are phosphorylated efficiently by
means of virus TK but not by cell-coded kinases. Once formed
the nucleoside monophosphates are converted to the triphosphates
by cellular, (and in some cases virus) enzymes and may then
interact with virus DNA-polymerase. Other compounds which are
not activated by virus TK include ara-A, aphidicolin, PAA, PFA
and more recently, hydroxyphosphonylmethoxypropyl adenine.
These are all thought to interact with herpes DNA-pol directly.
Thus the herpesvirus DNA-pol is the ultimate target for most of
the important antiherpes agents described to date.

Drug Resistance in Herpesvirus TK and DNA-Pol

The study of herpesvirus drug-resistance started in the
1960's with reference to the drugs bromodeoxyuridine and then
idoxuridine. This provided early evidence that the elevation in
TK activity in HSV-infected cells resulted from the expression

of a virus-coded enzyme (Dubbs & Kit, 1964). Subsequently many
other herpesviruses were found to express similar enzyme
activity including several neurotropic animal herpesviruses
(Kit, 1985) and EBV (Littler *et al*, 1986).

The discovery of acyclovir aroused much renewed interest in
the role of TK and DNA-pol in relation to antiviral drugs. It
was soon apparent that resistance to this drug was associated
with two loci (TK and DNA-pol). Circumstantial evidence (Field
et al, 1980) was confirmed by the more elegant genetic mapping
studies of Coen & Schaffer, (1980) and Schnipper & Crumpacker,
(1980). The next stage in this progression of work was to show,
using isolated enzymes from wild type and drug-resistant
strains, that the reduction in sensitivity of the mutants
reflected precise biochemical changes in the properties of the
respective enzymes (Furman *et al*, 1981). It is clear that TK-
defectiveness confers resistance to all the TK-mediated drugs
(Field *et al*, 1981) while subtle changes in the enzyme can
result in resistance to particular drugs such as acyclovir,
bromovinyldeoxyuridine (Larder *et al*, 1983) and
methoxymethyldeoxyuridine (Veerisetty & Gentry, 1983). The
enzymology of such mutants including their patterns of cross-
sensitivity has been extensively studied and the subject was
reviewed by Larder & Darby (1984). Similarly HSV DNA-pol has
been shown to be capable of acquiring resistance (or
hypersensitivity) to ara-A , phosphonoacetic acid or
phosphonoformic acid, aphidicolin, acyclovir and
dihydroxypropoxymethyl guanosine, and dihydroxybutyl guanosine.
This is a subject which has been thoroughly reviewed by Coen et
al, (1986).

The most recent progress has been to obtain the nucleotide
sequence from normal and resistant strains of virus and to show
the precise changes leading to the expression of an enzyme with
altered properties; these experiments are discussed below.

Drug Resistance in Other Members of the Herpetoviridae

Strains of VZ have been isolated whose resistance to
acyclovir has been located in either the TK or DNA-pol (Shiraki,
1983; 1986a; 1986b) and a TK-defective isolate of VZ has
recently been isolated from an acyclovir-treated patient (S.
Straus, personal communication). Murine CMV (unlike the human
virus) is very sensitive to acyclovir and acyclovir- resistant
variants have been obtained which appear to result from changes
in the DNA-pol (Sandford *et al*, 1985). Human CMV is naturally
fairly insensitive to acyclovir but is inhibited by
dihydroxypropoxymethyl guanosine and a resistant variant was
described recently by Biron *et al* (1986). A coding sequence for
human CMV TK has not been identified;yet the data of Biron *et al*
(1986) suggests a block in the production of the phosphorylated
analogue in the mutant-infected cells; these observations remain
unresolved.

Defining Functional Changes in TK and DNA-pol By Means of Nucleotide Sequencing

The HSV TK protein comprises two identical subunits
(Jamieson & Subak-Sharpe, 1974) each containing 376 amino acids
(McKnight, 1980; Wagner *et al* 1981). Recently Darby *et al.*

(1986) published the sequence of three mutants of HSV-1 derived
from the wild type strain SC16. Two were resistant to acyclovir
and one to bromovinyldeoxyuridine. They had previously been
shown to have TKs with altered substrate specificity and wild
type DNA-pol. The nucleotide sequences revealed a single base
substitution in each case leading to

Table 3. Amino acid substitutions in TK-substrate
specificity mutants of HSV predicted from
sequence analysis.

Virus	Drug ED50 (ug/ml)		TK Induction (%w/t)	Amino Acid residue No.	Predicted change
	ACV	BVDU			
SC16(w/t)	0.1	0.03	100	-	-
S1	10	0.5	30	336	CYS → TYR
B3	0.05	30	110	168	ALA → THR
Tr7	22	0.4	130	176	ARG → GLU

(Adapted from Darby *et al*, 1986; Larder *et al*, 1983b)

one amino acid change (Table 3) . The changes occurred at
residues 176, 168 and 336; the first two coming from a region
highly conserved between HSV-1 and HSV-2. It was noted by these
workers that the change in the BVDU-resistant mutant (SC16-B3)
which led to high resistance to the drug with relatively little
effect on dThd or ATP binding may resemble the HSV-2 enzyme
which is naturally resistant to this analogue (Cheng *et al*,
1981) which could reflect a block in the thymidylate kinase
activity of the enzyme (Fyfe, 1982) which may be important in
the conversion of bromovinyldeoxyuridine monophosphate to the
diphosphate. Kit *et al* (1987), using an acyclovir-resistant TK
substrate specificity mutant which had been isolated from a
clinical case, detected a change from arginine to histidine at
residue 223.

Using a similar approach a number of DNA-pol mutants of SC-
16 were analysed (Larder *et al*, 1987). These mutants had been
selected in HSV-TK-transformed cells by passage in the presence
of acyclovir. (This method avoids the ready isolation of TK⁻
defective mutants, Table 2). The DNA-pol contains some 1235
amino acid residues and it may be seen (Table 4) that the
changes occurred in 6 residues. It was noted that the only base
changes observed were those which resulted in an amino acid
change and that all but one mutant showed a single change.
Except for the mutant with a second mutation all the changes
occurred in an amino acid conserved between different
polymerases. One very interesting observation drawn from this
study was that the pattern of sensitivity to the drugs was more
complex than could be explained by the single substitutions and
one mutant revealed no change. The mutants were all checked for
changes in TK and none were found. This implies that other loci
outside TK and DNA-pol are involved in resistance to acyclovir.

Evidence has already been forthcoming that the HSV major DNA-binding proteins may be involved in the acquisition of resistance
(Honess et al, 1984; Chiou et al, 1985) and this is one likely
candidate. Other potential loci for resistance to herpes
inhibitors include the ribonucleotide reductase and the pathways
involved in glycosylation; we are not aware that there are any
proven examples of resistance in these functions to date.

Table 4. Amino acid changes predicted from nucleotide
sequence changes in DNA-Pol mutants of HSV.

Virus	Fold-resistance		Amino acid residue No.	Predicted change
	ACV	PAA		
SC16 (w/t)	1	1	-	-
RSC-26	10	8	597	GLU → ASP[+]
TP2.4	19	10	719	ALA → VAL
TP2.5	47	20	355 + 724	GLY → ASP+ SER → ASN
TP1.3	3	1.3	none	-
TP2.7	29	2.6	841	GLY → SER
TP4.4	233	0.6	815	ASN → SER
TP3.2	59	0.18	815	ASN → SER
TP4.1	187	0.24	815	ASN → SER

(Adapted from Larder et al, 1987).

The Relative Importance of TK and DNA-pol for Virus Replication

A functional DNA-pol is essential for herpesvirus
replication. It follows that mutant viruses must induce a
protein which has only minor and subtle differences from the
wild type enzyme. Even so the mutants charactarised to date
show that the patterns of cross-resistance can be very complex
(reviewed by Field, 1986). The studies of Larder & Darby,
(1986) on the cross-resistance of HSV mutants to a variety of
compounds illustrates this point. These workers propose that
co-resistance to phosphonoacetic acid, phosphonoformic acid and
ara-A is likely and that resistance to these compounds varies
inversely to aphidicolin to which they may be hypersensitive.
It is also clear from the work of Darby & Larder (1986) that
resistance to a single analogue (for example, acyclovir) can
occur without any obvious change in the response to a variety of
different analogues.

In contrast to DNA-pol, the herpesvirus TK is not essential
for virus replication and mutants exist which do not induce a
functional protein. Such mutants which have no demonstrable
enzyme activity have been described as TK- defective and they
are readily isolated by passage in the presence of any TK-
mediated inhibitor. The mutation usually results in premature
termination of translation and a truncated polypeptide (Summers
et al, 1975). Some read- through of the termination mutation
may account for a low level of enzyme activity which is commonly

observed (Cremer *et al*, 1975). An alternative explanation which
may be very important in practice is that the mutant stock is
contaminated with a low concentration of TK-inducing virus
(Field, 1982). Finally, low TK-activity may represent residual
activity in a mutant expressing full-length TK polypeptide with
altered substrate specificity in relation to its natural
substrate (Darby *et al*, 1984). Mutants which express low
levels of TK through changes in TK regulation have also been
described (Post *et al*, 1981) but have not been isolated as a
result of drug selection. However, mutants with subtle changes
in the TK polypeptide analogous to those described for DNA-pol,
have been isolated. In this case all or some of the normal
enzyme functions are retained but with acquired resistance to
one or more nucleoside analogues. Such mutants have been termed
"substrate specificity mutants" in TK and several of these have
been described above. This variety of different types of mutant
which is becoming apparent has an important bearing on the
pathogenicity of the mutant strains and inevitably this point
was not fully appreciated during some of the earlier work on
resistant strains *in vivo*.

Effects of Mutations in TK and DNA-pol on Pathogenesis

The clearest observation to date is that viruses with
defective TK-induction have a profound reduction in
neuropathogenicity. This means that TK-defective strains have a
significantly reduced capacity to cause neurological signs and
death in experimentally infected animals (Reviewed by Field,
1985; 1986). This difference is even more marked with the
extremely neuropathogenic pseudorabies virus where TK-defective
viruses were also found to be attenuated (Tenser *et al*, 1982).
Furthermore similar observations have been obtained with TK-
defective vaccinia (Buller *et al*, 1985).

A more controversial question is the effect of TK-
defectiveness on the establishment of neural latency. While all
agree that TK-defective viruses less readily establish and
reactivate from latent infections there is disagreement as to
whether the block is absolute. Sears *et al* (1985) could not
show a relationship between the maintenance of latency and the
level of TK-induction using genetically engineered "control"
mutants, however Tenser & Edris (1986) showed subsequently that
the production of TK by these mutants was higher under different
conditions making the animal data more difficult to interpret.
This highlights the difficulty associated with these studies.
It is clear to us that TK- defective virus can be reactivated
from latently infected animals (Field, 1982; Field & Lay, 1984)
but in association with wild type virus. This has recently been
proven absolutely using mixtures of wild type and a TK-defective
deletion mutant which can be distinguished by means of its
hybridization pattern in a Southern blot (S. Efstathiou,
personal communication). Furthermore it appears that true TK
defective virus can establish latency and although it may not
reactivate in a normal system, reactivation can be induced by
superinfection of explanted ganglia using wild type virus. These
observations may help to explain many of the conflicting results
obtained in previous studies and of course have important
implications with regard to the clinical use of TK- mediated
drugs.

HSV mutants with altered substrate specificity have also been shown to be similar to wild-type virus in murine infection models suggesting they are potentially pathogenic and similar results have been obtained with DNA-pol mutants. However, recently data have been obtained in two different laboratories which suggest that DNA-pol mutants may have changes consistent with altered pathogenicity (Larder et al, 1986; Field & Coen, 1987). We have also obtained some evidence that such mutants may have different patterns of replication in differentiated cells. For example we have examined a mutant which appears to have lost the capacity to replicate in neural tissue within the CNS, even on direct intra-cerebral inoculation; yet the retinal neurons appear to remain extremely sensitive and the virus multiplies to a high titre in the murine eye (Anderson & Field, 1982). The full significance of these findings and their molecular basis in terms of virus-host cell interactions remain mysterious.

Clinical Resistance in HSV and VZ

The clinical trials of acyclovir including prophylaxis and therapy using a variety of regimens for different manifestations of disease have yielded large numbers of clinical isolates. These have not yet shown any general trend towards resistance (Lehrman et al, 1986; Svennerholm et al, 1985; Gold & Corey, 1987) although it has been suggested that approximately 6% of HSV-2 clinical isolates from persons not receiving acyclovir have exhibited resistance in vitro (ED_{50}=>3 ug/ml) (McClaren et al, 1983). The lack of resistant virus among clinical isolates from drug trials contrasts markedly with the ready development of resistance in tissue culture.

There have been a relatively small number of resistant clinical isolates which resemble the different classes of laboratory resistant mutants. As mentioned above, both HSV and VZ which express TKs showing altered substrate specificity have been recovered from treated patients. Until recently there had not been an account of a DNA-pol resistant mutant from a clinical case, but one has been isolated from a recipient of a bone marrow transplantation with an unresolving herpes infection (P. Collins, personal communication).

These occasional resistant strains still represent a tiny proportion of the total number of post-therapy isolates although clearly the potential for resistance exists. Given the high mutation rate and the large number of virions produced in a single lesion this is perhaps surprising compared with the ready isolation of resistance in tissue culture. One possible explanation is that one or more compensating mutations are required in other sites before viruses with resistance in TK or DNA-pol are able to become truly pathogenic and the majority of potentially resistant viruses may have a selective disadvantage.

The immunocompromised host is a special case and the different types of acyclovir-resistant virus have been isolated from such patients (Field, 1985). This includes many examples of TK-defective viruses which in some cases undoubtedly were contributing to the patient's disease. The isolation of TK-defective virus from immunocompromised patients can readily be modelled using infected mice as discussed above. It is of interest that using plaque- autoradiography or the characterization of single plaques it has been shown that clinical isolates are mixtures of TK- defective and "normal" strains (Martin *et al*, 1985; Christophers *et al*, 1987). The occurrence of mixtures of viruses with TK-defective and wild-type phenotype was also observed in experimentally infected, acyclovir-treated mice (Field, 1982) when virus was passaged in animals undergoing chemotherapy. It should be noted that such mixtures could have considerable clinical significance and may be overlooked in screening assays (Harmenberg *et al*, 1986) especially where isolates are examined under high multiplicity conditions using methods such as the dye-uptake method.

It seems inevitable that with the continued use of acyclovir and other antiherpes drugs in the human population, and with transmission of infection among individuals, resistance will become more prevalent. We believe that it is certain to become a significant problem in the future.

Drug-Resistance in Poxviruses

In comparison with the enormous amount of published work concerning inhibitors of the herpetoviridae there is much less known about the poxviruses. However, at the advent of virus chemotherapy smallpox was still a clinical problem and some very good work was done early on and there has been a gradual and continued interest in this group of viruses.

Thiosemicarbazones. Mutants of vaccinia resistant to isatin-β-thiosemicarbazone were found to arise after as few as two passages in tissue culture in the presence of the drug (Appleyard & Way, 1966) or three passages in mice undergoing chemotherapy. The same workers noted that resistant viruses retained their pathogenic properties when re-inoculated into mice. More recently there has been renewed interest in thiosemicarbazone derivatives as inhibitors of herpesviruses and an extensive survey of the antiviral activity of a large series was published recently (Shipman *et al*, 1986). The likely mode of action is thought to involve the virus- specified ribonucleotide diphosphate reductase (Turk *et al*, 1986) but it is perhaps worth noting that we were unable to develop resistance in bovine herpesvirus-1 to 2-pyridyl ketone thiosemicarbazone, the parent compound in the group (Field & Reading, 1987). However, Katz & Margalith (1984) reported the isolation of variants of HSV with acquired resistance to 3-substituted triazinoindole.

Several nucleoside analogues are active against the poxviruses and it is of interest that idoxuridine-resistant strains were shown to have reduced pathogenicity in animals (Ferrari *et al*, 1965; Nakamura *et al*, 1967). Although the mutants were not characterised biochemically we can conclude that they were probably TK-defective strains analogous to those produced by HSV.

It was mentioned above that phosphonoacetic acid- resistant mutants of vaccinia have been used to study the virus DNA-polymerase (Moss & Cooper, 1982) and it can be mentioned in passing that resistance to aphidicolin has also been identified in the DNA-polymerase locus (DeFilippes, 1984).

Rifampicin has been shown to inhibit the maturation of poxviruses at a discrete step in envelope formation (Moss *et al*, 1969) and a rifampicin-resistant mutant has been characterised (Tartaglia & Paoletti, 1985). A DNA fragment which encompasses the target for resistance was sequenced by these workers and found to contain a base transition giving rise to a single amino acid change in the gene product.

The demise of smallpox led to a marked reduction in research effort directed to the poxviruses. However, other members of this family of viruses are widespread among man and animals and recently there has been renewed interest in the poxviruses as vectors for immunogenic proteins expressed by recombinant virus strains. The poxvirus virus is large and extremely complex with numerous targets for antiviral interactions and we may expect that drug-resistance will form a major part of future work with these and similar viruses.

DRUG RESISTANCE IN ORTHOMYXOVIRUSES

Amantadine and Rimantadine

Amantadine (1-aminoadamantane hydrochloride) selectively inhibits influenza A viruses (Davies *et al*, 1964). Many clinical trials have been conducted which show amantadine to be effective both prophylactically and therapeutically against influenza A (but not B which is naturally insensitive) in man (Oxford & Galbraith, 1980; Galbraith, 1985). The drug was licensed in 1976 for use against all influenza A subtypes although its use has been limited except in the Eastern Block. Rimantadine (alpha-methyl-1-adamantane methylamine hydrochloride) is as effective as amantadine but with fewer side effects (Dolin *et al*, 1982) and this drug has been used extensively in the USSR (Zlydnikov *et al*, 1981).

Amantadine-resistant variants can readily be obtained from tissue culture (Cochran *et al*, 1965) and from laboratory animals (Oxford *et al*, 1970) following passage in the presence of the drug. The study of such mutants has been particularly valuable in the elucidation of the mode of action of amantadine and related inhibitors. Amantadine has been found to exert two different actions depending on concentration (Hay & Zambon, 1984; Hay *et al*, 1986).

Inhibition by High Concentrations of Amantadine. Inhibition by concentrations of amantadine, of the order of 0.5mM, is not confined to influenza viruses; other enveloped RNA viruses which enter cells by endocytosis are also susceptible (Wallbank *et al*, 1966; Helenius *et al*, 1980). Similar effects are achieved using a variety of lysosomotropic amines (Helenius *et al*, 1982) which cause a rise in pH of endosomes. Resistance to high concentrations of amantadine maps to the haemagglutinin (HA) gene (Hay *et al*, 1986). Upon virus entry by endocytosis the HA normally undergoes a conformational change triggered by the low endosomal pH resulting in fusion of the virus and endosomal membranes and release of the nucleocapsid into the cytoplasm. The rise in pH associated with the presence of amantadine blocks this event. Resistant mutants may be explained by a change in the structure of HA which allows fusion to proceed at a higher pH. The nature and location of such mutations suggests a lowering of the energy barrier to the conformational change leading to the fusion active state (Daniels et al, 1985).

Inhibition by Low Concentrations of Amantadine. This effect is specific to influenza A viruses with optimum inhibition varying from 0.3 μM for 'Weybridge' H7N7 to 5 μM for other strains (Hay et al, 1986). Human influenza A viruses for example, A/Singapore/1/57 (H2N2) appear to be inhibited at an early stage since drug must be present prior to infection and primary transcription is prevented (Hay & Zambon, 1984). This is in contrast with data obtained from two avian strains 'Rostock' H7N1 and 'Weybridge' where early events are not affected and virus inhibition occurs when the drug is added later suggesting that a late stage of virus replication, probably assembly, is affected (Hay & Zambon, 1984; Hay et al,1986).

The stage at which amantadine inhibits viral replication seems to depend upon the HA. Substitution of the HA of "Weybridge" (late inhibition) by that of Singapore (early inhibition) conferred on the reassortants the characteristic sensitivity of the Singapore virus (Hay et al, 1986) Genetic analysis of reassortants between sensitive and resistant viruses and sequenceing studies indicate that resistance maps to the M2 gene and amino acid substitutions fall within the hydrophobic amino terminal region of about 20 amino acids which is suggested to be the membrane-spanning domain of the M2 protein which is expressed at the surface of infected cells (Hay et al, 1986; Lamb et al, 1985). The structural consequences of these changes are unknown but it is possible that they may affect the membrane association of the protein or specific intermolecular interactions involved in its function.

All amantadine-resistant variants isolated following passage in 5μM amantadine contained a mutation in their M gene leading to a single amino acid substitution in the M2 protein whereas less than a quarter of such mutants had a change in the amino acid sequence of the HA (Hay et al, 1985). Resistance is consistently transferred with M2 and the optimum inhibitory concentration reflects the parental M gene. However, the involvement of HA in drug action is indicated by the following: HA affects the degree of susceptibility and influences which stage (early, or late) is inhibited. Amantadine treatment of "Rostock"-infected cells, greatly reduced the membrane expression and antigenicity of HA; an effect which is relieved in resistant mutants. The fact that resistance to low concentrations of amantadine maps to M2 rather than HA suggests that the involvement of HA in drug action may be indirect and that amantadine may interfere with an HA/M2 protein interaction (Hay et al, 1986).

Norakin

Norakin (triperidin) or 1-tricyclo-(2,2,1,0)-heptyl (2)- 1-phenyl-3-piperidine-propanol HCl is a compound which is used as an anti-Parkinson agent. The discovery of its activity against influenza stemmed from the study of the anti- Parkinson effects of amantadine (Presber et al, 1984; Schroeder et al, 1985). It appears to be a selective inhibitor of ortho- and paramyxoviruses *in vitro*. All influenza A strains have been found to be sensitive while the susceptibility of influenza B is variable. Resistant variants can be isolated from tissue cultures infected in the presence of the drug and the study of reassortants derived from crosses between norakin-sensitive and -resistant fowl plague virus has mapped resistance to the HA gene (Ghendon *et al*, 1986). The evidence suggests that norakin acts by inhibiting the conformation change in HA at low pH (Ghendon *et al*, 1986). (In other words there seems to be a direct interaction with the virus protein rather than an indirect effect by raising endosomal pH, see above). Indeed, norakin- and rimantadine-resistant mutants exhibit very little cross- resistance and it appears that in combination the two drugs have an antagonistic interaction since they exert less than an additive effect (Schroeder *et al*, 1985).

The Outlook for Influenza Drug Resistance in Man

Since the mechanism of action of these drugs is not entirely clear it is difficult to predict from the study of laboratory-selected mutants what will be, if any, the clinical significance of drug resistance. So far their is little information available on the characteristics of resistant clinical isolates. However, recent data show that amantadine resistant strains can be isolated from the human population, furthermore, the incidence of resistant isolates may be increasing (Pemberton *et al*, 1986). In a striking series of experiments using flocks of chickens, Webster *et al* (1986) showed that amantadine and rimantadine resistant variants of a highly virulent influenza A virus could arise in infected birds receiving drug in their drinking water leading to unsuccessful therapy. A key factor appeared to be the transmission of infection among the flock during chemotherapy. The mortality associated with the drug-resistant variants was reduced by combining chemotherapy with vaccination. This kind of experiment and other animal models should be a very useful approach to assessing the implications of widespread use of amantadine and similar drugs in man and development of sensible strategies to minimise this potential problem.

Ribavirin

Ribavirin (3-β-D-ribofuranosyl-4-hydroxypyrazole-5-carboxamide), a synthetic nucleoside resembling guanosine, appears to have a very broad spectrum of antiviral activity *in vitro* and *in vivo*. It has been reported to be active against a wide variety of RNA and DNA viruses (reviewed by Canonico, 1983). Particular attention has been paid to the ortho- and paramyxoviruses. Ribavirin in aerosol form has been employed in the treatment of human influenza and respiratory syncytial virus infections in infants (Hall *et al*, 1983; Hall *et al*, 1985; Galbraith, 1985). It is important to point out that to date we know of no reports of ribavirin-resistant variants although they have been actively sought. The mode of action of the drug is unclear and it may have multiple targets requiring a number of mutations in several sites to confer measurable resistance. Ribavirin is a close structural analogue of guanosine being converted to its respective nucleotides by cellular enzymes. Ribavirin monophosphate is a potent inhibitor of inosine monophosphate dehydrogenase, which is involved in GMP synthesis and the drug may act in this way with a consequent effect on guanosine nucleotide pools. Alternatively there is evidence to suggest that ribavirin triphosphate, the principle intracellular form of ribavirin, inhibits the process of 5'-methyl capping of mRNAs (Canonico, 1983). It remains to be determined which if any of the theoretical targets are crucial to the mode of action. This drug remains as an interesting example where no resistance can be readily obtained and its specific antiviral selectivity must therefore remain in question.

DRUG RESISTANCE IN THE PICORNAVIRUSES

Guanidine

Guanidine is an inhibitor of human enterovirus RNA replication (Crowther & Melnick, 1961). Guanidine-resistant variants of poliovirus were among the first drug resistant viruses to be isolated from tissue culture and experimentally infected monkeys (Melnick et al, 1961). Guanidine dependent viruses were also isolated (Loddo et al, 1962). Early studies revealed that resistant mutants retained neurovirulence in monkeys (Carp, 1964) although the guanidine-dependent mutants were apathogenic (Loddo et al, 1962). These earlier workers were struck by the apparent ease with which resistance developed in their experimental systems and it was thought at that time to be a significant factor in the failure of chemotherapy in experimentally infected, guanidine-treated monkeys (Barera-Oro & Melnick, 1961).

Genetic recombination studies between guanidine-resistant and sensitive strains identified the region encoding the coat protein as the important site for mutations leading to resistance (Cooper, 1968; Cooper et al, 1970). Korant (1977) showed that resistant variants differed from the original sensitive strains in the electrophoretic and chromatographic behaviour of their structural proteins only. However, recent findings are not compatible with these early reports. Recombination studies using a neutralizing monoclonal antibody-resistant variant of the polio type 1 strain of the Sabin vaccine with a guanidine-resistant variant of the Mahoney strain indicated that guanidine-resistance mapped to a region of the genome specifying the non-structural proteins (Emini et al, 1984). Anderson-Sillman et al (1984) carried out peptide mapping and isoelectric focussing studies which demonstrated that 75% of guanidine-resistant mutants had modifications in the 37 kD protein known as 2C (formerly pX) which is a non-structural protein of unknown function. Sequence analysis of interstrain guanidine resistant and antibody resistant variant recombinants (Pincus et al, 1986) showed that 6 independently isolated guanidine-resistant mutants each contained a mutation in 2C at amino acid 179 which resulted in a change of either ASP to ALA or ASP to GLY. Guanidine-dependent mutants were also found to contain changes within 2C. Pincus & Wimmer (1986) produced resistant and dependent mutants from cloned cDNA demonstrating that mutations in polypeptide 2C are directly responsible for altered sensitivity to guanidine.
Similarly, guanidine-resistance mutations in foot and mouth disease virus have been mapped by isoelectric focusing, peptide

mapping and recombination studies to p34, the counterpart of poliovirus 2C (Saunders & King, 1982; Saunders *et al*, 1985). A comparison of the amino acid sequence of the 2C-equivalent protein in polio, foot and mouth, rhinovirus type 2 and 14 and encephalomyelocarditis virus reveals strong homology over a region of 115 residues in length and all the mutations associated with guanidine resistance appear to lie within this conserved region (Pincus *et al*, 1986). It may be that this region is involved in interaction with other viral proteins required for RNA replication. Membrane bound replication complexes contain 2C and guanidine could inhibit viral RNA replication by interfering with this interaction between 2C and replication proteins.

HBB (2-alpha-hydroxybenzyl) benzimidazole)

This is also an inhibitor of enterovirus RNA replication and, as for guanidine, resistant and dependent mutants have been isolated (Eggers & Tamm, 1961; Eggers & Tamm, 1963a) HBB has very similar antiviral activities to guanidine but its precise site of action is unknown. Combined guanidine and HBB treatment was synergistic suggesting that different sites are involved (Eggers & Tamm, 1963b). However, the existence of some cross-resistance and dependence between guanidine and HBB has been shown (Eggers & Tamm, 1963a).

While these two drugs are unlikely to be considered suitable for use in man these studies provide a good illustration of how virus inhibitors can be very useful tools in unravelling the complexity of virus structure and function.

Arildone and Rhodanine

Arildone, (4-[6-(2-chloro-4-methoxy)phenoxy] hexyl-3,5-heptandione) selectively inhibits a number of RNA and DNA viruses including polio. Arildone blocks poliovirus replication at the stage of uncoating of the virion. It has no effect on adsorption or penetration of the virus nor on virus synthetic processes (McSharry *et al*, 1979). The drug acts directly with virions so to stabilise the capsid *in vitro* against the effects of heat or high pH (Caliguiri *et al*, 1980). Both arildone-resistant and -dependent viruses have been isolated but in both cases the drug interacts with the mutants resulting in increased thermal stability suggesting that the binding of the drug to the virion does not directly account for the antiviral effect (Schram *et al*, 1982). The rapid metabolism of this drug in man makes it unlikely to be useful as a chemotherapeutic agent (Eggers, 1985).

Rhodanine (2-thio-4-oxothiazolidine) is a highly selective antiviral agent. It inhibits echovirus 12 but appears to have no effect on other viruses including other enteroviruses. It seems to act in a similar way to arildone by inhibiting uncoating as a result of stabilisation of the virus capsid. Resistant variants have been isolated (Eggers, 1977).

Dichloroflavan, Chalcone, Flavones, RMI and Enviroxime

4',6-dichloroflavan, the chalcone, Ro 09-0410 (4'-ethoxy-2'-hydroxy-4,6'-dimethoxy chalcone) the flavone Ro 09-0179

(4',5-dihydroxy-3,3',7 trimethoxyflavone), RMI-15,731 (1-[5-tetradecyloxy-2-furanyl]-ethanone) and enviroxime (anti-6-[(hydroxyamino)-phenylmethyl]-1-[(1-methylethyl)sulphonylimidazol-2-amine]) are all potential antirhinovirus agents. Different serotypes of human rhinovirus vary in their susceptibility to chalcone, dichloroflavan and RMI but enviroxime and flavone are equally active against all serotypes tested. Similar to arildone and HBB discussed above the inhibition of virus by dichloroflavan, RMI and chalcone appears to be associated with the binding of these agents to the virions. Infectivity of the virus may be restored by extraction of the agents with chloroform (Ninomiya et al, 1985; Ishitsuka et al, 1986). The three agents also stabilise the virus to inactivation by heat or acid (Tisdale & Selway, 1984). The binding of tritiated chalcone to rhinovirus type 2 is inhibited by unlabelled chalcone dichloroflavan and RMI but not flavone or enviroxime. Rhinovirus mutants selected for resistance to either dichloroflavan or RMI show cross-resistance to chalcone and vice versa. However, mutants resistant to these three agents are not cross-resistant to flavone or enviroxime (Ninomiya et al, 1985).

These data suggest that dichloroflavan, RMI and chalcone all act in a similar manner and interact with the virus particle at close or identical sites. It seems likely that they all stabilise the virions, preventing the normal process of uncoating although it has been reported that dichloroflavan inhibits an early stage of virus replication rather than adsorption or uncoating (Tisdale & Selway, 1983). By contrast, attempts to isolate resistant mutants to flavone and enviroxime have not been successful (Ishitsuka et al, 1986) and it appears that these drugs have a different mode of action as yet to be determined.

Unfortunately none of the antirhinovirus compounds discussed above have yet been demonstrated to have marked therapeutic potential (Phillpotts 1986) and therefore the role of drug resistance has been relevant only to the study of the virus and modes of action of the drugs. Whether resistance could influence clinical outcome remains an academic question for the time being.

HEPADNA AND RETROVIRUSES

Several agents have been used in clinical trials for hepatitis B therapy including interferon, ara-A and acyclovir (Thomas & Scully, 1985). The persistent nature of the infection in man means that protracted periods of chemotherapy are employed with these agents and this is also likely to be necessary with future compounds and this may increase the chance of resistant virus emerging. None of the antiviral strategies tested to date have been particularly successful and the question of resistance has yet to be addressed.

The emergence of AIDS has stimulated research on compounds with anti-retrovial properties. The reverse transcriptase enzyme has been a focus of attention for selective attack. The dideoxynucleoside analogue 3'-azido- 3'-deoxythymidine (AZT) has shown some promise in early trials for AIDS chemotherapy and there is also interest in the possibility of using dideoxycytidine. AZT is phosphorylated by cellular kinases to the nucleoside triphosphate which can compete with TTP and serve as a chain-terminating inhibitor of HIV reverse transcriptase(Mitsuya & Broder, 1987) . Unlike the situation with the herpesviruses there is no virus deoxypyrimidine kinase to allow the virus to adopt a similar strategy to herpesvirus TK-resistance. Resistance would most likely have to result from changes in the reverse transcriptase enzyme itself; some early evidence for this is presented in the recent paper of Vrang *et al*, (1987). Resistance in HIV is currently being sought and although we know of no published reports to date we anticipate that resistant mutants will be isolated. Similar to hepatitis B lengthy periods of chemotherapy are required to suppress the replication of HIV *in vivo* and if AZT has a true virus-specific mode of action in man we anticipate the likely emergence of drug-resistant variants.

Other inhibitors of reverse transcriptase include the pyrophosphate analogues and strains of retrovirus resistant to phosphonoformate have been selected by laboratory passage of the virus in the presence of the drug (Muratore *et al*, 1984). Another agent, coumermycin A1, a novobiocin analogue, has been shown to inhibit murine retroviruses and resistant variants can be isolated although the mode of action of this drug is not yet known (Varnier *et al*, 1985). These few reports clearly demonstrate that retroviruses have the capacity to develop resistance to a variety of inhibitors. This will doubtless provide useful tools for the study of new antiretrovirus compounds but will pose problems for the development of effective chemotherapy. The question of drug-resistance among retroviruses is likely to form a much more substantial part of any future review of this field.

POST SCRIPT

Drug-resistance has been demonstrated in many virus families in response to a wide variety of different drugs. In several cases the resistant strains have been shown to have a single point mutation in a gene coding for a protein involved in the mode of action of that drug. As mentioned above viruses can show a high rate of point mutations and this, coupled with the enormous replication potential in an infected individual host suggests that resistance could develop rapidly. In practice this may not be the case. An important feature of viruses in this respect may be that the virus genome itself and its products are often multifunctional. This, coupled with their complex host-parasite relationships may put considerable constraints on the development of new biological traits such as drug resistance. It is likely that several mutations in different sites could be required to give a constellation of changes in the virus to produce a "successful" drug-resistant strain. However, the "field" experiments of Webster *et al* (1986) clearly demonstrated that there is a potential for a rapid development of amantadine- resistant strains of influenza in birds and the occurrence of a small but significant number of cases of acyclovir-resistant herpesviruses in man suggest that drug resistance must be taken seriously. We have already mentioned that there may be a small proportion of naturally drug-resistant variants in the virus population from a single isolate (which may be difficult to detect). It remains to be seen whether this will be a significant factor in the development of clinical resistance. The observations to date with different agents and viruses, alluded to in this review suggest that it may be difficult to find general principles which apply equally to all viruses and drugs. It seems likely that each new drug, and clinical situation will require an individual approach. It will be important to establish suitable experimental models which will help to predict the future role of drug-resistance in virus chemotherapy.

REFERENCES

Allaudeen, H.S. (1985) Distinctive properties of DNA polymerase induced by herpes simplex virus type-1 and Epstein-Barr virus. Antiviral Res. 5, 1-12.

Allen, L.B. & Fingal, C.M. (1977) Failure of type 1 herpesvirus to develop resistance to ribavirin. Antimicrob. Ag. Chemother. 12, 120-121.

Anderson, J.R. & Field, H.J. (1982) The development of retinitis in mice with non-fatal herpes simplex encephalitis. Neuropath. Appl. Neurobiol. 8, 277-287.

Anderson-Sillman, K., Bartal, S. & Tershak, D.R. (1984) Guanidine-resistant poliovirus mutants produce modified 37-kilodalton proteins. J. Virol. 50, 922-928.

Appleyard, G. & Way, H. (1966) Thiosemicarbazone-resistant rabbitpox virus. Brit. J. Exp. Pathol. 47, 144-151.

Barrera-Oro, J.G. & Melnick, J.L. (1961) The effect of guanidine: (1) on the experimental poliomyelitis induced by oral administration virus to cynomolgus monkeys; (2) on naturally occurring enteroviruses of cynomolgus monkeys. Tex. Rep. Biol. Med. 19, 529-539.

Barry, D.W., Lehrman, S.N. & Ellis, M.N. (1986) Clinical and laboratory experience with acyclovir-resistant herpes viruses. J. Antimicrob. Chemother. 18, Suppl. B. 75-84.

Biron, K.K., Fyfe, J.A., Stanat, S.C., Leslie, L.K., Sorrell, J.B., Lambe, C.U. & Coen, D.M. (1986) A human cytomegalovirus mutant resistant to the nucleoside analog 9- {[2-hydroxy-1(hydroxymethyl)ethoxy] methyl} guanine (BW B759U) induces reduced levels of BW B759U triphosphate. Proc. Nat. Acad. Sci. USA 83, 8769-8773.

Buller, R.M.L., Smith, G.L., Cremer, K., Notkins, A.L. & Moss, B. (1985) Decreased virulence of recombinant vaccinia virus expression vectors is associated with a thymidine kinase- negative phenotype. Nature 317, 813-815.

Caliguiri, L.A., McSharry, J.J. & Lawrence, G.W. (1980) Effect of arildone on modifications of poliovirus in vitro. Virol. 105, 86-93.

Canonico, P.G. (1983) Ribavirin - a review of efficacy, toxicity and mechanisms of antiviral activity. In 'Antibiotics Vol. VI, Modes and Mechanisms of Microbial Growth Inhibitors', Ed. F.E. Hahn, Springer Verlag, Berlin

Carp, R.I. (1964) Studies on the guanidine character of poliovirus. Virol. 22, 270-279.

Cheng, Y-C., Dutschman, G., De Clercq, E., Jones, A.S., Rahim, S.G., Verhelst, G. & Walker, R.T. (1981) Differential affinities of 5-(2-halogenovinyl)-2'-deoxyuridines for deoxythymidine kinases of various origins. Molec. Pharmacol. 20, 230-233.

Chiou, H.C., Weller, S.K. & Coen, D.M. (1985) Mutations in the herpes simplex virus major DNA binding protein gene leading to altered sensitivity to DNA polymerase inhibitors. Virol. 145, 213-226

Christophers, J. & Sutton, R.N.P. (1987) Characterization of acyclovir-resistant and sensitive clinical isolates of herpes simplex virus from an immunocompromised patient. J. Antimicrob. Chemother. in press.

Cochran, K.W., Massaab, H.R., Tsunoda, H.F. & Berlin, B.S. (1965) Studies on the antiviral activity of amantadine hydrochloride. Ann. N.Y. Acad. Sci. 130, 432-439.

Coen, D.M. (1986) General aspects of virus drug resistance with special reference to herpes simplex virus. J. Antimicrob. Chemother. 18, Suppl. B, 1-10.

Coen, D.M. & Schaffer, P.A. (1980) Two distinct loci confer resistance to acycloguanosine in herpes simplex virus type 1. Proc. Natl. Acad. Sci. USA 77, 2265-2269.

Coen, D.M., Chiou, H.C., Fleming, H.E. Jr., Leslie, L.K. & Retondon, M.J. (1984) Drug resistant and hypersensitive herpes simplex virus mutants: isolation and application to dissection of the pol locus. In 'Herpesvirus' Ed. F. Rapp, Alan R. Liss, New York, 373-385.

Coen, D.M., Fleming, H.E. Jr., Leslie, L.K. & Retondo, M.J. (1985) Sensitivity of arabinosyladenine-resistant mutants of herpes simplex mutants to other antiviral drugs and mapping of drug hypersensitivity mutations to the DNA polymerase locus. J. Virol. 53, 477-488.

Coleman, V.R., Tsu, E. & Jawetz, E. (1968) 'Treatment resistance' to idoxuridine in herpetic keratitis. Proc. Soc. Exp. Biol. Med. 129, 761-765.

Cooper, P., Wentworth, B. & McCahon, D. (1970) Guanidine inhibition of poliovirus: a dependence of viral RNA synthesis on the configuration of the structural protein. Virol. 40, 486-493.

Cooper, P.D. (1968) A genetic map of poliovirus temperature sensitive mutants. Virol. 35, 584-596.

Cremer, D.K., Bodemer, M., Summers, W.P., Summers, W.C. & Gesteland, R.F. (1979) *In vitro* suppression of UAG and UGA mutants in the thymidine kinase gene of herpes simplex virus. Proc. Natl. Acad. Sci. USA 76, 430-434.

Crowther, D. & Melnick, J. L. (1961) Studies on the inhibitory action of guanidine on poliovirus multiplication in cell cultures. Virol. 15, 65-74.

Crumpacker, C.S., Schnipper, L.E., Marlowe, S.I., Kowalsky, P.N., Hershey, B.J. & Levin, M.J. (1982) Resistance to antiviral drugs of herpes simplex virus isolated from a patient treated with acyclovir. New Eng. J. Med. 306, 343-346.

Daniels, R.S., Downie, J.C., Hay, A.J., Knossow, M., Skehel, J.J., Wang, M.L. & Wiley, D.C. (1985) Fusion mutants of the influenza virus haemagglutinin glycoprotein. Cell, 40, 431- 439.

Darby, G. Larder, B.A. & Inglis, M.M. (1986) Evidence that the 'active centre' of the herpes simplex virus thymidine kinase involves an interaction between three distinct regions of the polypeptide. J. Gen. Virol. 67, 753-758.

Darby, G., Churcher, M.J. & Larder, B.A. (1984) Cooperative effects between two acyclovir resistance loci in herpes simplex virus. J. Virol. 50, 838-846.

Darby, G., Field, H.J. & Salisbury, S.A. (1981) Altered substrate specificity of herpes simplex virus thymidine kinase confers acyclovir-resistance. Nature 289, 81-83.

Datema, R., Ericson, A-C, Field, H.J., Larsson, A. & Stenberg, K. (1987) Critical determinants of antiherpes efficacy of buciclovir and related acyclic guanosine analogs. in press.

Davies, W.L., Grunert, R.R., Haff, R.M., McGahen, J.W., Neumayer, E.M. & Paulshock, M. (1964) Antiviral activity of 1- adamantanamine (amantadine). Science 144, 862-863.

De Clercq, E. (1982) Specific targets for antiviral drugs. Biochem. J. 205, 1-13.

De Filippes, F.M. (1984) Effect of aphidicolin on vaccinia virus: isolation of an aphidicolin-resistant mutant. J.Virol. 52,474-482.

Dolin, R., Reichman, R.C., Madore, H.P., Maynard, R., Linton, P.N. & Webber-Jones, J. (1982) A controlled trial of amantadine and rimantadine in the prophylaxis of influenza A infection. New Eng. J. Med. 307, 580-584.

Dubbs, D.R. & Kit, S. (1964) Mutant strains of herpes simplex deficient in thymidine kinase-inducing activity. Virol. 22, 493-502.

Eggers, H.J. (1977) Selective inhibition of uncoating of echovirus 12 by rhodanine. A study of virus-cell interactions. Virol. 78, 241-252.

Eggers, H.J. (1985) Antiviral agents against picornaviruses. Antivir. Res. Suppl. 1, 57-65.

Eggers, H.J. & Tamm, I. (1961) Spectrum and characteristics of the virus inhibitory action of 2- (alpha-hydroxybenzyl)- benzimidazole. J. Exp. Med. 113, 657.

Eggers, H.J. & Tamm, I. (1963a) Drug dependence of enteroviruses: variance of Coxsackie A9 and ECHO 13 viruses which require 2-(alpha-hydroxybenzyl) benzimidazole for growth. Virol. 20, 62-74.

Eggers, H.J. & Tamm, I. (1963b) Synergistic effect of 2- (alpha-hydroxybenzyl) benzimidazole and guanidine on picornavirus replication. Nature 199, 513-514.

Ellis, M.N., Martin, J.L., Lobe, D.C., Johnsrude, J.D. & Barry, D.W. (1986) Induction of acyclovir-resistant mutants of herpes simplex virus type 1 in athymic nude mice. J. Antimicrob. Chemother. 18, Suppl. B, 95-101.

Emini, E.A., Leibowitz, J., Diamond, D.C., Bonin. J. & Wimmer, E. (1984) Recombinants of Mahoney and Sabin strain poliovirus type 1: analysis of *in vitro* phenotypic markers and evidence that resistance to guanidine maps in the non-structural proteins. Virol. 137, 74-85.

Ferrari, W., Gessa, G.L., Loddo, B. & Schivo, M.L. (1965) Decreased pathogenicity for the rabbit skin of IDU-resistant vaccinia virus. Virol. 26, 154-155.

Field, H.J. (1982) Development of clinical resistance to acyclovir in herpes simplex virus-infected mice receiving oral therapy. Antimicrob. Ag. Chemother. 21, 744-752.

Field, H.J. (1985) Resistance and latency. Brit. Med. Bul. 41, 345-350.

Field, H.J. (1986) Resistance to acyclovir. In 'Human Herpes Infections' Ed. C. Lopez & B. Roizman. Raven Press, New York, 177-192.

Field, H.J. & Coen, D.M. (1986) Pathogenicity of herpes simplex virus mutants containing drug resistance mutations in the viral DNA polymerase gene. J. Virol. 60, 286-289.

Field, H.J. & Darby, G. (1980) Pathogenicity in mice of strains of herpes simplex virus which are resistant to acyclovir *in vitro* and *in vivo*. Antimicrob. Ag. Chemother. 17, 209-216.

Field, H.J., Darby, G. & Wildy, P. (1980) Isolation and characterization of acyclovir-resistant mutants of herpes simplex virus. J. Gen. Virol. 49, 115-124.

Field, H.J. & Lay, E. (1984) Characterization of latent infections in mice after inoculation with herpes simplex virus which is clinically resistant to acyclovir. Antivir. Res. 4, 43-52.

Field, H., MacMillan, A. & Darby, G. (1981) The sensitivity of acyclovir-resistant mutants of herpes simplex virus to other antiviral drugs. J. Inf. Dis. 143, 281-285.

Field, H.J. & Neden, J. (1982) Isolation of bromovinyldeoxyuridine-resistant strains of herpes simplex virus and successful chemotherapy of mice infected with one such strain by using acyclovir. Antiviral Res. 2, 243-254.

Field, H.J. & Reading, M.J. (1987) The inhibition of bovine herpesvirus-1 by methyl 2-pyridyl ketone thiosemicarbazone and its effects on bovine cells. Antiviral Res. in press.

Foster, D.A., Hantzopoulos, P. & Zubay, G. (1982) Resistance of adenoviral DNA replication to aphidicolin is dependent on the 72-kilodalton DNA-binding protein. J. Virol. 43, 679- 686.

Furman, P.A., Coen, D.M., St. Clair, M.H., & Schaffer, P.A. (1981) Acyclovir-resistant mutants of herpes simplex virus type 1 express altered DNA polymerase or reduced acyclovir phosphorylating activities. J. Virol. 40, 936-941.

Fyfe, J.A. (1982) Differential phosphorylation of (E)-5-(2-bromovinyl)-2'deoxyuridine monophosphate by thymidylate kinases from herpes simplex viruses types 1 and 2 and varicella zoster virus. Mol. Pharmacol. 21, 432-437.

Galbraith, A.W. (1985) Influenza - recent developments in prophylaxis and treatment. Brit. Med. Bull. 41, 381-385.

Gauri, K.K. (1979) Antiherpes polychemotherapy. Adv. Ophthalmol. 38, 151-163.

Ghendon, Y., Markushin, S., Heider, H., Melnikov, S. & Lotte, V. (1986) Haemagglutinin of influenza A virus is a target for the antiviral effect of Norakin[R]. J. Gen. Virol. 67, 1115-1122.

Gold, D. & Corey, L. (1987) Acyclovir prophylaxis for herpes simplex virus infection. Antimicrob. Ag. Chemother. 31, 361-367.

Grinsted, J. (1986) Evolution of transposable elements. J. Antimicrob. Chemother. 18, Suppl. C, 77-83.

Hall, C.B., McBride, J.T., Gala, C.L., Hildreth, S.W. & Schnabel, K.C. (1985) Ribavirin treatment of respiratory syncytial viral infection in infants with underlying cardio pulmonary disease. J. Am. Med. Assoc. 254, 3047-3051.

Hall, C.B., McBride, J.T., Walsh, E.E., Bell, D.M., Gala, C.L., Hildreth, S., Eyck,L.G.I. & Hall, W.J. (1983) Aerosolized ribavirin treatment of infants with respiratory syncytial viral infection. A randomized double blind study. N. Eng. J. Med. 308, 1443-1447.

Hall, M.J. (1986) Anti-herpes virus combinations in relation to drug resistance. J. Antimicrob. Chemother. 18, Suppl. B, 165-176.

Harmenberg, J., Abele, G. & Wahren, B. (1985a) Nucleoside pools of acyclovir-treated herpes simplex type 1 infected cells. Antiviral Res. 5, 75-81.

Harmenberg, J., Abele, G. & Malm, M. (1985b) Deoxythymidine

pools in animal and human skin with reference to antiviral drugs. Arch. Dermatol. Res. 277, 402-403.

Harmenberg, J., Sundqvist, V-A., Gadler, H., Leven, B., Brannstrom, G. & Wahren, B. (1986) Comparative methods for detection of thymidine kinase-deficient herpes simplex virus type 1 strains. Antimicrob. Ag. Chemother. 30, 570-573.

Harmenberg, J., Wahren, B., Sundqvist, V-A., Leven, B. (1985c) Multiplicity dependence and sensitivity of herpes simplex virus isolates to antiviral compounds. J. Antimicrob. Chemother. 15, 567-573.

Hay, A.J. & Zambon, M.C. (1984) Multiple actions of amantadine against influenza viruses. In 'Antiviral Drugs and Interferon: The Molecular Basis of their Activity' Ed. Y. Becker, Martinus Nijhoff, Boston, USA.

Hay, A.J., Wolstenholme, A.J., Skehel, J.J. & Smith, M.H. (1985) The molecular basis of the specific anti-influenza action of amantadine. EMBO J. 4, 3621-3624.

Hay, A.J., Zambon, M.C., Wolstenholme, A.J., Skehel, J.J. & Smith, M.H. (1986) Molecular basis of resistance of influenza A viruses to amantadine. J. Antimicrob. Chemother. 18, Suppl. B, 19-29.

Helenius, A., Kartenbeck, J., Simons, K. & Fries, E. (1980) On entry of Semliki Forest virus into BHK-21 cells. J. Cell Biol. 84, 404-420.

Helenius, A., Marsh, M. & White, J. (1982) Inhibition of Semliki Forest virus penetration by lysosomotropic weak bases. J. Gen. Virol. 58, 47-61.

Herrmann, E.C. Jr. & Herrmann, J.A. (1977) A working hypothesis - virus resistance development as an indicator of specific antiviral activity. Ann. N. Y. Acad. Sci. 284, 632- 637.

Hill, B.T. (1986) Resistance of mammalian tumour cells to anticancer drugs: mechanisms and concepts relating specifically to methotrexate and vincristine. J. Antimicrob. Chemother. 18, Suppl. B, 61-73.

Honess, R.W., Purifoy, D.J.M., Young, D., Gopal,R., Cammack, N. & O'Hare, P. (1984) Single mutations at many sites within the DNA polymerase locus of herpes simplex viruses can confer hypersensitivity to aphidicolin and resistance to phosphonoacetic acid. J. Gen Virol. 65, 1-17.

Honess, R.W. & Watson, D.H. (1977) Herpes simplex virus resistance and sensitivity to phosphonoacetic acid. J. Virol. 21, 584-600.

Howell, C.L. & Miller, M.J. (1984) Rapid method for determining the susceptibility of herpes simplex virus to acyclovir. Diagn. Microb. Inf. Dis. 2, 77-84.

Ishitsuka, H., Ninomiya, Y. & Suhara, Y. (1986) Molecular basis of drug resistance to new antirhinovirus agents. J. Antimicrob. Chemother. 18, Suppl. B, 11-18.

Jamieson, A.T. & Subak-Sharpe, J.H. (1974) Biochemical studies on the herpes simplex virus-specified deoxypyrimidine kinase activity. J. Gen. Virol. 24, 481-492.

Jawetz, E., Coleman, W.R., Dawson, C.R. & Thygeson, P. (1970) The dynamics of IUDR action in herpetic keratitis and the emergence of IUDR-resistance in vivo. Ann. N.Y. Acad. Sci. 173, 282-291.

Katz, E. & Margalith, E. (1984) Antiviral activity of SK & F 21681 against herpes simplex virus. Antimicrob. Ag. Chemother. 25, 195-200.

Katz, E., Margalith, E., Winer, B. & Lazar, A. (1973) Characterization and mixed infections of three strains of vaccinia virus: wild type, IBT-resistant and IBT-dependent mutants. J. Gen. Virol. 21, 469-475.

Kerridge, D. & Nicholas, R.O. (1986) Drug resistance in the opportunistic pathogens candida albicans and candida glabrata. J. Antimicrob. Chemother. 18, Suppl. B, 39-49.

Kit, S. (1985) Thymidine kinase. Microbiol. Sci. 2, 369-375.

Kit, S., Sheppard, M., Ichimura, H., Lehrman, N.S. & Otsuka, H. (1987) Nucleotide sequence changes in the thymidine kinase gene of herpes simplex virus type 2 clones from a patient treated with acyclovir. In the press.

Klein, R.J. (1975) Isolation of herpes simplex virus clones and drug resistant mutants in microcultures. Arch. Virol. 49, 73-80.

Korant, B.D. (1977) Poliovirus coat protein as the site of guanidine action. Virol. 81, 17-28.

Lamb, R.A., Zebedee, S.L. & Richardson, C.D. (1985) Influenza virus M2 protein is an integral membrane protein expressed on the infected cell surface. Cell, 40, 627-633.

Larder, B.A., Cheng, Y-C. & Darby, G. (1983a) Characterization of abnormal thymidine kinases induced by drug-resistant strains of herpes simplex virus type 1. J. Gen. Virol. 64, 523-532.

Larder, B.A. & Darby, G. (1984) Virus drug-resistance: mechanisms and consequences. Antiviral Res. 4, 1-42.

Larder, B.A. & Darby, G. (1986) Susceptibility to other antiherpes drugs of pathogenic variants of herpes simplex virus selected for resistance to acyclovir. Antimicrob. Ag. Chemother. 29, 894-898.

Larder, B.A., Derse, D., Cheng, Y-C. & Darby, G. (1983b) Properties of purified enzymes induced by pathogenic drug-resistant mutants of herpes simplex virus. J. Biol. Chem. 258, 2027-2023.

Larder, B.A., Derse, D., Cheng, Y-C. & Darby, G. (1983b) Properties of purified enzymes induced by pathogenic drug-resistant mutants of herpes simplex virus. Evidence for virus variants expressing normal DNA polymerase and altered thymidine kinase. J. Biol. Chem. 285, 2027-2033.

Larder, B.A., Kemp, S.D. & Darby, G. (1987) Related functional domains in virus DNA polymerase. EMBO J. 6, 169- 175.

Larsson, A., Brannstrom, G.,& Oberg, B. (1983) Kinetic analysis in cell culture of the reversal of antiherpes activity of nucleoside analogs by thymidine. Antimicrob. Ag. Chemother. 24, 819-822.

Lehrman, S.N., Hill, E.L., Rooney, J.F., Ellis, M.N., Barry, D.W. & Strauss, S.E. (1986) Extended acyclovir therapy for herpes genitalis: changes in virus sensitivity and strain variation. J. Antimicrob. Chemother. 18, Suppl. B. 85-94.

Littler, E., Zeuthen, J., McBride, A.A., Sorensen, E.T., Powell, K.L., Walsh-Arrand, J.E. & Arrand, J.R. (1986) Identification of an Epstein-Barr virus-coded thymidine kinase. EMBO J. 5, 1959-1966.

Loddo, B., Ferrari, W., Spanedda, A.,& Brotzu, G. (1962) In vitro guanidine resistance and guanidine dependence of polioviruses. Experientia 18, 518-519.

Martin, J.L., Ellis, M.N., Keller, P.M., Biron, K.K., Lehrman, S.N., Barry, D.W. & Furman, P.A. (1985) Plaque autoradiographic assay for the detection and quantitation of thymidine kinase-deficient and thymidine kinase-altered mutants of herpes simplex in clinical isolates. Antimicrob. Ag. Chemother. 28, 181-187.

McKnight, S.L. (1980) The nucleotide sequence and transcript map of the herpes simplex virus thymidine kinase gene. Nuc. Acids Res. 8, 5949-5964.

McLaren, C., Corey, L., Dekker, C. & Barry, D.W. (1983) In vitro sensitivity to acyclovir in genital herpes simplex virus from acylovir-treated patients. J. Inf. Dis. 148, 868- 875.

McLaren, C., Ellis, M.N. & Hunter, G.A. (1983) A colorimetric assay for the measurement of the sensitivity of herpes simplex virus to antiviral agents. Antiviral Res. 3, 223-234.

McSharry, J.J, Caliguiri, L.A. & Eggers, H.J. (1979) Inhibition of uncoating of poliovirus by arildone a new antiviral drug. Virol. 97, 307-315.

Melnick, J.L., Crowther, D. & Barrera-Oro, J. (1961) Rapid development of drug resistant mutants of poliovirus. Science 134, 551.

Mitsuya, H. & Broder, S. (1987) Strategies for antiviral therapy in AIDS. Nature 325, 773-778.

Moss, B. & Cooper, N. (1982) Genetic evidence for vaccinia virus-encoded polymerase: isolation of phosphonoacetate-resistant enzyme from the cytoplasm of cells infected with mutant virus. J. Virol. 43, 673-678.

Moss, B., Rosenblum, E.N., Katz, E. & Grimley, P.M. (1969) Rifampicin: a specific inhibitor of vaccinia virus assembly. Nature 224, 1280-1284.

Muratore, O., Varnier, O.E. Raffanti, S.P. & Schito, G.C. (1984) In vitro recovery of resistant retrovirus isolates after exposure to phosphonoformate. Eur. J. Clin. Microbiol. 3, 447-449.

Nakamura, Y., Bolloli, A., & Varaldi, V. (1967) Ricerche sulla patogenicita e sul potere immunizzante di un ceppo di virus vaccinico resistente alla 5-iodo-2'-deossiuridina. Boll. Inst. Sieroterapico Milanese 46, 281-287.

Ninomiya, Y., Aoyama, M., Umeda, I, Suhara, Y. & Ishitsuka, H. (1985) Comparative studies on the modes of action of the antirhinovirus agents RO 09-0410, RO 09-0179, RMI-15,731, 6- dichloroflavan, and enviroxime. Antimicrob. Ag. Chemother. 27, 595-599.

O'Brien, T.F. (1986) Resistance to antibiotics at medical centres in different parts of the world. J. Antimicrob. Chemother. 18, Suppl. C, 243-253.

Oxford, J.S. & Galbraith, A. (1980) Antiviral activity of amantadine: a review of laboratory and clinical data. Pharm. Therapeut. 11, 181-262.

Oxford, J.S., Logan, I.S. & Potter, C.W. (1970) In vivo selection of an influenza A2 strain resistant to amantadine. Nature 226, 82-83.

Parris, D.S. & Harrington, J.E. (1982) Herpes simplex virus variants resistant to high concentrations of acyclovir exist in clinical isolates. Antimicrob. Ag. Chemother. 22, 71- 77.

Patterson, S. & Oxford, J.S. (1986) Analysis of antigenic determinants on internal and external proteins of influenza virus and identification of antigenic sub-populations of virions in recent field isolates using monoclonal antibodies and immunogold labelling. Arch. Virol. 88, 189-202.

Pemberton, R.M., Jennings, R., Potter, C.W. & Oxford, J.S. (1986) Amantadine resistance in clinical influenza A (H3N2) and (H$_1$N$_1$) virus isolates. J. Antimicrob. Chemother. 18, Suppl. B, 135-140.

Phillips, I. (1986) Resistance as a cause of treatment failure. J. Antimicrob. Chemother. 18, Suppl. C, 255-260.

Phillpotts, R.J. & Tyrrell, D.A.J. (1985) Rhinovirus colds. Brit. Med. Bull. 41, 386-390.

Pincus, S.E., Diamond, D.C., Emini, E.A. & Wimmer, E. (1986) Guanidine-selected mutants of poliovirus: mapping of point mutations to polypeptide 2C. J. Virol. 57, 638-646.

Pincus, S.E. & Wimmer, E. (1986) Production of guanidine-resistant and -dependant poliovirus mutants from cloned cDNA: mutations in polypeptide 2C are directly responsible for altered guanidine sensitivity. J. Virol. 60, 506-514.

Post, L.E., Mackem, S. & Roizman, B. (1981) Regulation of genes of herpes simplex virus: expression of chimeric genes produced by fusion of thymidine kinase with alpha-gene promotors. Cell 24, 555-565.

Presber, H.W., Schroeder, C., Hegenscheid, B., Heider, H., Reefschlager, J. & Rosenthal, H.A. (1984) Antiviral activity of Norakin[R] (triperidin) and related anticholinergic anti- Parkinson drugs. Acta Virol. 28, 501-507.

Purifoy, D.J.M., Lewis, R.B. & Powell, K.L. (1977) Identification of the herpes simplex virus DNA polymerase gene. Nature 269, 621-623.

Rapazzo, G., Grillo, A., Biondi, O., Sammartano, F. & Pignatti, P.F. (1986) Resistance of HSV-1 growth to aphidicolin in two aphidicolin resistant cell lines. Microbiol. 9, 381-386.

Reed, S.E., Craig, J.W. & Tyrrell, D.A.J. (1976) Four compounds active against rhinovirus: comparison in vitro and in volunteers. J Inf. Dis. 133, Suppl.A, 128-135.

Sandford, G.P., Wingard, J.R., Simons, J.W., Staal, S.P., Saral, R. & Burns, W.H. (1985) Genetic analysis of the susceptibility of mouse cytomegalovirus to acyclovir. J. Virol. 53, 104-113.

Saunders, K. & King, A.M.Q. (1982) Guanidine-resistant mutants of aphthovirus induce the synthesis of altered non-structural polypeptide, p34. J. Virol. 42, 389-394.

Saunders, K., King, A.M., McCahon, D., Newman, J.W., Slade, W.R. & Forss, S. (1985) Recombination and oligonucleotide analysis of guanidine-resistant foot and mouth disease virus mutants. J. Virol. 56, 921-929.

Schinazi, R.F., Chou, T-C., Scott, R.T., Yao, X. & Nahmias, A.J. (1986) Delayed treatment with combinations of antiviral drugs in mice infected with herpes simplex virus and application of the median effect method of analysis. Antimicrob. Ag. Chemother. 30, 491-498.

Schnipper, L.E. & Crumpacker, C.S. (1980) Resistance of herpes simplex virus to acycloguanosine: role of viral thymidine kinase and DNA polymerase loci. Proc. Natl. Acad. Sci. USA 77, 2270-2273.

Schram, M., Laffin, J.A., Evans, B., McSharry, J.J & Caliguiri, L.A. (1982) Isolation of poliovirus variants resistant to and dependent on arildone. Virol. 122, 492- 497.

Schroeder, C., Heider, H., Hegenscheid, B., Schoffel, M.,

Bubovich, V.I. & Rosenthal, H.A. (1985). The anticholinergic anti-Parkinson drug Norakin[R] selectively inhibits influenza virus replication. Antiviral Res. Suppl. 1, 95-99.

Sears, A.E., Meignier, B. & Roizman, B. (1985) Establishment of latency in mice by herpes simplex virus 1 recombinants that carry insertions affecting regulation of the thymidine kinase gene. J. Virol. 55, 410-416.

Shipman, C.Jr., Smith, S.H., Drach, J.C. & Klayman, D.L. (1986) Antiviral activity of 2-acetylpyridine thiosemicarbazones against herpes simplex virus, Antimicrob. Ag. Chemother. 19, 682-685.

Shiraki, K., Ogino, T., Yamamoto, T., Yamanishi, K. & Takahashi, M. (1986a) Susceptibilities of phosphonoacetic acid and acyclovir resistant varicella-zoster virus mutants to 9-beta- arabinofuranosyladenine and 1-beta-arabinofuranosylcytosine. Biken J. 29, 11-17.

Shiraki, K., Ogino, T., Yamanishi, K. & Takahashi, M. (1986b) Thymidine kinase with altered substrate specificity of acyclovir resistant varicella-zoster virus. Biken J. 29, 7- 10.

Shiraki, K., Ogino, T., Yamanishi, K. & Takahashi, M. (1983) Isolation of drug resistant mutants of varicella-zoster virus: cross resistance of acyclovir resistant mutant with phosphonoacetic acid and bromodeoxyuridine. Biken J. 26, 17-23.

Simon, E.H., King, S., Koh, T.T. & Brandman, P. (1976) Interferon-sensitive mutants of mengovirus. I Isolation and biological characterization. Virol. 69, 727-736.

Summers, W.P., Wagner, M. & Summers, W.C. (1975) Possible peptide chain termination mutants in thymidine kinase gene of a mammalian virus, herpes simplex virus. Proc. Nat. Acad. Sci. USA 72, 4081-4084.

Svennerholm, B., Valne, A., Lowhagen, G.B., Widell, A. & Lycke, E. (1985) Sensitivity of HSV strains isolated before and after treatment with acyclovir. Scand. J. Inf. Dis. Suppl. 47, 149-154.

Tamm, I. & Eggers. H.J. (1963) Unique susceptibility of enteroviruses to inhibition by 2-(alpha-hydroxybenzyl)-benzimidazole and derivatives. In: 2nd. International Symposium of Chemotherapy, Part II. Ed. H.P. Kuemmerle, P. Preziosi, & P. Rentchnick, Karger, Basel, New York, 88.

Tartaglia, J. & Paoletti, E. (1985) Physical mapping and DNA sequence analysis of the rifampicin resistance locus in vaccinia virus. Virol. 147, 394-404.

Tenser, R.B. & Edris, W.A. (1986) Thymidine kinase (TK) activity in herpes simplex virus type 1 recombinants that carry insertions affecting regulation of the TK gene. Virol. 155, 257-261.

Tenser, R.B., Ressel, S.J., Fralish, F. A. & Jones, J.C. (1983) The role of pseudorabies virus thymidine kinase expression in trigeminal ganglion infection. J. Gen. Virol. 64, 1369-1373.

Tisdale, M. & Selway, J.W.T. (1983) Inhibition of an early stage of rhinovirus replication by dichloroflavan (BW683C). J. Gen. Virol. 64, 795-803.

Tisdale, M. & Selway, J.W.T. (1984) Effect of dichloroflavan (BW683C) on the stability and uncoating of rhinovirus type 1B. J. Antimicrob. Chemother. 14, Suppl. A, 97- 105.

Thomas, H.C. & Scully, L.J. (1985) Antiviral therapy in

hepatitis B infection. Brit. Med. Bull. 41, 374-380.

Tolskaya, E.A., Romanova, L.A., Kolesnikiva, M.S. & Agol, V. (1983) Intertypic recombination in poliovirus: genetic and biochemical studies. Virol. 124, 121-132.

Turk, S.R., Shipman, C.Jr. & Drach, J.C. (1986) Selective inhibition of herpes simplex virus ribonucleotide diphosphate reductase by derivatives of 2-acetylpyridine thiosemicarbazone. Biochem. Pharmacol. 35, 1539-1545.

Van-Dyke, R.B. & Connor, J.D. (1985) Uptake of [^{125}I] iododeoxycytidine by cells infected with herpes simplex virus: a rapid screening test for resistance to acyclovir. J. Inf. Dis. 152, 1206-1211.

Varnier, O.E., Muratore, O., Raffanti, S.P. & Schito, G.C. (1985) Antiviral activity of coumermycin: identification of resistant and sensitive retrovirus strains. Microbiologica 8, 283-287.

Veerisetty, V. & Gentry, G.A. (1983) Alterations in substrate specificity and physicochemical properties of deoxythymidine kinase of a drug-resistant herpes simplex virus type 1 mutant. J.Virol. 46, 901-908.

Vrang, L. Bazin, H., Remaud, G., Chattopadhyaya, J. & Oberg, B. (1987) Inhibition of the reverse transcriptase from HIV by 3'-azido-3'-deoxythymidine triphosphate and its threo analogue. Antiviral Res. 7, 139-149.

Wade, J.C., Newton, B., McLaren, C., Flournoy, N., Keeney, R.E. & Meyers, J.D. (1982) Intravenous acyclovir to treat mucocutaneous herpes simplex virus infection after bone marrow transplantation. Ann. Intern. Med. 96, 265-269.

Wagner, M.J., Sharp, J.A. & Summers, W.C. (1981) Nucleotide sequence of the thymidine kinase gene of herpes simplex virus type 1. Proc. Nat. Acad. Sci. USA. 78, 1441-1445.

Wallbank, A.M., Matter, R.E. & Klinowski, N.G. (1986). 1-Adamantane hydrochloride inhibition of Rous and Esk sarcoma viruses in cell culture. Science 152, 1760-1761.

Webster, R.G., Kawoaka, Y. & Bean, W.J. (1986). Vaccination as a strategy to reduce the emergence of amantadine and rimantadine resistant strains of A/chick/Pennsylvania/83 (H5N2) influenza virus. J. Antimicrob. Chemother. 18, Suppl. B, 157-164.

Zlydnikov, D.M., Kubar, O.I., Kovaleva, T.P. & Kamforin, L.E. (1981). Study of rimantadine in the USSR. Review of the Literature. Rev. Inf. Dis. 3, 408-421.

INTERFERON AS AN ANTIVIRAL AGENT

H. Schellekens

Primate Center
P.O.Box 5815
2280 HV Rijswijk
The Netherlands

INTRODUCTION

Interactions that occur when an organism is infected with two or more different types of viruses simultaneously are called viral interference. In general, it manifests itself as tissue immunity. That means that a tissue infected with a virus is resistent to infection with other types of viruses. This phenomenon was already described in the beginning of this century, not only between antigenically related viruses such as yellow fever viruses (1) and herpes viruses (2), but also between antigenically unrelated viruses, such as yellow fever and vaccinia (3,4) on the one hand, and mumps and Western equine encephalomyelitis virus on the other hand (4). Especially the latter indicates that a non-immune mechanism was responsible. Although the most simple explanation for tissue immunity is a direct interaction between the different viruses, it has been assumed for a long time that a non-viral agent was responsible for this effect (3,5).

The cause of tissue immunity was discovered in 1957 by Isaacs and Lindenmann (6) when they were studying the interfering capacity of inactivated influenzavirus. They incubated the inactivated virus with fragments of chick chorio-allantoic membranes and expected that, because the virus was absorbed by the chick cells, the interfering capacity of the preparation would decrease. To their surprise the reverse happened, and they were able to show that the interfering activity was caused by a non-viral agent. They named the agent interferon. Interferon appeared to be produced by quite a variety of species including fish and reptiles after induction by a variety of viruses, both alive and inactivated, and also an impressive number of viruses appeared to be sensitive to interferon (7).

This broad antiviral spectrum and the initially supposed lack of toxicity made interferon a potential useful antiviral agent for clinical application. And most of the research devoted to interferon since 1957 had as goal the use of interferon to treat viral infections in man. During the last 10 years, however, the anti-cancer use of interferon has dominated research and the antiviral application has become of lower priority.

The main obstacle for clinical evaluation of interferons was its production. It is produced only in minute amounts by induced cells. The scarcity of material has hampered to a great extent the development of interferon research. In the years following its discovery the enthusiasm for its direct clinical application waned; not only because of the disappointing results of the first trials obtained in man (8) by interferon produced in monkey cells but especially because of the discovery of the so-called non-viral inducers (9,10) that were supposed to act as alternatives for interferon.

By the mid seventies it became clear that the inducers were not the answer. The main problem of inducers is their refractoriness when used repeatedly. After a limited number of applications the host response disappears (11).

Concomitantly with the waning interest in interferon inducers the need was felt for the large-scale production of human interferons. The anticancer activity of interferon in experimental animals which was show in those years by the pioneering activity of Ion Gresser (12,13) means an extra stimulus for setting up production facilities. The main producer of human interferon until the introduction of recombinant DNA technology has been Kari Cantell in Helsinki (14,15). He and his associates of the Finnish Red Cross developed a large-scale production system based on human leukocytes derived from pooled donor buffy coats, which are a byproduct of processing whole blood. These leukocytes were infected with Sendai virus after pretreatment with low doses of interferon. The interferon was purified by potassium thiocyanate at low pH and re-solubilized into ethanol by increasing pH. With this interferon preparation most of the clinical trials in the seventies have been performed. Cantell and his coworkers reported their year production to be around 400 bilion units in 1980. Because donor blood was considered too a limited source for human interferon, alternative production systems were developed. Wellcome Laboratories, among others, set up a production unit based on transformed B cells or lymphoblastoid cells (16). These cells have an indefinite lifespan and can be grown in suspension in large fermentation tanks. They produce relatively large amounts of human interferon after challenge with Sendai or Newcastle disease virus (NDV). The most commonly used cell line is the Namalwa line derived from a patient with Burkitt's lymphoma. The rigorous purification schedules used for these so-called lymphoblastoid interferons makes them safe products to use and the product made by Wellcome has been licensed (16). The production of human fibroblast interferon for clinical evaluation became possible after the development of superinduction schemes by Vilcek et al. (17). The large scale production of this type of interferon was pioneered by Billiau and his colleagues at the Rega Institute in Leuven (18). They used human diploid cells induced by poly(I).poly(C), a synthetic double-stranded RNA, cycloheximide and actinomycin D for superinduction. This type of interferon is commercially produced by Rentschler and licensed in West Germany for the treatment of severe viral infections; yet, it has not become very popular. It is quite expensive to produce fibroblast interferon because the anchorage-dependent cells cannot be grown in large fermentation tanks but only in systems that need many manipulations and/or expensive equipment. Fibroblast interferon is also less stable than leukocyte interferon, and the pharmacokinetic behavior has been considered as disadvantageous although its clinical significance is still unclear (19).

The availability of relative large amounts of human interferon did not only make it possible to carry out limited clinical trials but also

provided the material for complete purification and physico-chemical characterization. By the end of the seventies both leukocyte and fibroblast
interferon were purified to homogeneity and partial amino acid sequences
were elucidated (20-22). The main result from these studies, which was
later confirmed by cloning of the genes, was that in a single species a
whole family of different interferon (sub)types can be detected. The different members of the human interferon family are shown in Table 1. There
are three types of interferon designated alpha, beta, and gamma. Initially the typing was based on antigenic properties, but now the genes for
the interferons are known, the nomenclature is based on structural properties. HuIFN-α is the type of interferon produced by viral induced leukocytes or lymphoblastoid cells. It is a mixture of at least 15 different
subtypes which show a 80 % amino acid homology. The subtypes differ in
biological activity in vitro (28). In contrast with earlier reports, the
majority of the human alpha interferons are not glycosylated. The official nomenclature of fibroblast interferon is HuIFN-β. This interferon is
glycosylated and consists of a single subtype. In the last years there
have been reports on a second subtype of HuIFN-β, called HuIFN-β_2 (23-
25). This subtype has recently been shown to be a B-cell-differentiation-
factor and not an interferon (26). HuIFN-α and HuIFN-β are structurally
related; yet, there is no homology with HuIFN-γ (29). The latter type of
interferon is produced by T-lymphocytes during immune reactions. There is
only a single subtype, which is glycosylated and consists of 146 amino
acids, 20 amino acids less than the other human interferons. In contrast
with the other human interferons, HuIFN-γ can be activated by pH 2 treatment. Also, because of its biological activities, HuIFN-γ should be considered a lymphokine rather than a classical interferon.

The introduction of recombination DNA technology has provided a major breakthrough in interferon research. Weissman and his associates were
the first to introduce in E.coli an interferon gene, later termed the Hu-
IFN-α_2 gene (30). That was quite an achievement because hardly anything

Table 1.

Official nomen-clature	Former names	Natural source	Structure
HuIFN-α (1,2, ...)	leukocyte interferon lymphoblastoid interferon type I interferon	leukocytes lymphoblastoid cells	At least 15 different sub-types with 80 % homology in amino acid composition Molecular weight 18,500-22,000 Consists of 166 amino acids
HuIFN-β	fibroblast interferon type I interferon	diploid fibroblast	One single glycosylated subtype of 166 amino acids Molecular weight 23,000
HuIFN-γ	immune interferon type II interferon acid-labile interferon	activated T-lymphocytes	One glycosylated subtype of 146 amino acids Molecular weight 20,000-25,000

was known about the protein or gene structure at that time and they had to follow an indirect and laborious way to find the gene. They isolated the messenger RNA pool from induced leukocytes, transcribed this in cDNA, introduced this cDNA into plasmids, and transformed E.coli cells with these plasmids. To identify transformants harboring cDNA of interferon genes the plasmids DNA were hybridized with the original mRNA pool. The mRNA that hybridized was eluted and injected in Xenopus laevis oocytes. The oocytes were grinded and tested for interferon activity. Because interferon is only produced in minute amounts by induced cells, also the amount of interferon mRNA is minimal and the number of transformants containing the interferon cDNA is extremely small. Several hundred thousands E.coli clones were tested before an interferon containing oocyte was found. The plasmid that had hybridized with the mRNA resulting in the positive oocyte therefore contained at least part of an alpha interferon gene. With this clone other IFN cDNA containing clones could be identified. One of these clones produced small amounts of interferon. Subsequently the gene was further manipulated resulting in an E.coli clone that produced enough HuIFN-α_2 for in vivo studies (31).

Other groups have subsequently succeeded in cloning HuIFN-β and Hu-IFN-γ (29,32) and at present all human interferon (sub)types can be produced in unlimited amounts. Also purification methods have been developed in which affinity chromatography employing monoclonal antibodies play an important role and all human interferons are now available in essentially pure form (33).

MODE OF ACTION

As I mentioned earlier, the antiviral activity of interferon has been the main theme of research for years. The paramount question was : how is the viral replication inhibited in interferon-treated cells and in what stage ? The lack of pure interferon preparations and the fact that all investigators studied their own virus-cell combinations resulted in many different answers to this question. Virtually all stages of viral replication have been reported to be affected by interferon (7). The discovery that in cell-free systems the antiviral state is activated by double-strands RNA (34), using virus-infected cells (35), has generated the consensus that two enzyme systems play an important role in the antiviral state: an oligo-A synthetase and a protein kinase. Oligo A synthetase leads to the formation of an oligonucleotide known as 2'-5'A (36). This 2'-5'A activates a cellular endonuclease which degrades viral mRNA (37). The protein kinase is supposed to inactivate by phosphorylation initiation factors needed by the virus (38,39). Both enzyme systems block viral replication at the stage of viral translation, but this is not the complete explanation of the antiviral activity of interferon. Some viruses seem to be blocked at other stages; e.g. retroviruses can be inhibited at the level of budding (40). For influenza virus the inhibitory effect is virus-specific, and under control of the Mx gene. This dominant Mx gene was discovered during crossbreeding of mouse strains that were resistant or sensitive to influenza virus (41). The resistance is directed towards all orthomyxoviruses but not other viruses. In the mouse the Mx gene is located on chromosome 16 and it codes for a protein of 75,000 dalton (42). The Mx gene can be activated only by IFN-α or -β and not by IFN-γ. The 75K protein is located in the nucleus. The human Mx protein can also be found in the cytoplasma. The Mx gene has been cloned and transfection experiments with the gene have proven its specific anti-influenza activity (43). How the Mx gene product functions, however, is still not solved (44).

Some "interferonologists" have in the past considered interferon a selective inhibitor of viral replication, eager as they were to develop it as the ideal antiviral agent (45). However, in the years following its discovery other activities of interferon such as inhibition of cell multiplication were reported (46). Because of the lack of purified material these effects could be attributed to the impurities.

When pure interferon preparations became available, it was shown that interferons indeed have multiple biological effects (47). By now more than 200 different effects of interferons have been reported. These effects are often referred to as the non-antiviral activities of interferon (48), which is quite misleading because many of these effects may well contribute to the antiviral activity of interferon in vivo.

Some of these effects are listed in Table 2. Interferons have been reported to inhibit DNA, RNA and protein synthesis in cells (49). It is clear that such an effect contributes to the inhibition of viral replication. Interferons play an important role in the activity of natural killer cells and macrophages (50). IFN-α/β plays an important regulatory role in the activity of the NK cells while IFN-γ plays a crucial role in macrophage activity. Antigens of the major histocompatibility complex are sometimes necessary for recognition of infected cells by activated T cells. Interferons can induce MHC antigens and IFN-γ may well be the most important inducer of class II antigens (50). Interferons regulate their own production. In cells pretreated with small doses of interferon, the interferon production is enhanced (48). In cells that have been in contact with high concentrations, the interferon response is blocked (49). At an early stage of the viral infection the interferon production is amplified. There also is a mechanism that prevents that too much interferon is produced. Interferon enhances the production of prostaglandins and histamine that are active in the aspecific inflammatory response (51). Specific immune reactions such as cytotoxicity of T cells and antibody production are also stimulated (50). A number of effects on the cell membrane, such as increase in rigidity, have been reported (51) and these changes may also contribute to the antiviral effects of interferon in vivo.

Table 2. Some biological effects of interferon relevant to their antiviral activity

---

Inhibition of viral replication
Inhibition of macromolecular synthesis
Activation of NK cells and macrophages
Induction of MHC expression
Priming and inhibition of interferon production
Enhancement of antibody production
Enhanced synthesis of prostaglandins and histamine
Enhanced expression of Fc receptors
Enhanced cytotoxicity of T cells
Increase in cell membrane rigidity

ANTIVIRAL STUDIES IN ANIMALS

The animal studies on the antiviral efficacy of interferon will not be discussed in detail. We reviewed the animal studies performed since 1980 in a recent review (52) and we have also described experiments carried out in non-human primates (53). I will only discuss some general problems here. As stated earlier, the number of published animal studies is disappointingly low (54) and this may explain some of the uncertainties concerning the clinical application of interferon. By 1982 the total number of studies reported in experimental animals did not exceed 60 while there were 2500 original papers published on interferon in general. In comparison, when acyclovir was licensed in 1981, 38 out of the 147 studies reported on this drug were animal studies. Initially, interferons for animal studies had to be produced in cell culture systems, and this did not provide enough material for the contemplated work. Recently, a number of murine, rat and bovine interferons produced by recombinant DNA technology have become available and all necessary studies can now be performed (55).

Animal studies are essential for establishing the efficacy of agents with multiple biological effects because in vitro assays only measure a single isolated effect. Often the in vitro effect cannot be extrapolated to the in vivo situation. We have shown that human interferons are active against vaccinia virus infections in rhesus monkeys while they have no activity against this virus in vitro (56). In the same animals we have shown that a recombinant DNA interferon that was ten times more active than natural HuIFN-α in vitro was much less active in vivo, and a hybrid recombinant DNA interferon that had no activity in vitro was active in vivo (57). Sun and colleagues (58) found that two HuIFN-α subtypes that were equally active against VSV in vitro had quite different in vivo effects in cotton rats infected with the same virus. Sasaki and coworkers (59) treated mice influenza A virus-infected with mouse interferon and human hybrid interferon alpha. Although these interferons were equally active in vitro, in vivo only the mouse interferon proved active.

Now that recombinant DNA interferon preparations for rodent studies are available, toxicity studies should have high priority. Such studies, apart from some investigations in newborn mice and rats, have hardly been done (60). Toxicity studies with human interferons in experimental animals are hardly feasible. In laboratory animals such as rats, mice and guinea pigs, human interferons are not active. In higher primates such as rhesus monkeys and African green monkeys in which human interferons have an efficient antiviral activity, they do not exhibit any toxicity (61). The only animal in which the side effects of human interferons have been reproduced is the chimpanzee (62). This animal is, however, scarce and some studies are for ethical reasons impossible. This means that toxicity data have to originate from rodents treated with species-specific interferon.

CLINICAL STUDIES

In Table 3 are listed the viral infections in humans in which interferon has shown activity. They are divided in established indications which are beyond doubt, and possible indications, for which the clinical significance remains to be proven. The viral infection in which interferon has led to the most dramatic response, is juvenile laryngeal papillomatosis (JLP) (63). It is a wartlike disease which occurs commonly around the vocal cords and leads to considerable morbidity and mortality, especially when the disease starts early in life. It is caused by human pa-

Table 3. Use of interferon in viral infections

Established indications	Juvenile laryngeal papillomatosis (JLP)
	Condylomata acuminata
	Herpes keratitis (combined with TFT or acyclovir)
Possible indications	Prevention of common cold
	Prevention of herpes virus infection
	Chronic hepatitis B virus infections

pillomavirus type 6 or 11 and the treatment before interferon consisted of surgery or laser therapy. In general the papillomas recur making multiple excisions up to hundreds necessary. Interferon has been shown to have an effect in the majority of cases and even complete remission has been achieved. If a remission can be achieved over a period of 1 or 1.5 years, the disease will not recur when the treatment is stopped. This is in contrast with hairy cell leukemia in which interferon therapy has to be given continuously to keep the disease under control.

Also condylomata acuminata (63) or genital warts are associated with human papillomavirus types 6 and 11. For this disease, as for JLP, there was no consistent therapy before interferon was proven to be effective. Complete or partial regression is achieved by different types of interferon injected systemically or locally. Interferon has also been reported to be effective in other wart-like diseases (63), and thus papillomavirus infections can be considered as sensitive to interferon therapy.

A combination of high titer interferon with either trifluorothymidine (TFT) or acyclovir has a significant better effect in dendritic herpes keratitis than chemotherapy alone (64). A single drop of interferon is enough to reduce the average healing time from 6 to about 3 days. This is well before the immune mechanisms get involved in the disease. Such immune mechanisms are held responsible for the destruction of the cornea that may result from recurrent herpes keratitis. Recently human interferon has been licensed in W.-Germany for combination therapy of dendritic keratitis (65).

Other indications for the antiviral use of interferon in man are less clear. In several studies interferon has been shown to be effective in preventing the common cold, both in experimental infections and naturally acquired infections in family settings (66). The practical consequence is doubtful because treatment schedules longer than a week cause side effects which are similar to the symptoms of the common cold. Interferon has been reported to reduce herpes labialis reactivation by surgery for trigeminal neurology, the spread of herpes zoster in cancer patients and cytomegalovirus disease in immunosuppressed kidney transplant recipients (67). Herpes labialis has been reported to be activated by interferon therapy; also, interferon treatment for herpes zoster has to start within 24 hours after appearance of the first lesions and interferon treatment of immunosuppressed patients is compounded by considerable toxicity. There are a number of reports on the effect of interferon in chronic hepatitis B virus infection (66). Although interferon undoubtedly affects viral replication, the clinical significance remains dubious. There are some reports indicating that interferon combined with antiviral chemotherapy is beneficial to selected groups of chronic HBV hepatitis patients, and there are numerous case studies and small open trials with interferon in various viral infections. The beneficial effects of interferon in these infections remain to be confirmed by properly controlled trials.

THE FUTURE

The major problem in the application of interferon in the treatment of viral infections are its side effects (Table 4). The side effects are generally comparable with an influenza-like syndrome (68). They diminish after prolonged treatment, are completely reversible after stopping the treatment and tend to be more severe in older patients. Yet, they are too severe to be acceptable for the treatment of a trivial viral infection such as the common cold. The most frequent viral infections in man are situated in the upper respiratory tract, and, of course, a treatment with side effects that are equal to, or greater than, the symptoms caused by the disease itself is not likely to become popular.

Table 4. Toxic side effects of human interferon

Frequent	Fever Pain and local inflammation at injection site
Sometimes	Malaise Chills Fatigue Muscle pain
Infrequent	Lumbago Paresthesia Dry mouth Herpes simplex reactivation Arthritis Anorexia Vomiting Tremors Pale hands Pain behind eyes Neck stiffness Nephrotic syndrome Renal insufficiency Diarrhea Hair loss Drop in blood pressure
At extremely high dose	Convulsions Coma
Abnormal laboratory values	Leukopenia (frequent) EEG changes (infrequent) Transaminase increase (rare) Antibodies to IFN (rare) Proteinuria (rare) Serum creatine increase (rare) Serum urea increase (rare)

It will be necessary to develop a "second-generation" interferon with less side effects by manipulating the genes of the naturally occuring interferons. There is evidence that the side effects can be dissociated from the antiviral activity. For instance in rhesus monkeys 50,000 HuIFN-α U/kg body weight causes a significant antiviral effect, while even at 200 x 10⁶ U/kg no toxic side effects are noted (61). Also, the different biological effects of interferon seem to be associated with different parts of the molecule (69). It is important that in vitro assays with interferon do not bear on its in vivo activity. This means that all new interferon molecules have to be evaluated in vivo.

An alternative may be the use of gamma interferon as an antiviral agent. This type of interferon (or lymphokine) was initially considered as a potential antitumor or immunomodulating agent and its antiviral activity has been overlooked. It has been shown in a number of studies to have an antiviral activity that is at least as great as the antiviral activity of the alpha interferons (52). However, the toxicity profile of gamma interferon differs from that of alpha interferon. There is also a considerable synergism in the in vivo antiviral activity of alpha and gamma interferons (52,57,65). To be of practical usefulness, however, these combinations should not lead to synergism in the toxic side effects.

REFERENCES

1. M. Hoskins, A protective action of neurotropic against viscerotropic yellow fever virus in Macaca rhesus, Am. J. Trop. Med. 15:675 (1935).
2. F. Magrassi, Studii sul'infezione e sul'immunita del virus erpetico, Z. Hyg. Infektionskr. 117:573 (1935).
3. E.H. Lennette, and H. Koprowski, Interference between viruses in tissue culture, J. Exp. Med. 83:195 (1945).
4. A. Vilches, and G.K. Hirst, Interference between neurotropic and other unrelated viruses, J. Immunol. 57:125 (1947).
5. W. Henle, Interference phenomena between animal viruses: a review, J. Immunol. 64:203 (1950).
6. A. Isaacs, and J. Lindenmann, Virus interference. I. The interferon, Proc. R. Soc. B147:258 (1957).
7. W.E. Stewart II, The Interferon System. Springer-Verlag, New York (1979).
8. Scientific Committee on Interferon, Experiments with interferon in man, Lancet 1:505 (1965).
9. J.S. Youngner, and W.R Stinebring, Interferon production in chickens injected with Brucella abortus, Science 144:1022 (1964).
10. A.K. Field, G.P. Lampson, A.A. Tytell, M.M. Nemes, and M.R. Hilleman, Inducers of interferon and host resistance, IV. Double-stranded replicative form RNA (MS2-RF-RNA) from E.coli infected with MS2 coliphage, Proc. Natl. Acad. Sci. USA 58:2102 (1967).
11. H.B. Levy, W. London, D.A. Fuccillo, S. Baron, and J. Rice, Prophylactic control of simian hemorrhagic fever in monkeys by an interferon inducer, polyriboinosinic-polyribocytidylic acid-poly-l-lysine, J. Infect. Dis. 133:A256 (1976).
12. I. Gresser, C. Bourali, J.P. Lévy, D. Fontaine-Brouty-Boyé, and M.T. Thomas, Increased survival in mice inoculated with tumor cells and treated with interferon preparations, Proc. Natl. Acad. Sci. USA 63:51 (1969).

13. I. Gresser, and C. Bourali, Antitumor effects of interferon preparations in mice, J. Natl. Cancer Inst. 45:365 (1970).
14. K. Cantell, S. Hirvonen, H.-L. Kauppinen, and G. Myllylä. Production of interferon in human leukocytes from normal donors with the use of Sendai virus, in: "Interferons". Part A. Methods in Enzymology, vol. 78, S. Pestka, ed., Academic Press, New York, p. 29 (1981).
15. K. Cantell, S. Hirvonen, and V. Koistinen, Partial purification of human leukocyte interferon on a large scale, in: "Interferons". Part A. Methods in Enzymology, vol. 78, S. Pestka, ed., Academic Press, New York, p. 499 (1981).
16. N.B. Finter, and K.H. Fantes, The purity and safety of interferons prepared for clinical use: the case for lymphoblastoid interferon, in: "Interferon 2", I. Gresser, ed., Academic Press, London, p. 65 (1980).
17. J. Vilcek, and E.A. Havell, Stabilization of interferon messenger RNA activity by treatment of cells with metabolic inhibitors and lowering of the incubation temperature, Proc. Natl. Acad. Sci. USA 70:3909 (1973).
18. J. Van Damme, and A. Billiau, Large-scale production of human fibroblast interferon, in: "Interferons". Part A. Methods in Enzymology, vol. 78, S. Pestka, ed., Academic Press, New York, p. 101 (1981).
19. J. Treuner, and D. Niethammer, Studies with human fibroblast (β) interferon preparations in patients with cancer, in: "Interferon, vol. 4, In Vivo and Clinical Studies, N.B. Finter, and R.K. Oldham, eds., Elsevier Science Publishers, Amsterdam, p. 281 (1985).
20. G. Allen, and K.H. Fantes, A family of structural genes for human lymphoblastoid (leukocyte-type) interferon. Nature 287:408 (1980).
21. E. Knight Jr., M.W. Hunkapiller, B.D. Korant, R.W.F. Hardy, and L.E. Hood, Human fibroblast interferon: amino acid analysis and amino terminal amino acid sequence. Science 207:525 (1980).
22. S. Stein, C. Kenny, H.-J. Friesen, J. Shively, U. Del Valle, and S. Pestka, NH_2-terminal amino acid sequence of human fibroblast interferon, Proc. Natl. Acad. Sci. USA 77:5716 (1980).
23. J. Weissenbach, Y. Chernajovsky, M. Zeevi, L. Shulman, H. Soreq, U. Nir, D. Wallach, M. Perricaudet, P. Tiollais, and M. Revel, Two interferon mRNAs in human fibroblasts: in vitro translation and Escherichia coli cloning studies, Proc. Natl. Acad. Sci. USA 77:7152 (1980).
24. J. Van Damme, M. De Ley, H. Claeys, A. Billiau, C. Vermylen, and P. De Somer, Interferon induced in human leukocytes by concanavalin A: isolation and characterization of γ- and β-type components, Eur. J. Immunol. 11:937 (1981).
25. P.B. Sehgal, and A.D. Sagar, Heterogeneity of poly(I).poly(C)-induced human fibroblast interferon mRNA species, Nature 288:95 (1980).
26. A. Billiau, BSF-2 is not just a differentiation factor, Nature 324:415 (1986).
27. T. Taniguchi, N. Mantei, M. Schwarzstein, S. Nagata, M. Muramatsu, and C. Weissmann, Human leukocyte and fibroblast interferons are structurally related, Nature 285:547 (1980).
28. M. Evinger, M. Rubinstein, and S. Pestka, Antiproliferative and antiviral activities of human leukocyte interferons, Arch. Biochem. Biophys. 210:319 (1981).
29. P.W. Gray, D.W. Leung, D. Pennica, E. Yelverton, R. Najarian, C.C. Simonsen, R. Derynck, P.J. Sherwood, D.M. Wallace, S.L. Berger, A.D. Levinson, and D.V. Goeddel, Expression of human immune interferon cDNA in E.coli and monkey cells, Nature 295 : 503 (1982).
30. S. Nagata, H. Taira, A. Hall, L. Johnsrud, M. Streuli, J. Ecsödi, W. Boll, K. Cantell, and C. Weissmann, Synthesis in E.coli of a polypeptide with human leukocyte interferon activity, Nature 284:316 (1980).

31. C. Weissmann, The cloning of interferon and other mistakes, in: "Interferon 3", I. Gresser, ed., Academic Press, London, p. 101 (1981).

32. R. Derynck, J. Content, E. De Clercq, G. Volckaert, J. Tavernier, R. Devos, and W. Fiers, Isolation and structure of a human fibroblast interferon gene, Nature 285:542 (1980).

33. W.G. Lewis and K.H. Fantes, The purification of interferons: recent developments, in: "Interferon. Vol. 1. General and Applied Aspects", A. Billiau, ed., Elsevier Sci. Publ., Amsterdam, p. 251 (1984).

34. I.M. Kerr, R.E. Brown, and L.A. Ball, Increased sensitivity of cell-free protein synthesis to double-stranded RNA after interferon treatment, Nature 250 :57 (1974).

35. R.M. Friedman, D.H. Metz, R.M. Esteban, D.R. Tovell, L.A. Ball, and I.M. Kerr, Mechanism of interferon action: inhibition of viral messenger ribonucleic acid translation in L-cell extracts, J. Virol. 10:1184 (1972).

36. I.M. Kerr, R.E. Brown, and A.G. Hovanessian, Nature of inhibitor of cell-free protein synthesis formed in response to interferon and double-stranded RNA. Nature 268:540 (1977).

37. C. Baglioni, M.A. Minks, and P.A. Maroney, Interferon action may be mediated by activation of a nuclease by pppA2'p5'A2'p5'A, Nature 273:684 (1978).

38. B. Lebleu, G.C. Sen, S. Shaila, B. Cabrer, and P. Lengyel, Interferon, double-stranded RNA, and protein phosphorylation, Proc. Natl. Acad. Sci. USA 73:3107 (1976).

39. A. Zilberstein, P. Federman, L. Shulman, and M. Revel, Specific phosphorylation in vitro of a protein associated with ribosomes of interferon-treated mouse L cells, FEBS Lett. 68:119 (1976).

40. A. Billiau, V.G. Edy, H. Sobis, and P. De Somer, Influence of interferon on virus-particle synthesis in oncornavirus-carrier lines. II. Evidence for a direct effect on particle release, Int. J. Cancer 14:335 (1974).

41. O. Haller, Inborn resistance of mice to orthomyxoviruses, Curr. Top. Microbiol. Immunol. 92:25 (1981).

42. P. Staeheli and O. Haller, Interferon-induced human protein with homology to protein Mx of influenza virus-resistant mice, Mol. Cell. Biol. 5:2150 (1985).

43. P. Staeheli, O. Haller, W. Boll, J. Lindenmann, and C. Weissmann, Mx protein: constitutive expression in 3T3 cells transformed with cloned Mx cDNA confers selective resistance to influenza virus, Cell 44:147 (1986).

44. R.M. Krug, M. Shaw, B. Broni, G. Shapiro, and O. Haller, Inhibition of influenza viral mRNA synthesis in cells expressing the interferon-induced Mx gene product, J. Virol. 56:201 (1985).

45. J.A. Sonnaben and R.M. Friedman, Mechanism of interferon action, in: "Interferons and Interferon Inducers", N.B. Finter, ed., North-Holland Publ. Co., Amsterdam, p. 201 (1973).

46. K. Paucker, K. Cantell, and W. Henle, Quantitative studies on viral interference in suspended L cells. III. Effect of interfering viruses and interferon on the growth rate of cells, Virology 17:324 (1962).

47. I. Gresser, J. De Maeyer-Guignard, M.G. Tovey, and E. De Maeyer, Electrophoretically pure mouse interferon exerts multiple biologic effects, Proc. Natl. Acad. Sci. USA 76:5308 (1979).

48. W.E. Stewart II, L.B. Gosser, and R.Z. Lockart Jr., Priming: a nonantiviral function of interferon, J. Virol. 7:792 (1971).

49. W.E. Stewart II, Varied biologic effects of interferon, in: "Interferon 1", Academic Press, London, p. 29 (1979).

50. E. De Maeyer, Interferons and the immune system, in: "Interferon. Volume 1. General and Applied Aspects", A. Billiau, ed., Elsevier Sci. Publ., Amsterdam, p. 167 (1984).

51. J. Taylor-Papadimitriou, Effects of interferon on cell growth and function, in: "Interferon. Volume 1. General and Applied Aspects", A. Billiau, ed., Elsevier Sci. Publ., Amsterdam, p. 139 (1984).

52. H. Schellekens, K. Nooter, and P.H. van der Meide, Efficacy and toxicity of interferons in experimental animals, in: "Interferon 8", Academic Press, London, in press (1987).

53. H. Schellekens, K. Nooter, and P.H. van der Meide, in: "Clinical Aspects of Interferon ", Revel and Becker, eds., Martinus Nijhoff Publ., Dordrecht, in press (1987).

54. H. Schellekens and P.H. van der Meide, Animal studies with interferon, in: "The Biology of the Interferon System 1983", E. De Maeyer and H. Schellekens, eds., Elsevier Sci. Publ., Amsterdam, p. 409 (1983).

55. H. Schellekens and P.H. van der Meide, in: "The Interferon System: A Current Review", S. Baron et al., eds., 3rd ed., The University of Texas Press, in press (1987).

56. H. Schellekens, W. Weimar, K. Cantell, and L. Stitz, Antiviral effect of interferon in vivo may be mediated by the host, Nature 278 :742 (1979).

57. H. Schellekens, A. De Reus, and P.H. van der Meide, How to achieve the more general use of interferon as an antiviral agent in man, in: "The Biology of the Interferon System 1985", W.E. Stewart II and H. Schellekens, eds., Elsevier Sci. Publ., Amsterdam, p. 365 (1986).

58. C.-S. Sun, P.R. Wyde, S.Z. Wilson, and V. Knight, Efficacy of aerosolized recombinant interferons against vesicular stomatitis virus-induced lung infection in cotton rats, J. Interferon Res. 4:449 (1984).

59. O. Sasaki, T. Karaki, and J. Imanishi, Protective effect of interferon on infections with hand, foot and mouth disease virus in newborn mice, J. Infect. Dis. 153:498 (1986).

60. I. Gresser, Can interferon induce disease ?, in: "Interferon 4", I. Gresser, ed., Academic Press, London, p. 95 (1982).

61. H. Schellekens, The toxicity of human interferons in nonhuman primates, in: "Interferons", T.C. Merigan and R.M. Friedman, eds., UCLA Symposia, vol. 25, p. 387 (1982).

62. H. Schellekens and P.H. van der Meide, J. Med. Primatol. 13:235 (1984).

63. P.K. Weck, J.L. Brandsma, and J.K. Whisnant, Interferons in the treatment of human papillomavirus diseases, Cancer Metast. Rev. 5:139 (1986).

64. R. Sundmacher, The role of interferon in prophylaxis and treatment of dendritic keratitis, in: "Herpes Simplex Infections of the Eye". Contemporary Issues in Ophthalmology", vol. 1, F.C. Blodi, ed., Livingstone, New York, p. 129 (1984).

65. R. Sundmacher, A. Mattes, D. Neumann-Haefelin, G. Adolf, and B. Kruss, The potency of interferon-alpha 2 and interferon-gamma in a combination therapy of dendritic keratitis. A controlled clinical study, Current Eye Res. 6:273 (1987).

66. G.M. Scott and D.A.J. Tyrrell, Antiviral effects of interferon in man, in: "Interferon. Volume 4. In vivo and Clinical Studies", N.B. Finter and R.K. Oldham, eds., Elsevier Sci. Publ., Amsterdam, p. 181 (1985).

67. K.L. Hartshorn and M.S. Hirsch, Interferons, in: "The Antimicrobial Agents Annual 2", P.K. Peterson and J. Verhoef, eds., Elsevier Sci. Publ., Amsterdam, p. 339 (1987).

68. J.M. Bottomley and J.L. Toy, Clinical side effects and toxicities of interferon, in: "Interferon. Volume 4. In Vivo and Clinical Studies", N.B. Finter and R.K. Oldham, eds., Elsevier Sci. Publ., Amsterdam, p. 155 (1985).

69. M. Cebrián, E. Yagüe, M.O. de Landázuri, M. Rodiguez-Moya, M. Fresno, N. Pezzi, S. Llamazares, and F. Sánchez-Madrid, Different functional sites on rIFN-α2 and their relation to the cellular receptor binding site, J. Immunol. 138:484 (1987).

RISK-BENEFIT ANALYSIS IN DRUG RESEARCH

Paul Janssen

Janssen Pharmaceutica
B-2340 Beerse, Belgium

*"When a man has anything to tell in this world, the difficulty is
not to make him tell it, but to prevent him from telling it too
often."*
(Bernard Shaw).

In preparing this presentation on risk-benefit analysis in drug research,
I found it exceedingly difficult, as you will experience in the minutes
to come, to do really more than to merely repeat the classical statements
on this difficult and controversial subject that can be found in the
abundant scientific and other literature of the last decades.

"It is not enough to speak, but to speak true."
(William Shakespeare).

I'll therefore try to follow Edmund Burke's advice :

*"Show the thing you contend for to be reason, show it to be common
sense, and then I am content to allow it what dignity you please."*
(Edmund Burke).

Keeping in mind the ancient Chinese proverb that says :

I hear and I forget, I read and I remember, I do and I understand.

I'll restrict my remarks to the design of drug-related clinical trials
for the simple reason that I have had no personal experience in other
types of clinical trials.

That there is a need for well-designed and informative clinical trials
goes without saying :

*"The symptoms of a common cold, if treated vigorously, will go away
within seven days, whereas if left alone they disappear over the
course of a week."*
(Common Medical Adage).

It is equally obvious that the subject of "risk-benefit analysis in drug research and therapy of human patients" is hotly debated, not only among scientists, but also in the news media. In the classical struggle between rational arguments and irrational convictions, the latter often prevail. As Mark Twain would have said :

"In matters of opinion our adversaries are all insane."
(Mark Twain).

In his most interesting book "Le refus du réel", Maurice Tubiana, the famous French oncologist, provides us with a lucid analysis of the psychological state of mind of those of us who refuse to accept obvious facts, but rather prefer to believe in a wide variety of old myths. He concludes by saying :

"Il faut aider l'homme à retrouver un équilibre. L'accroissement des connaissances doit y contribuer."
(Maurice Tubiana).

Let us hope that those of us who do not believe that solid scientific endeavour is essential for the advancement of medical welfare and health will sooner or later understand the meaning of Barbara Tuchman's words in her excellent book "The March of Follies".

"Experience of the failure of your policy should shake your believe in its essential excellence.
If pursuing disadvantage after the disadvantage has become obvious is irrational, then rejection of reason is the prime characteristic of folly."
(Barbara Tuchman).

"Once an effective treatment in an area of medicine has been discovered, the requirements for the next advance in that area will be more difficult, because no one is asking to achieve the same level, but something that is better than that."
(C.T. Dollery, 1980).

Clinical trials can be designed to answer such questions. Reliable measurement is the only source of knowledge in pharmacotherapy.

"Measurement is the process of mapping or assigning numbers to objects or observations. The kind of measurement which is achieved is a function of the rules under which the numbers are assigned."
(Sidney Siegel).

In designing a clinical trial it is essential to keep in mind that there are at least 4 levels of measurement

1. nominal or classificatory scale   (weakest)
2. ranking or ordinal scale
3. interval scale
4. ratio scale   (strongest)

All scales have certain formal properties. The operations and relations employed in obtaining the scores define and limit the manipulations and operations which are permissible in handling the scores :

Scale	Defining relations	
1. Nominal	$=$	equivalence
2. Ranking	$= >$	greater than
3. Interval	$= > \Delta x/\Delta y$	known ratio of 2 intervals
4. Ratio	$= > \Delta x/\Delta y \; x/y$	known ratio of 2 scores

Any system of diagnostic groups constitutes a nominal scale. In a ranking scale the carat > generally means "greater than" or " is preferred to", "is higher than", "is more difficult than", etc., depending on the nature of the relation that defines the nominal scale. Body temperature is usually measured on an interval scale (Celsius or Fahrenheit), the zero point being arbitrary. We measure mass and weight in a ratio scale with a true zero point. The ratio of 2 scores (grs or kg) is known.

The central topic of modern statistics is statistical inference,
which is concerned with 2 types of problems :
- estimation of population parameters, and
- tests of hypotheses.

In the development of modern statistical methods, the first techniques of inference which appeared were those which made a good many assumptions about the nature of the population from which the scores were drawn. Since population values are "parameters", these statistical tests are called parametric tests.

More recently we have seen the development of a large number of techniques which do not make numerous or stringent assumptions about parameters. These newer nonparametric techniques result in conclusions which require fewer qualifications. It is obvious that the fewer or weaker are the assumptions that define a particular model, the less qualifying we need to do about our decision arrived at by the statistical test associated with that model: the fewer or weaker are the assumptions, the more general are the conclusions.

In contemporary statistical decision theory, the procedure of adhering rigidly to the classical but arbitrary level of significance of 0.05 has been rejected in favor of making decisions in terms of loss functions, utilizing such principles as the principle of minimizing the maximal loss, the so-called minimax principle. Although the desirability of such a technique for arriving at decisions is clear, one should keep in mind that different people can hold different views as to the relative importance of the "losses" and "gains" involved.

What I am trying to make clear is that even the best designed drugrelated clinical trial, well carried out and analysed with the best statistical techniques, can only lead to a rational statement about the probability level at which the null hypothesis (e.g. no difference between drugs A and B) can be rejected.

If the difference in the incidence of desirable or undesirable effects caused by the two drugs is very large, say 0% versus 100%, and if these effects are clinically obvious and predictable from animal data, and if these large differences are confirmed by several independent groups of clinical investigators enjoying a good reputation, the scientific community tends to regard the conclusions of the clinical trials as "established facts".

Statistically significant ($P < 0.05$), but small and clinically apparently unimportant differences in clinical efficacy between various treatments tend to be looked upon with scepticism and to be the source of controversy and heated debate among parties with different interests and points of view.

However a single case report about an unanticipated undesirable event occurring during treatment with a relatively new drug, tends to attract great interest, even in the lay press, and the inevitable question will be asked almost automatically: "Did the drug do it?"

In a recent issue of British Medical Journal (July 26, 1986) Professor M.J.S. Langman concludes that

> *"Determining the risk of drug-induced disease does not get easier; indeed, it is getting progressively more costly and taking more time and effort. The public very reasonably wants safer medicines and expects close scrutiny of those on the market. The methods of investigation are, however, only partially effective. If new approaches are to be tried we shall need to examine them critically and to decide whether they are cosmetic or effective."*
> (M.J.S. Langman, 1986).

In their interesting book "Risk Watch, the Odds of Life", John Urquhart and Klaus Heilmann, point out that

> *"Risk and fear are twin themes in life. Fear drives us to shun risk, yet risk accompanies any action, however trivial. To grasp opportunity, one must act, and, in acting, one incurs risk: opportunity and risk cannot be separated, and no goal can be attained without accepting risk. When fear paralyses action, opportunity fades. To understand risk is to balance fear and opportunity. How we try to balance these - as individuals and as a society - reveals strange paradoxes which show how we misunderstand risk."*
> (J. Urquhart and K. Heilmann)

Throughout medical history risks of unknown magnitude have been taken in quest for progress.

In 1720 at the start of a smallpox epidemic in Boston, Zabdiel Boylston took the tremendous risk of inoculating 247 volunteers, including his own son, with pus from the smallpox lesions of others, in an effort to produce a mild case of the disease and thus obtain life-saving immunity in subsequent epidemics. When the smallpox epidemic in Boston, brought in by the crew of a visiting ship, had finally run its course, a careful count of victims revealed the following: 5759 people had developed smallpox, 844 of whom had died - 1 in 7. Six in 247, or 1 in 41, died in the inoculated group ($X^2 = 29.14$; $P = < 0.001$). In subsequent years, others practiced the inoculation procedure, but always in a limited way and rarely without controversy.

70 years later, in 1792, Edward Jenner discovered that the almost innocuous disease called cowpox provided a safe means to the same end as Boylston's obviously hazardous inoculation with smallpox.

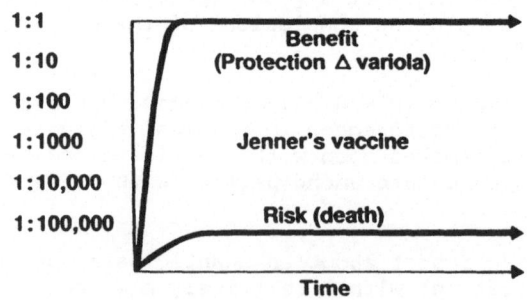

Today, almost 200 years later, we know that the risk of death due to cowpox vaccination is approximately 1 in 100,000.

Risk of death
No vaccination   1 : 7
Boylston's method   1 : 40   smallpox
Jenner's method   1 : 100,000   cowpox

Fortunately, much time passed after Jenner's discovery before the risk of cowpox vaccination became evident, so that Jenner's method did not suffer the same fate as Boylston's.

The conclusion of all this is unescapable: one of the great thriumphs of medicine, i.e. the total eradication of smallpox, was achieved using a very effective but not completely riskfree vaccine.

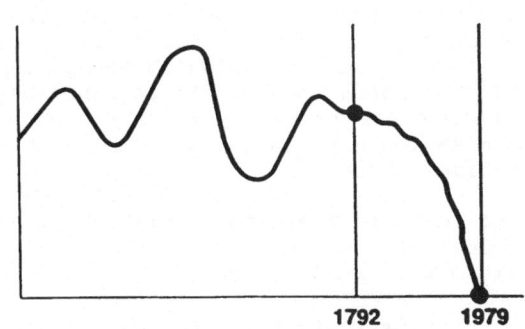

World-wide incidence of smallpox

Jenner's vaccine, in retrospect, prevented many thousands of smallpox deaths but killed quite a few people that would probably not have died as quickly without the vaccine.

In a recent paper (Brit. Med. J. 293, 363, 1986) R. Dahan et al. (Paris and Sheffield) concluded that the ethically required informed or written consent procedure biases the results of controlled clinical trials and might affect their general applicability.

These clinical investigators designed a randomised, observer blinded trial in order to examine whether informed consent might lead to biased sampling of the population and influence response to placebo treatment in hospitalized patients suffering from insomnia.

Of the 189 patients, 103 were excluded before randomisation for the following reasons :
- psychiatric disorders (12),
- use of drugs that commonly cause insomnia or cause sleep (28),
- insomnia due to pain, fever, dysuria and other non-specific medical problems (25),
- insomnia due to nursing procedures (12),
- sensorial or cognitive dysfunction (20),
- other reasons (7).

```
                    103/189 excluded
                    2 pairs of 43 included
                    26/43 refused written consent
                     ⌈ 18/27   females ≧ 60
                     ⌊  8/29   others

                    "Hypnotic" effect of placebo:
                    written consent      yes      no
                    excellent             3        6
                    good                  2        7
                    mediocre             13       11
                    very bad             12        6
                                         ⌊_____⌋
                                          p < 0.05

                    side-effects        4/30     0/30
```

R. Dahan et al., Br. Med. J. 293, 363 (1986)

The remaining 86 patients were first paired according to sex, age
(>60, ≧ 60 years old) and hospital environment (single or double
room). The first patient of each pair was then assigned randomly to the
control group or to the group that was asked to give written informed
consent after having read a document in which the patient was asked to
participate in a trial designed to evaluate the therapeutic potential of
a new drug versus placebo. The side-effects of benzodiazpines were listed
and the use of randomisation explained, after which 26 out of 43 patients
declined to give informed consent :

    12 were reluctant to participate in a trial,
    10 had fear of side-effects and
     4 wished to take a well known drug.

The 30 remaining pairs of patients all received one placebo tablet one to
two hours before going to bed. The next morning each patient evaluated
sleep subjectively on a 4 point scale (excellent, good, mediocre or very
bad) and reported any side-effects to an independent observer. The data
were analysed using two adequate non-parametric statistical methods.

Women over 60 were significantly ($p < 0.02$) less disposed than others
to give informed consent :

    Females ≧ 60   18/27 refused consent
    Females < 60   1/ 8 ⌉
    Males   ≧ 60   3/12 | 8/29
    Males   < 60   4/ 9 ⌋

The "hypnotic activity" of the placebo tablet was scored as follows in
the two groups :

    written consent   yes   no
    ⌈ excellent   3   6
    | good   2   7
    | mediocre   13   11
    ⌊ very bad   12   6
    Side-effects   4/30   0/30

Statistically (Wilcoxon's signed rank test) the placebo had a better
"hypnotic" effect in the control group than in those patients who had
given written consent ($p < 0.05$).

In this study obtaining informed consent had several obvious disadvantages :

1. it increased the duration of the trial,
2. it modified the characteristics of the population included in the trial,
3. it modified the hypnotic response obtained with placebo treatment,
4. anxiety was induced by the experiment and
5. guilt feelings in patients who refused to give consent,
6. it induced some patients who had given consent to complain about side-effects.

I would like to conclude this presentation with the same general remarks I made in London some 6 years ago on a similar occasion :

> *"Let us take two generally held principles that form the basis of civilized society, principles that are taught to our children at a very early age :*
> *(1) Don't do unto others as you wouldn't have them do unto you.*
> *(2) Don't expect others to do what you would refuse to do yourself.*

In applying these general principles to the field of drug research, this would mean that
(a) It is ethically unacceptable to expect a patient to receive a drug that you yourself would not take in a similar situation.
(b) It is unreasonable to expect others to work under research conditions and constraints that are unacceptable to yourself.

And to make it even more concrete one must add that weighing the risks of the administration of a drug against its anticipated benefits is a matter of conscience which a doctor faces continually and from which he cannot escape.

It is essential for a doctor under all circumstances to treat his patient as if he himself were the patient. If there is no problem with the diagnosis, or if he is dealing with a well-known and rather common disease or syndrome, the treatment for which is readily available, effective and safe, then the doctor will find no major obstacles in carrying out his job. In such cases, the doctor's professional skill is his main asset. But if it is a rare condition of unknown aetiology and the treatment option ill-defined, then common sense will be more valuable than book lore and the doctor will have to take a calculated risk. In extreme cases, when a patient's life is at stake, his decision must come quickly and decisively.

Surgeons in particular are often confronted with such agonizing problems. Doctors are of course individuals and there is no way of avoiding different doctors placing different interpretations on the same basic situation. Different interpretations will result in different courses of action, but always with the same objective   patient care. Thus a doctor may be expected to keep abreast of medical progress and be prepared day and night to take action to the best of his ability.

And what can be reasonably expected from those who are engaged in the search for better medicines? As much as possible, of course, But then, what is possible? What is reasonable or practicable, and where does Utopia begin? To what frontiers does science reach and where lie the limits of our economic possibilities, of the socially and economically acceptable? On the one hand, our world of 1986 witnesses the existence and development of a technologically oriented science, capable of quickly and efficiently unravelling and solving problems traditionally regarded as inextricable, and on the other hand it fosters a bureaucracy created and supported by public opinion and intended to protect society from unexpected and unwanted dangers which modern technology seems to entail.

Fear of the unknown is a particularly strong sociological motive which politicians experience day after day. So where lies the real interest of the citizen in general and of the diseased in particular?

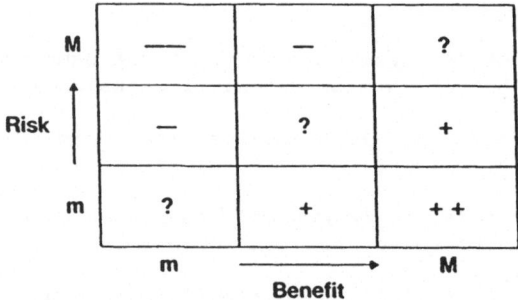

It is not my intention to give you a lengthy and possibly sterile analysis of the problem, but I would like to contribute to what may be a small but, in my opinion, important step forward. The benefits of any new drug must always be weighed against possible risks. The benefit-risk ratio will vary according to the condition to be treated. Thus in the case of common, self-limiting minor ailments, the benefits must always significantly outweigh the risks, as is also the case when effective and safe treatments are already available. However, in rare diseases, where the prognosis is poor and no effective therapy exists, it is justifiable to take great risks, but always with due consideration to the hazards of remaining untreated.

In the interest of the unfortunate patient who is suffering from a rare and serious, spontaneously irreversible and, so far, intractable disease, I propose that all those who are trying to find an adequate therapeutic solution for such a patient be well encouraged in their endeavours rather than demotivated by all sorts of bureaucratic red tape. The legislator who has set up the bureaucratic control systems of new drugs ought to finally be aware that there is a considerable difference in the risk-benefit analysis between a toothache and sleeping sickness. It is a matter for regret that too little attention is paid to diseases of rare occurrence and to typically tropical diseases that cause great damage to the Third World but are seldom diagnosed in thriving countries.

The main cause of this shortcoming is usually sought in merely economic considerations. It is quite obvious, indeed, that the financial risks of drug research in these two areas bear no acceptable relation to the potential market. From a purely economic point of view, these problems are very difficult indeed. Yet there are sufficient people in this world who would be willing to take that risk. Fortunately, there are sill some idealists, even in the pharmaceutical industry. The government ought to encourage these people. It would be wise to deliver them from the bureau-cratic yoke under which they are currently bending their necks. This deliverance would suffice to give drug research in the field of rare and tropical diseases new and vigorous impetus. And if, besides, the govern-ments of many developing countries were able to set up a fair and sound patent legislation, this would, in itself, mark a considerable step forward. Whether there should be any direct support by the government remains a matter of controversy.

To conclude with a more general remark. Risk-benefit analysis in drug research is, first and foremost, the responsibility of the drug researchers themselves, and afterwards, it is a problem which the attending physician will have to solve in each individual case. In these matters, public authorities can play only a minor role for the protection of public health. They must, of course, track down any evil-doers and fight their ill effect upon society. But they do certainly not serve the interest of the public in general or the needy patient in particular by creating a dictatorial bureaucracy which rapidly loses sight of the real aims of its existence and unintentionally imposes a check on the real progress of drug research.

Creative people dislike an exaggerated bureaucracy and are, to a great extent, demotivated by it. To them nothing can be worse than the obligation to complete piles of paper with insipid data. That deadly dull sort of work will finally result in a loss of the creative capacity itself. Useless work drives out useful work, and who may benefit from that?

SAFETY ASSESSMENT OF ANTIVIRAL DRUGS

Elaine C. Esber, Robert, C. Nelson and Norma Jean Browder

Office of Biologics Research and Review
Center for Drugs and Biologics, FDA
Rockville, Maryland 20857, USA

INTRODUCTION

Over the past two decades, accumulating knowledge of viral replication and viral/host cell interaction has made it possible to define virus specific events that are unique and to identify compounds that have the potential to interfere with these events. The available new antiviral agents have already impacted upon the treatment of several viral illnesses and more are under development. With the advent of compounds that interrupt or interfere with events at the genomic level, questions have been raised regarding the safety of these compounds as well as the adequacy of the available test systems to evaluate safety. Just as the field of antiviral drug therapy is young and evolving, so too are the methods used to assess their associated risk.

The scope of this presentation will include only those issues relevant to a safety evaluation of antiviral agents. Although efficacy testing will not be discussed, therapeutic use information is considered as integral part of the information necessary for guiding the safety evaluation. Additionally, comments will be restricted to the true antivirals, thereby excluding non-specific virucidal agents and immunomodulators. The following is a listing of true antiviral compounds approved for marketing in the United States: amantadine, idoxuridine, vidarabine, acyclovir, ribavirin, zidovudine, and trifluridine.

The existing U.S. statutes and regulations provide the basic requirements for determining the safety and efficacy of products. The process includes the development of safety and efficacy data during the investigational or IND phase, and the submission of these data in support of a marketing application - the New Drug Application (NDA). As a basic principle, the sponsor of a new drug or biologic bears the responsibility for identifying and characterizing actual and potential safety issues associated with the use of the product that is proposed for market approval.

With a novel class of agents, such as the antiviral, the series of tests that will need to be performed to fulfill that responsibility can only be partially defined at the onset. In a rapidly evolving field as this one, there can be no firm guidelines, rather judgements should be made based on accumulating information. Therefore, close cooperation and communication between the sponsor and the FDA is necessary to arrive at a relevant and agreed upon series of tests to address evolving issues.

During new drug development, in vitro tests, animal safety and toxicity tests, and appropriate animal models of drug activty are extremely valuable for predicting potential observations in humans, but the ultimate method to determine if a drug is safe and effective is the test in humans. In the following discussion, we will present the issues and concepts that are considered in assessing the safety of these novel antiviral agents. Since the field is rapidly evolving and individual compounds may present unique questions, contact should be made directly with our Division of Anti-Infective Drug Products for consultation on specific compounds. At present, a case-by-case approach is still most prudent.

The format of this paper is as follows : in part I: Safety Assessment of a New Drug in the U.S., a brief overview of the multistaged and interrelated process will be presented to provide a basis for understanding the "usual" process. Part II: An Antiviral Drug Risk Analysis Illustration - Acyclovir will provide a practical and tutorial example of the new and unique safety issues addressed during the development of the various dosage forms of acyclovir for its various clinical indications. In part III: Current State-of-the-Art and Future Needs, as modified by the lessons learned thus far, will be explained. Areas of need and future challenges will also be addressed because components of this testing process have considerable room for improvement. These will be highlighted and the challenges of adapting to the rapid advances in medical technology will be discussed.

PART I: SAFETY ASSESSMENT OF A NEW DRUG IN THE U.S.

The evaluation of a new drug during its development and throughout the regulatory review process involves a continual risk analysis which is later coupled with a benefit analysis. These culminate with the ultimate risk/benefit determination which takes the form of a decision to allow or not allow marketing of the drug product and if marketed the basis for that decision appears in the labeling that is approved for the product.

Simplistically, these risk analyses flow along a continuum, from in vitro testing (e.g. mutagenicity) to preclinical animal testing (e.g. toxicity and carcinogenicity), through the three phases of premarketing clinical testing (in humans), to Phase IV studies and post-marketing surveillance efforts. In actuality there is a coordinated interaction between specific sets of preclinical tests and each phase of clinical testing.

Thus far, we've used the word safety and risk interchangeably and that may lead to misunderstanding. Safety is the reciprocal of risk. In order to judge a drug's safety one must identify, characterize and assess (quantitatively and qualitatively) its risk. Then a conceptual conversion to safety needs to be made.

To further clarify terms, we use risk analysis as a bi-compartment concept which includes risk assessment and risk management. Risk assessment includes the identification and characterization of the hazards. Risk management involves a decision based upon the risk assessment. A benefit analysis is similar in structure. The risk-to-benefit decision making requires the balancing of the results from both forms of analyses.

This section of the presentation will describe the safety assessment of the "average or typical" new drug to provide a background for understanding the challenges faced with the true antivirals.

Before any new drug is administered to humans, it must undergo a series of preclinical tests. What are the reasons, the expectations and the limitations of this strategy ?

The reasons are multiple :

- detect over toxicity
- establish dose-response estimation of pharmacological and toxic effects
- assess drug distribution to organ systems
- identify metabolic, kinetic and elimination pathways
- assess carcinogenicity
- assess reproductive toxicity and teratogenic potential
- direct clinical safety assessments

Drugs make up the only group of chemicals that are deliberately used or are administered at doses intended to have an effect on the body. By definition, a drug is biologically active and, therefore, can be expected to be toxic at some dose. Ideally, a drug's wanted effects are separable by dose-range from its unwanted effects.

Drug toxicity, no different from that of other types of chemicals, may derived from one or more of a variety of biological responses. Toxicity may result from exaggerated pharmacological activity, altered systemic physiology or a direct action of the compound, or its metabolites, on one or more organ systems or tissues. Manifestations of toxicology may include functional and/or morphological changes which may or may not be correctable or reversible.

In drug development, toxicology and other types of preclinical studies precede and then interdigitate with the clinical investigation. These toxicology data are not intended to assure or establish safe human use, nor are they typically used for de facto prediction of probable human risk. Instead, the primary purpose of the preclinical toxicity studies is to provoke and maximize the drug's effects in two or more species of animals, and to determine the dose and time relationship of these effects under various experimental conditions. This characterization of the drug action may identify certain toxic effects or narrow margins of safety which pose an unacceptable risk for the proposed patient population or for humans in general. More commonly, however, the toxicology information serves to direct attention to potential adverse effects that may warrant particularly rigorous clinical monitoring or patient selection until actual assessment of the drug effects in humans is known.

An exception to this lies in the use of data pertaining to certain types of irreversible toxicity seen in animals, such as teratogenicity or carcinogenicity, which cannot be evaluated in human investigation. But even in this case, such data from animals are seldom extrapolated directly to human risk decisions without considering many qualifying factors such as the dose and time response relationships and other information from the preclinical studies, the dosage, extent and duration of intended or actual human use, the severity of illness being treated and the availability of more acceptable therapeutic alternatives. Such drugs, if approved for marketing, have full disclosure of these animal data in the package insert. Generally, however, the more extensive or chronic the intended clinical use and the less incapacitating or life-threatening assurance of a considerable margin of safety as based on clinical and on animal study results.

Acceptable toxicity for drugs is not based on a mathematically-derived upper limit of acceptable exposure, but rather on an acceptable separation of toxic and beneficial doses that vary according to the particular disorder being treated, the therapeutic benefit sought, and the extent to which the particular type of toxicity can be detected early and reversed in humans.

Over the years the types and amount of preclinical toxicity testing considered necessary for drug development and marketing have evolved and expanded in a continuing effort to avert unacceptable and unexpected human hazard as has occurred in the past, or which may be perceived as a possibility from current scientific knowledge. The purpose of the US FDA Investigational New Drug Regulations[2] is to assure that no compound is tested as a human drug without appropriate safeguards. To rephrase, the major function of preclinical toxicology data is to provide an integrated background of basic information about the drug to support and guide the clinical investigations.

Specific requirements for animal toxicology studies for INDs and NDAs generally depend upon the extent and duration of inteded or actual human use, the dosage form to be used, the severity of the illness being treated and the availability of acceptable therapeutic alternatives. The sequence of these tests proceed in lock-step with the successive phases of clinical investigation. In brief, the three phases of premarketing clinical investigation are as follows: Phase I - clinical pharmacology usually in normal volunteers with attention to pharmacokinetics, metabolism and both single dose and dose-range safety. Phase II - limited sized closely monitored investigations designed to assess efficacy and relative safety. Phase III - Full scale clinical investigations designed to provide an assessment of safety, efficacy, optimum dose and more precise definition of drug-related adverse effects in a given disease or condition.

An illustration of the interplay between the preclinical testing and the phases of clinical trials appear in Table 1. Toxicity testing is based upon anticiptated human exposure in the developmental phases and the ultimate projected human exposure. The guiding principle behind the preclinical testing is to have exposure durations exceed the anticiptated duration of human exposure in the clinical trials. The durations of preclinical testing in each developmental phase are also a function of the anticipated use; thus, the closer one is to the anticipation of multiple dose exposures in large numbers of patients (Phase III or pre-NDA), the longer the toxicity studies would be to support the relative safety of that duration of human exposure. Further relative assurances of safety are built into the toxicity trials by using multiple doses, the highest dose generally producing measurable toxicty. The data generated from toxicity tests are compared to those which produce the desired pharmacological effect. This can be done by comparing acute and, if possible, subacute ratios of ED_{50} to LD_{50} doses. Of course, it is also possible to learn about the problems of toxicity by comparing the slopes of the dose-effect curves for pharmacological and toxic effects, their steepness, and degree of overlap.

Toxicity assessment is performed in two or more species for several reasons. More than one species is used to evaluate the relative and "absolute" vulnerability of a second or a third species to a toxic effect that may have been observed in rodents. The relative vulnerability, which is expressed as a dose ratio of desirable to toxic effects or a ratio of median effective to toxic doses, is important as a general principle as it suggests the predictability of the ratio of desirable to undesirable effects likely to be encountered in the human. Compounds which have variable effects across the phylogenetic scale should be introduced into clinical trials with great caution; i.e., low starting doses with careful escalations in the dose run-up.

Table 1. General Outline for Animal Toxicity Studies for New Drugs

Category	Duration of human administration	Phase	Subacute or chronic toxicity
Oral or or parenteral	several days	I,II,III,DNA	2 species; 2 weeks
	up to 2 weeks	I	2 species; 2 weeks
		II	2 species; up to 4 weeks
		III,NDA	2 species; up to 3 months
	up to 3 months	I,II	2 species; 4 weeks
		III	2 species; 3 months
		NDA	2 species; up to 6 months
	6 months to unlimited	I,II	2 species; 3 months
		III	2 species; 6 months or longer
		NDA	2 species; 12 months (nonrodent) 18 months (rodent)

The US FDA requires that a drug be adequately characterized with respect to its pharmacology and toxicity before marketing. This includes an evaluation of the pharmacological tests which predict or correlate with clinical activity as well as the characterization of responses in other systems; e.g., cardiovascular responses for a neuropharmacological drug. The toxicological work-up for a new drug includes an evaluation of the acute toxicity, including the biological and statistical issues surrounding this phenomenon, subacute and chronic toxicity (two week to one year studies) and toxicological testing for which human assessment is not possible (teratogenic, mutagenic and carcinogenic properties).

The parameters which are of interest in acute toxicity testing are the calculation of the LD_{50} (for certain classes of drugs) and its 95 % confidence limits, the toxic signs and the progression to death, and the time interval to death as a function of route of administration. Interferences about the absorption of a compound are made by comparing the relative toxicities and rapidity of death to a parenteral route.

The usual effects monitored in a multiple dose subacute or chronic toxicity trial are the pharmacotoxic signs, the preagonal signs (i.e. those occurring just prior to death), mortality, and changes "in life" during the study. These changes include but are not limited to body weight and food consumption, clinical chemistry, hematology, and urinalysis parameters, as well as physical examinations with measurement of physiological parameters if appropriate, ECGs, and ophthalmoscopic exams. Following sacrifice of the animals, organ weight and gross pathology are recorded; representative and unusual looking tissues from all major organ systems are prepared for histopathological analysis. Changes seen in the toxicity trials are used as a guide for monitoring in clinical studies.

Reproductive studies have traditionally been catalogued into three segments: segment I, fertility and reproductive performance, is an evaluation of the drug's effect on mating behavior and gametogenesis, segment II, teratology, is an evaluation of the drug's effect on organogenesis, and segment III is an evaluation of the drug on the peri-natal and post-natal behaviors, development and, the reproductive capacity of the FI generation.

Drug disposition studies are performed in preclinical evaluations to quantitate the A.D.M.E. - absorption, distribution, metabolism and excretion across species. This allows a comparison of qualitative and quantitative differences across species, including animal to human disposition comparisons. These types of data are used to explain species differences on many occasions.

Mutagenicity testing is an evolving area which presents new problems in interpretation, extrapolation (bacterial to mammalian), and regulatory significance, emphasis and policy formulation. Currently the US FDA requires the submission of all available data on the mutagenicity of a compound, recommends this type of testing for drug categories suspected of being possibly mutagenic (similar chemical type to a known mutagen), and has a labeling sub-section for drugs with positive mutagenic effects. No current battery of tests are recommended as being definitive for the detection of this type of effect.

As previously stated, upon reasonable assurance of relative safety from the preclinical testing, Phase I investigations can commence. These studies involve a small number of persons (normal volunteers and/or patients) and are conducted under carefully controlled circumstances by clinical investigators qualified to determine pharmacologic effects, toxicity, metabolism, absorption, elimination, and safe dosage ranges. At the end of Phase I the sponsor should have preliminary information on the drug's pharmacological effects in humans, and sufficient information on the short-term toxicity of the drug and its pharmacokinetics in humans to design Phase II trials. The second phase of clinical trials are well-controlled studies conducted on a limited number of patients to determine the drug's effectiveness for treatment, prevention, or diagnoses of a specific disease and to obtain additional data on short-term toxicity. Additional and/or longer duration toxicological studies performed concurrently on animals may be necessary to provide further support of safety. For example, reproduction studies would now be carried out for drugs to be used in women of child bearing potential. At the end of Phase II the sponsor should have sufficient information to determine whether the drug should be developed to the NDA stage. An end-of-Phase II conference may be held between the FDA and the sponsor to discuss the results to date and to identify the need for further studies. Phase III studies, involving more extensive clinical trials, are undertaken if the information obtained in the preceding phases demonstrates reasonable assurance of effectiveness for the intended use and if the drug has potential benefits outweighing the potential risks discovered up to that point. These studies are intended to assess the drug's safety on the basis of exposure to a reasonably large population group, to appraise effectiveness for the intended use in broader patient populations than those studies in Phase II, and to obtain the additional information needed to write a comprehensive package insert, i.e., to provide adequate directions for use. Phase III studies may include both controlled and open studies and may more closely approximate the use of the drug in medical practice than to Phase II studies.

The NDA is prepared following completion of the bulk of the Phase III studies. The clinical safety section of the NDA describes the extent of exposure to the drug, contains listings of adverse drug reactions with their incidence and comparison to that found in the placebo or control group, a summary of clinical laboratory data highlighting abnormal values, any evidence concerning the consequences of acute overdosage, and a description of and reason for the drop-out of patients.

The need for formal Phase IV safety studies are assessed prior to NDA approval, and are usually conducted to evaluate a residual safety concern. Observational post-marketing surveillance systems serve to monitor for drug safety in the general medical practice environment.

PART II: AN ANTIVIRAL DRUG RISK ANALYSIS ILLUSTRATION - ACYCLOVIR

Each new class of therapeutic drug products present the research community with a set of special challenges and concerns. In this section of the paper we plan to use the experiences surrounding the research and review of acyclovir[4] will illustrate the developmental process related to the risk analysis of nucleoside analogue antivirals.

Acyclovir was not the first drug in this class. However, it is especially interesting as an illustration because it was developed as a variety of dosage forms for a variety of therapeutic indications. Therefore, there were many product formulation (i.e. dosage form) specific safety concerns addressed and its relative safety was considered in concert with the severity and/or chronicity of a variety of underlying disease states.

It is important to clarify that in the U.S. drug approval process specific drug product formulations are approved for specific indications. Therefore, as we discuss acyclovir we are really discussing investigations leading to multiple NDAs.

To begin, we draw a distinction between the evaluation and the testing of a new drug. A proper initial evaluation will aid in directing the preclinical testing efforts. An initial evaluation should take into account the known structural, biochemical, and pharmacological properties as well as all information, experimental or epidemiological, which can be obtained from similar or related compounds. Any special testing requirements should be guided by what was learned.

The development of a comprehensive toxicological profile for antivirals was built upon that obtained from the few antiviral agents which preceded acyclovir. The first nucleoside analogues were used as anti-cancer agents (e.g., cytarabine) and the initial toxicological database was developed in that context. Then in 1976, the purine nucleoside, vidarabine was approved as an ophthalmic ointment for the treatment of acute keratoconjunctivitis and recurrent epithelial keratitis due to herpes simplex virus, types 1 and 2, and approved in 1978 as a parenteral suspension for I.V. infusion in the treatment of the life-threatening disease, herpes simplex encephalitis.

The true antivirals that are nucleoside analogues interact with the host cell genome and therefore it is assumed that they are cytotoxic and have mutagenic and carcinogenic potential. Most of these agents are also teratogenic and clastogenic (or capable of producing chromosome breaks).

Therefore, an in vitro mutagenicity study, the AMES plate test, was performed for this drug class. Unfortunately, this test appeared to be insensitive; all drugs in this class tested negative except for trifluridine.

In 1976, FDA requested in vitro tests using mammalian cells such as the mouse lymphoma assay and/or the Chinese hamster ovary cell assay to test for mutagenicity. The mouse lymphoma assay specifically evaluates a chemical's potential to induce forward mutations at the thymidine kinase (TK) locus. Also, in vitro cell transformation studies and in vivo tumor induction studies in syngeneic hosts were performed at that time in an attempt to obtain preliminary information on the carcinogenic potential of these drugs. Vidarabine tested positive in most of the tests in this new battery.

Clearly, the confirmation of theoretical concerns with these results called for a cautious and conservative approach toward any subsequent nucleoside analogue antiviral. In the spring of 1977, a working group of FDA scientists assembled to draft a set of proposed guidelines for the preclinical toxicology testing of antivirals. The scientific assessment set forward in these proposed guidelines is continuously updated and revised to retain the state-of-the-art relevance. In addition to the standard set of tests as described earlier, and the special mutagenicity and carcinogenicity screening battery just described, elucidation of the exact mechanism of action was also included in this assessment. Therefore, prior to the initiation of a Phase I trial with these drugs it is important to possess an understanding of the items listed above. While we request that the sponsor determine and specify the drug's exact mechanism of action, Phase I trials are not held up awaiting these results.

In late 1977 FDA reviewers began to address a company sponsored IND submitted for acyclovir ophthalmic ointment. In the preceding discussions with the sponsor, the test battery developed for vidarabine was reviewed for applicability in this case. The proposed Phase I clinical trial would include patients with herpes simplex virus infections of the eye who also had a documented history of herpes simplex keratitis.

Safety concerns were able to be focused appropriately because the mechanism of action of acyclovir was known. Acyclovir is a synthetic acyclic purine nucleoside analogue with in vitro inhibitory activity against herpes simplex virus, types 1 and 2, varicella-zoster virus, Epstein-Barr virus and cytomegalovirus. In cell cultures, the inhibitory activity of acyclovir for herpes simplex virus is highly selective. Cellular thymidine kinase does not effectively utilize acyclovir as a substrate. Herpes simplex virus-coded thymidine kinase, however, converts acyclovir into acyclovir monophosphate, a nucleotide analogue. The monophosphate is further converted into diphosphate by cellular guanylate kinase and into triphosphate by a number of cellular enzymes. Acyclovir triphosphate interferes with herpes simplex virus DNA polymerase and inhibits viral DNA replication. Acyclovir triphosphate also inhibits cellular alpha-DNA polymerase but to a lesser degree. In vitro, acyclovir triphosphate can be incorporated into growing chains of DNA by viral DNA polymerase and to a much smaller extent by cellular alpha-DNA polymerase. When incorporation occurs, the DNA chain is terminated. Acyclovir is preferentially taken up and selectively converted to the active triphosphate form by herpes virus-infected cells.

Thus, acyclovir is much less toxic in vitro for normal uninfected cells becase: 1) less is taken up; 2) less is converted to the active form; 3) cellular alpha-DNA polymerase is less sensitive to the effects of the active form.

In mid-1978, while the ophthalmic product was still in its early stages of review, a second IND was submitted for an intravenous preparation of acyclovir which was initially intended for use in immunocompromised patients at risk of developing potential life-threatening infections caused by herpes simplex virus and varicella-zoster virus. Therefore there were two different formulations for two very different indications of quite differing severity and prognoses.

Clinical safety concerns with the ophthalmic product included the production of drug resistant mutants, the extent of absorption and local irratation. The I.V. formulation raised the additional concerns of general systemic toxicity, nephrotoxicity secondary to renal excretion of unchanged drug by glomerular filtration and tubular secretion (the major route of acyclovir elimination), central nervous system toxicity, cytotoxicity to cells that turn-over rapidly such as bone marrow and sperm cells, as well as reproductive toxicity. Therefore, preclinical testing was conducted to assess these concerns in the appropriate in vitro and in vivo animal models. The formulation specific tests for the ophthalmic product included a 21 day eye irritation study in rabbits, a seven day acute parenteral toxicity study to examine the drug effect on the visual system of rats, ocular penetration studies in rabbits and subcutaneous teratology studies in rats.

The proposed use of high systemic doses in an intravenous formulation increased the importance of an extensive addition to the already lengthy list of preclinical tests. These included formulation specific ADME and pharmacokinetic studies including measurement of excretion rates and creatinine clearance; in vivo immunosuppressant screening studies, intraperitoneal dominant lethal mutagenicity study (performed on male rodents to detect chromosomal mutations produced in germinal cells which would present as fetal wastage in first generation embryos); clastogenesis assays in the form of both in vitro chromosomal aberration studies in human lymphocytes and in vivo bone marrow chromosomal aberration studies in rodents; an in vitro cell transformation assay for the rapid assessment of the potential to induce neoplastic transformations in mouse fibroblasts (if positive results are found, the transformed cells are transplanted into syngeneic animals for tumor induction); two year carcinogenicity studies in mice and rats; I.V. teratology studies in rabbits; and the full three segment reproduction studies in rats.

The first IND for the topical formulation was received in October 1978. The proposed Phase I clinical trials were in patients with mucocutaneous herpes simplex and localized varicella zoster virus infections in both non-immunocompromised and immunocompromised patients. The eventual indication would also include initial and recurrent genital herpes simplex virus, type 2 infections. The clinical concerns with this dosage form included local irritation, cutaneous absorption and the development of drug resistant mutants. Additional toxicological tests performed for the topical formulation included: a 21 day dermal irritation study on the intact and abraded skin of guinea pigs; evaluation of drug absorption and plasma levels in guinea pigs; dermal sensitization studies; and, an evaluation of dermal penetration through isolated human skin or mucosa.

At this point in time many, if not most, of the toxicological tests performed for the earlier dosage forms of acyclovir had been completed. The drug was qualitatively positive in some of the tests for mutagenicity, clastogenicity and cell transformation; the carcinogenicity screening was negative. A decision was made to permit the early clinical trials for the ophthalmic and topical preparations. Trials using the I.V. preparation in the life-threatening infections were also begun. However, these were conducted with a slow I.V. infusion in well hydrated patients, as crystallization of drug in the renal tubules was noted from preclinical studies. ADME studies with radio-labelled drug and confirmatory post-mortem examination also revealed substantial distribution to the central nervous system.

In early 1980, an IND was submitted for the oral capsule product. The porposed Phase I trials were in immunocompromised patients with herpes simplex virus, types 1 or 2 or varicella zoster virus infections, using a dosage regimen of 200 mg given every four hours for five days. Early bioavailability and kinetic tests on this dosage form raised some concern. Only 15-20 % of the oral dose was absorbed. Additionally, a flat dose-response curve was seen because of an apparent plateau in absorption at around a single dose of 400 mg. Clinical concern was related to the unknown effect of the non-absorbed dose on the rapidly turning over epithelial cells of the intestinal tract. Furthermore, the concern over gonadal toxicity arose again. Toxicity had been seen with another nucleoside analogue under review by the FDA. Therefore, this aspect was reappraised and explored further.

Chronic 12 month oral toxicity studies in dogs; oral LD_{50} studies in two species; oral ADME and pharmacokinetics studies; a two year carcinogenicity study in two species (mice and rats); and a two generation oral fertility study in mice were performed as formulation specific tests. These studies gained an increased level of importance when the IND for the oral capsule was amended to include studies for both initial and recurrent episodes of genital herpes in immunocompetent persons. Suppression of frequent episodes of genital herpes for up to six months was a subsequently added regimen.

The use of acyclovir for chronic suppression of episodes, though highly effective, is not without concern and controversy. The areas of clinical concern are: a theoretically increased probability for the production of drug resistant strains, gonadal and reproductive toxicity, excessive drug accumulation, and post-therapy rebound. Therefore, a series of long-term safety studies were agreed upon by FDA and the sponsor some for immediate conduct and some for the Phase IV or post-approval period. They include both clinical and animal toxicology studies. In general, studies conducted in Phase IV. They are designed to explore a residual concern noted pre-approval.

Parallel to the review of the oral product was a further amendment to the IND for the I.V. formuation to include use in the non-life threatening condition of severe initial clinical episodes of herpes genitalis in patients who are not immunocompromised. The expanded indications for the oral capsule and the I.V. infusion followed the general recognition that based on current knowledge acyclovir probably does not produce major toxicity to the genome, that is, based upon clinical experience the drug did not appear to be causing readily detectable effects on the host.

To recap the sequence of acyclovir products as they were submitted for US FDA review; initially, there were the ophthalmic, topical for external use and the I.V. for a life-threatening condition. Then, when greater confidence in the safety profile was attained, the drug was tested for use systemically for the non-life threatening conditions, and for longer durations.

What was learned about the safety of acyclovir from this substantial battery of preclinical tests and subsequent clinical trials ? Overall, acyclovir appears to have a relatively good safety profile. It has a remarkably specific antiviral effect, and it is that high degree of selectivity which may explain the lack of major effect to the host cells as experienced thus far. The most significant feature appears to be that cellular thymidine kinase does not effectively utilize acyclovir as a substrate. Therefore, most of the theoretical safety concerns which could be related to the nucleoside analogues were not actualized with acyclovir. However, we must

be reminded that safety is not an absolute term. While the overall interpretation of the major toxicological issues was favorable to acyclovir,
there were positive mutagenicty and gonadal toxicity findings in certain
models at certain usually very high doses. Additionally, given the inexpe-
rience that the scientific community has with the interpretation of these
newer techniques their relevance to human toxicity is still uncertain. The
professional labeling (i.e. the package insert) for the products marketed
in the U.S. clearly outline the fact that there are still unanswered questions in this regard.

Regarding the production of drug resistant mutants; to date all of the
less sensitive viruses obtained from patients have been the result of alterations in the viral thymidine kinase and were not the result of changes in
the viral DNA polymerase.

The major systemic toxicities verified during the Phase III clinical
trials were associated with I.V. administration. Abnormal renal function
(decreased creatinine clearance) occurred in 4.6 % of patients even when
administered by slow infusion in well hydrated patients. If administered by
bolus injection the rate exceeds 10 % and precipitation of acyclovir crystals in renal tubules can occur. Approximately 1 % of patients receiving
I.V. acyclovir have manifested encephalopathic changes characterized by
either lethargy, tremors, confusion, hallucinations, agitation, seizures or
coma.

PART III: CURRENT STATE-OF-THE-ART AND FUTURE NEEDS

Just as the initial evaluation of a new drug guides its preclinical
testing course, the preclinical results guide the conduct of the clinial
trials, and so too, the cumulative experience should guide any Phase IV
studies and serve to focus the post-marketing surveillance efforts. Safety
assessment is a continual process.

The dominant role assumed by the preclinical testing in the acyclovir
illustration highlights the need for constant communication and coordina-
tion between the pharmacologists and the medical officers both within the
FDA and within the companies.

During the extensive review of the acyclovir products there was some
trial and error, serious discussions over what tests should be conducted,
some disagreement over the interpretation of their results, and as a result
much was learned and much was refined. With acyclovir, the theoretical safety concerns and their actuality were distinguished with a firm degree of
confidence as a result of the safety testing process. In the new drug product approval process, empirical data are key. NDAs are approved on the
basis of facts, and not theory. That holds for efficacy as well as for safety.

Clearly, new classes of drugs, including those products derived from
biotechnology, must satisfy all the usual expectations of the regulatory
process, plus some additional requirements generated by their special nature. While the US FDA does not consider it prudent to issue guidelines in
areas of rapidly advancing technology, the need for general guidance
exists. We have met this general need by constructing documents entitled
"Points to Consider". An example is the "Points to Consider" document for
the safety evaluation of antiviral drugs for non-life threatening diseases
prior to Phase I clinical studies. At present, a case-by-case approach is
the most prudent mechanism for interaction between our regulatory agency

and antiviral drug sponsors. Early and continuous communication leads to an efficient development and review process.

With that principle in mind, Table 2 contains, in an ordered manner, the basic categories of toxicological tests used for new antiviral nucleoside analogues. Recall that the length and type of tests within most of these categories are dependent upon the intended route and duration of administration, and the severity of the disorder to be treated. This table repeats the standard battery of tests as outlined in the first part of this paper, then presents the additional studies used to assess safety for the antiviral drugs.

Table 2. Outline for the Preclinical Toxicity and Pharmacogenetic Evaluation of Antiviral Drugs

Standard Test Categories
- Acute toxicity
- Multidose toxicity
- Carcinogenicity
- Formulation specific
- Drug disposition (A.D.M.E.)
- Reproduction studies (3 segments)

Additional test categories
- Mechanism of action
- Mutagenicity assay series
- Clastogenesis assays
- Cell transformation assays
- Gonadal toxicity tests
- Transplacental carcinogenicty assay
- Immunosuppressant screening tests

A full battery of tests, such as the ones outlined above, need to be performed for each new antiviral drug product before substantial exposure in humans in order to minimize the potential for harm. The time of the conduct of these studies may vary with the phase of the clinical time depending on the severity of the underlying disease process. Based upon our current knowledge, acyclovir appears to be relatively safe. However, it's profile does not assure that the theoretical risks associated with the nucleoside analogues will not be actualized with subsequent compounds which may, for example, have a different level of selectivity.

In conclusion, the antiviral drugs characterized by the nucleoside analogues represent a relatively new and increasingly important class of therapeutic agents. Recent breakthroughs in knowledge about viruses, especially the retroviruses, will without doubt open the door to a new round of antiviral drug development. The challenge to the scientists at the US FDA is to remain at the cutting edge of this new knowledge, and to maintain their rigor in new drug safety assessment and at the same time to maintain reasonable flexibility in an evolving field.

ACKNOWLEDGMENTS

Grateful acknowledgments to Jose Canchola, M.D., Vera Glocklin, Ph.D., Edward Tabor, M.D., Frank Vocci, Jr., Ph.D. and John Davitt for their review and comments and to Margaret Cunningham for her clerical support.

REFERENCES

1. V. Glockin, Toxicology in Drug Development, Presented at the American Chemical Society Symposium on Toxicology and Risk Assessment (1982).
2. New Drug, Antibiotic, and Biologic, Drug Product Regulations, Federal Register, Vol. 52, No. 53, pp. 8798-8847, March 19, 1987.
3. E.I. Goldenthal, Current Views on Safety Evaluation of Drugs, FDA Papers, May 1968 (GPO: 1970-396-012175).
4. Zovirax[R], Burroughs Wellcome Co., Research Triangle Park, North Carolina, United States.

ANTIVIRAL AGENTS: A NEW BEGINNING

George J. Galasso

National Institutes of Health

Bethesda, Maryland 20892

INTRODUCTION

Review of the statements I made at the last meeting four years ago
showed that the major change that has occurred since then is not so much
one of increased knowledge on antiviral drugs, but more importantly, one
of attitude. The prevailing skepticism of the past has given way to
optimism and determination. This determination is partially due to
recent successes in the field, but more likely results from the acquired
immunodeficiency syndrome (AIDS) epidemic.

When this group met, we still had not realized the severity of
AIDS. The causative agent was just being identified as a virus. Since
then, an impressive amount of knowledge has been gained about the
disease, the virus and antiviral agents. It is most unfortunate that a
disease such as AIDS proved to be the needed stimulus to advance
antiviral research. During my remarks four years ago, I expressed the
need for the formation of multitalented groups for targeted antiviral
research (1). My words were meant to serve as a catalyst for the
establishment of such groups, but I did not have high hopes that this
would occur. Again, little did we know that AIDS would change that
equation.

The urgency of the AIDS epidemic was the much needed focal point to
bring together government, academic, and private resources for targeted
antiviral research. In the United States, the National Institutes of
Health (NIH) has become the logical institution to coordinate research
in AIDS.

Every approach to understanding and conquering this disease is
being undertaken and a major focus is the development of antiviral
agents. This includes not only the current methods of screening and
modifying compounds, but also the targeted approach. Molecular biology
has brought us many new techniques for studying the structure and
function of viruses and nucleic acids. We know more about the causative
agent of AIDS, the human immunodeficiency virus (HIV), than probably any
other virus or cell. We should be able to use this information to
tailor antiviral agents to HIV. This effort requires a
multidisciplinary approach, including scientists with diverse scientific

backgrounds such as virology, biochemistry, biophysics, structural chemistry and biology, organic chemistry, crystallography, computer analysis and pharmacology. The goal would be to identify essential components of the HIV which are specific for the virus and develop methods of blocking them. An example would be to isolate a particular protein and produce it in sufficient quantities to crystallize it for x-ray diffraction followed by computerized three dimensional molecular mapping. A drug then could be designed to block active sites on the protein. This is an approach that is not out of our grasp; similar work has been done with rhinovirus (2). NIH, with its broad range of talented scientists, is an excellent location to initiate such an activity, and it is indeed under way. Intramural scientists have been invited to submit proposals for a broad range of projects related to targeted antivirals. These will be coordinated with the hope of achieving the goal of controlling AIDS. It is understood that it will not be achieved rapidly, if at all, but the information gained in this approach will be invaluable to antiviral research in general. Funds have also been made available to support similar collaborative efforts in the extramural programs. Investigators throughout the United States also have been invited to submit proposals for creative research projects directed at targeted antivirals. They have been encouraged to assemble research teams involving several institutions. This collaboration of industry, nonprofit organizations, academia and government may yield a much needed antiviral agent for AIDS and lead the way for the development of antiviral agents for other diseases.

The intent of this final chapter is to address the state of the art of antiviral agents, present some of the problems, and address the future. The preceding chapters have discussed in great detail some of the specific compounds and diseases. This next portion of the chapter will give a general overview of the whole field.

HERPES INFECTIONS

Except for amantadine, rimantadine, and phosphonoformate, nearly all antiviral drugs approved by the U.S. Food and Drug Administration (FDA) or approved as investigational drugs, are nucleoside analogues for the treatment of herpesvirus infections or AIDS. In the past, drug development concentrated on the herpesvirus because a herpesvirus specific enzyme, thymidine kinase, was able to phosphorylate substrate analogues such as acyclovir (ACV), vidarabine, and bromovinyldeoxyuridine (BVDU) leading to an inhibition of DNA synthesis. In the case of ACV, the virus specific thymidine kinase phosphorylates ACV into acyclovir monophosphate. Cellular enzymes then phosphorylate acyclovir monophosphate into its activated triphosphate form which acts as a DNA chain terminator, thereby inhibiting viral replication. Unlike its parent, the nucleoside guanosine, ACV does not have a 3' hydroxyl group to which the next nucleoside is attached.

Herpes Keratitis

The eye continues to serve as an excellent model for antiviral studies because it is an isolated organ. Just as the first approved antiviral drug, idoxuridine (IDU) was developed for an eye infection, new antiviral approaches using drug combinations are being pursued in the eye model. The advantage is that drugs which would be toxic systemically, may be tolerated in the eye (e.g. IDU). Successful combination drug therapies have been designed for herpes keratitis using interferon and either ACV, vidarabine, or trifluorothymidine.

Interferon alone is inactive, but the synergistic effect of its combination with one of the other nucleoslides results in the more rapid healing of lesions than any drug alone.

Varicella-zoster

In the healthy individual, varicella-zoster virus infections are usually self-limiting. However, in the immunocompromised patient, the resulting diseases, either chicken pox or shingles, can be very serious and sometimes fatal. Recent clinical trials comparing ACV to vidarabine in the treatment of varicella-zoster in immunocompromised patients reported the prevention of cutaneous dissemination in all ten patients on ACV in contrast to five of the ten patients on vidarabine. Acyclovir also compared more favorably in other parameters such as defervescence, new lesion formation, time to decrease in pain, crusting and healing (3). Another study in Paris using immunocompromised patients with disseminated varicella-zoster found no significant difference in the efficacy of vidarabine and ACV (4). However, this may be due to the lower dose of ACV used in this study compared to others. Ease of administration, favorable results from ACV, and occasional neurological toxicity observed when using vidarabine, especially in children, (5) make acyclovir the drug of choice. Further controlled studies are in progress in the United States.

2'-Fluoro-5-iodarabinosylcytosine (FIAC) was also compared with vidarabine in immunosuppressed patients and reported to be more effective in reducing pain and time to cessation of new lesions (6). Other FIAC analogs (FIAU and FMAU) and BVDU are also being investigated for the treatment of this disease, but not in the United States. Although in vitro and animal studies have demonstrated antiherpetic properties, 2'-halogenated pyrimidine analogues have been found to be highly toxic in preliminary clinical trials.

Herpes Encephalitis

Herpes encephalitis is a rare but fatal infection with a mortality rate of 70% and serious neurological impairment in many who survive. In past studies, vidarabine was found to be effective in reducing the mortality rate by half. However, with the advent of ACV, comparison studies have shown ACV to be the preferred drug. In a study comparing vidarabine with ACV, vidarabine decreased the mortality rate to 54%, a rate much higher than earlier studies because of the inclusion of a larger number of elderly patients with a lower level of consciousness. On the other hand, ACV decreased the mortality rate to 28%. During a six month follow-up, 14% of patients on vidarabine were functioning normally compared with 38% of patients on ACV (7).

Early diagnosis and treatment of herpes simplex encephalitis is essential in reducing mortality and preventing debilitating sequelae. Since herpes simplex encephalitis is only one of many causes of encephalitis, an accurate rapid, noninvasive method of diagnosis is needed, but currently not available. Brain biopsy remains the best diagnostic tool. Therefore, in the light of the low toxicity of ACV and the imperative for early treatment, ACV is oftentimes administered before confirmed diagnosis of herpes simplex encephalitis. However, it must be emphasized that herpes encephalitis cannot be diagnosed clinically and viral etiology should be confirmed.

Neonatal Herpes

Like herpes simplex encephalitis, neonatal herpes is a rare but potentially fatal infection. Approximately one third of affected infants die and one third become severely retarded. In a placebo control study, mortality decreased from 75% with placebo to 38% with vidarabine. The prognosis is more optimistic in infants with infections localized to either the central nervous system or to the skin, eyes, or mouth rather than disseminated to multiple organs (8). Since ACV proved to be more effective in the treatment of herpes simplex encephalitis, clinical trials comparing the efficacy of vidarabine and ACV were recently completed (9). None of the infants with infections localized to the skin, eyes, and mouth died after one year on either treatment as in the previous study. Ninety-three percent of those on ACV and 85% of those on vidarabine were developing normally at one year. The mortality rate for neonates with central nervous system involvement was 8% using ACV and 13% using vidarabine with 34% of those on ACV and 37% of those on vidarabine developing normally after one year. In neonates with disseminated infection of the organs, the mortality rate was 65% using ACV and 50% using vidarabine with only 29% on ACV and 23% on vidarabine developing normally at one year. The results, therefore, indicate that both drugs are equally effective.

Oral-Genital Herpes

Until the advent of AIDS, the increasing incidence of oral and genital herpes was the principal reason for a change toward a less promiscuous life style. Herpes simplex infection is a serious problem causing considerable morbidity in the normal population and more serious effects in the immunocompromised patient. Recently, these infections have increased in importance because of their frequency as opportunistic infections among AIDS patients. Prior to identification of HIV, herpes simplex virus was the virus which lent itself most readily to antiviral therapy. However, a cure still has not been found. Oral treatment of genital herpes with ACV for initial episodes is effective in decreasing the number of new lesions, the duration of lesion formation, viral shedding, and pain. In recurrent episodes, ACV shortened the period of viral shedding and lesion formation, but no significant symptomatic relief such as decrease in pain and itching was reported. Acyclovir was effective in suppressing recurrence during a 120 or 125 day prophylactic trial using 200 mg ACV 2-5 times daily. Sixty-five to ninety percent of the patients were free of recurrence during treatment (10,11). A small double-blind, placebo-controlled study to determine a once daily dosage of ACV to suppress recurrence is currently in progress. Preliminary results suggest that 800 mg/day is adequate for suppression (Personal communication, Dr. S. R. Mostow). However, ACV neither prevented recurrence nor changed the natural history of recurrence once treatment had terminated. Latent virus infections were not eradicated.

Because of the expense of long term suppressive therapy and the potential for toxicity and drug resistance, the optimum frequency of ACV administration for suppression in the normal patient is now being determined. In a study to determine an optimum intermittent therapy, weekend treatment was found to be ineffective in suppressing recurrence. In a double-blind study using 200 mg oral ACV three times daily or 400 mg three times daily on Saturdays and Sundays only, significantly more individuals using weekend therapy developed recurrence (13/17 weekend therapy vs. 3/18 daily therapy). For the weekend group, recurrence was more frequent as the week progressed, suggesting some residual suppressive effect from the high dose on the weekend (12). More studies need to be done to determine an optimum intermittent treatment in normal patients.

In 1985, the Food and Drug Administration approved the prophylactic continuous use of oral ACV for up to six months in the treatment of genital herpes. A large clinical study to determine the toxicity of suppressive ACV therapy using high levels of ACV over a long period of time was recently reported. One thousand patients using 400 mg ACV three times a day for one year demonstrated no significant clinical or laboratory toxicity (13). Although suppressive ACV treatment significantly prevents recurrence, it does not promise freedom from transmission. Herpes simplex virus still can be transmitted during breakthrough and asymptomatic excretion of virus.

In patients undergoing bone marrow transplants, a once daily intravenous dose of ACV did not suppress recurrence indicating a need to maintain a constant serum level of ACV in this population. When compared with placebo, the once daily regimen demonstrated a longer, although not significant, time to onset of lesion. Nine of thirteen patients on placebo developed lesions during therapy as compared to four of fourteen patients taking ACV (14). Long-term treatment should be given with caution since drug-resistant viruses were isolated from patients receiving ACV.

Clinical trials using foscarnet (phosphonoformate) cream and arildone cream reported some clinical benefit (15,16). Patients on foscarnet appear to develop less new lesions with a higher proportion reporting no symptoms after one day of treatment than those on placebo. However, healing time did not improve significantly. Similarly, arildone significantly reduced the localized itching in men and the duration of viral shedding in women, but duration of pain and healing time were not significantly reduced. Studies are now being conducted using topical InterVir-A, an ointment containing two surfactants, p-diisobutylphenoxy polyethoxyethanol and polyoxyethylene-10-oleyl ether, and alpha interferon.

Cytomegalovirus Infection

Early studies demonstrated an in vitro effect of 9-[2-hydroxy-1-(hydroxymethyl)ethoxymethyl] guanine (DHPG, ganciclovir) against cytomegalovirus (CMV) with subsequent encouraging clinical studies. As with herpes simplex, CMV, which is always a problem in immunocompromised patients, became an even greater infectious disease problem because of its role in AIDS, causing serious opportunistic infections, particularly retinitis. Several studies were undertaken with beneficial results against a variety of CMV infections in AIDS patients (17,18). Results indicated that a therapeutic response to DHPG was dose-related. Dosage of 7.5 mg/kg/day demonstrated the best results in AIDS patients although neutropenia occurred in 27% of these patients. Nearly all patients with eye, lung, gastrointestinal or multiple organ involvement improved with 7.5 mg/kg/day of DHPG. Earlier studies using lower dosages (2.5 mg/kg/day and 5 mg/kg/day) demonstrated promising results, especially with retinitis, but the improvement was not as great as that seen using 7.5 mg/kg/day. Pneumonitis, the most fatal of CMV infections, appears to be the most resistant to DHPG therapy. More vigilant post-surgical monitoring for CMV in immunocompromised organ transplant patients to identify the need for therapy before pulmonary damages are irreversible, may improve the outcome for pneumonitis. Clinical and viral relapses occur frequently, sometimes within thirty days after cessation of treatment. Because of this recrudescence and the neutropenia that occurs at the high but successful dosage regimen, long-term clinical trials are needed to evaluate suppressive therapy. Development must begin on a more effective, less toxic regimen, perhaps a drug combination, for prophylactic use.

Preliminary results indicate foscarnet is potentially beneficial in treating CMV infections in allograft recipients (19). The results of this study are difficult to evaluate because the patients are severely ill and heavily medicated. Twelve of twenty-five patients died. Seventy percent showed some improvement at some time during treatment in one or more of the parameters such as eradication of CMV infection, resolution of fever, or improved laboratory values. Similar post-treatment recrudescense seen with DHPG treatment was observed with foscarnet. Other studies appear to support these results (20), however these preliminary investigations indicate that further carefully controlled trials may be warranted. The limited effects seen to date and the associated toxicity are not encouraging. It appears that CMV cannot be eradicated by foscarnet or DHPG, but they may suppress viral formation until the body's immune system can control the infection. In a small study, response from CMV-specific hyperimmune globulin therapy was reported beneficial in five of eight bone marrow transplant recipients with CMV pneumonitis (21). It would be of interest to examine the effects of treatment combining foscarnet and hyperimmune globulin therapy in a controlled study to determine if a more optimal clinical effect could be achieved.

Laboratory studies have found that difluoromethylornithine, a polyamine synthesis inhibitor, enhanced the virustatic effects of DHPG when used in combination to inhibit CMV infections (22). This synergistic combination shows promise for clinical use to reduce the neutropenia observed with high doses of DHPG. Several pyrrolopyrimidine compounds with interesting anti-CMV properties are also being studied (23).

Past attempts to treat CMV infections with ACV, vidarabine, interferon, and vidarabine with interferon were not effective. Interferon has had some beneficial effect when used prophylactically to prevent infections in the seropositive renal transplant recipient.

Epstein-Barr Virus Infection

The role of ACV in treating infectious mononucleosis is not clear. Thirty-one adults treated with intravenous ACV were shown to have a significant decrease in oropharyngeal viral shedding of the Epstein-Barr virus during treatment. This effect, however, was transient. Improvements in the clinical symptoms such as duration of fever, weight loss, tonsillar swelling, and pharyngitis were not significant unless multiple parameters were combined (24).

RESPIRATORY INFECTIONS

Common Cold

In the general population, the common cold is the most frequent acute infection causing significant economic loss due to absenteeism from work. The search for a cure for the common cold is faced with two unique problems. First, the common cold is an umbrella term for infections caused by a wide variety of viruses and strains. Second, the symptoms of a cold are usually mild and short-lived. Therefore, the antiviral drug must be broad-spectrum, free of toxicity, and easily administered at the earliest onset of symptoms. Ideally, the cure for the common cold should' be an over-the-counter product.

Rhinovirus causes about 30-50% of all acute respiratory illness. Recently, the possibility for treatment of colds caused by the rhinovirus is brightening with new information on the molecular structure of the virus and its specific cell binding sites. Although the rhinovirus has over 100 serotypes, there is now evidence that the virus attaches to a single cell receptor. Some of the isoxazole derivatives designed from information from X-ray crystallography may produce an effective, nontoxic antiviral drug (2). Rhinovirus receptor monoclonal antibodies are also being evaluated.

The first clinical study using interferon against rhinovirus was conducted in 1973. Many studies have since demonstrated intranasal interferon's prophylactic efficacy against the common cold. However, interferon administered for longer than seven days causes nasal mucosal bleeding and other nasal involvement. Often, the side effects also include symptoms mimicking a cold. A new strategy for administration was then designed where family members self-administered interferon nasal spray once a day for seven days as soon as a family index case was identified. These studies have shown that interferon is 80% effective in preventing laboratory documented rhinovirus infections and 40% effective in preventing general respiratory illness (25,26). The shorter time of interferon administration needed under these circumstances resulted in fewer, more tolerable side effects. Although the treatment has drawbacks such as limited efficacy against ongoing rhinovirus infections, short-term prophylactic use in family settings may have considerable benefit for individuals susceptible to cold-related complications such as otitis media, sinusitis, asthma, or bronchitis, or in a family situation where a cold may ruin an upcoming trip or major social event.

Colds caused by rhinovirus appear to be spread principally by aerosol transmission either through coughing, sneezing, or noseblowing. However, the transmission rate is surprisingly low, infecting only 50% of the individuals in a family setting (27). An interesting series of experiments was performed by Elliot Dick and his associates to control the aerosol transmission of the common cold by using virucidal facial tissue which had been treated with 9.1% citric acid, 4.5% malic acid, and 1.8% sodium lauryl sulfate (28). Citric acid was the major virucidal agent with malic acid added for synergism. Sodium lauryl sulfate was used as a surfactant to provide added activity against envelope viruses such as paramyxovirus. In a confined room, healthy male volunteers played poker for 12 hours with volunteers experimentally infected with rhinovirus type 16. A remarkable reduction in the transmission of a cold was reported when volunteers using virucidal facial tissues were compared with those using cotton handkerchiefs. Twenty-four controls used cotton handkerchiefs and were instructed to use them whenever they desired. Of the twenty-four, fourteen controls became infected. Whereas none of the 24 individuals using virucidal facial tissues became infected. This group, however, appeared to have more precise directions on the generous and careful use of virucidal tissues to catch all nasal effluent and smother coughs and sneezes. Although the results are impressive, the same instructions encouraging generous use and careful containment should have been conveyed to both groups for more conclusive results.

In another controlled clinical trial, zinc gluconate lozenges shortened the duration of a cold by seven days (29). These results are impressive but they may be unintentionally biased since the very unpalatable nature of the zinc lozenges may have encouraged patients to report recovery in order to terminate their use. A more palatable zinc lozenge would need to be developed to be of any use. However, a recent study appears to support the debatable effects of zinc on rhinoviruses. When examined in vitro, zinc salts had little effect on rhinoviruses, indicating that the disagreeable taste may have influenced the clinical study (30).

Influenza

Amantadine and rimantadine are therapeutically and prophylactically effective and relatively nontoxic in the treatment of influenza A. Both antiviral agents are well absorbed as oral formulations which make administration easier for outpatient use.

Rimantadine, an amantadine analog, was developed because of concerns for the infrequent but potential central nervous system toxicity of amantadine. Side effects of amantadine, although less than those with antihistamines, include insomnia, anxiety, and difficulty in concentrating. Rimantadine has no central nervous side effects and its side effects were indistinguishable from placebo in controlled studies. A study comparing the prophylactic benefits of amantadine and rimantadine in the treatment of influenza A reported amantadine reduced laboratory confirmed influenza by 91% and rimantadine by 85% (31). With the reported efficacy of rimantadine and amantadine being essentially comparable, rimantadine is the drug of choice because of its relative lack of central nervous system toxicity.

As with other antivirals with proven efficacy, the current research on rimantadine centers on the optimal treatment regimen and comparative studies with other accepted methods of influenza treatment. In a double-blind, placebo control trial using a 200 mg dose of rimantadine daily for five days, rimantadine was found to be well tolerated with significant therapeutic effectiveness against uncomplicated influenza A when administered within 48 hours after onset of symptoms (32).

Acetaminophen, a common drug prescribed to alleviate the symptoms of a cold, was recently compared with rimantadine in a controlled study using 63 children, 49 of whom had confirmed influenza A. Both drugs were well tolerated in this pediatric population. Although this study did not demonstrate superiority of either regimen due to the rapid resolution of influenza in these children, the decline in viral shedding in the rimantadine group was impressive. This has beneficial implications for the prevention of virus spread (33).

Amantadine has been approved by FDA for influenza A. Rimantadine has been submitted for FDA approval. Only in the Soviet Union is rimantadine used extensively for influenza A. In individuals for whom influenza may cause severe complications such as persons with cardiovascular, pulmonary, or immunodeficiency diseases or in situations where it is advantageous to reduce transmission as in a health care facility, amantadine should be given prophylactically for the duration of the outbreak of influenza A. This could be replaced by rimantadine once it is approved.

No consistent data are available on the prophylactic and therapeutic effectiveness of oral ribavirin on influenza A. However,

beneficial effects such as a significantly more rapid improvement in
signs and symptoms, decreased viral shedding, and diminished height and
duration of fever during influenza A (34) and B (35) were seen in a
hospital setting using aerosol administration of ribavirin for 12-20
hours per day. The relative merit of this treatment is overshadowed by
the fact that rimantadine is less toxic, easily administered, and
therapeutically superior to ribavirin in the treatment of uncomplicated
influenza A. Ribavirin's role may be in the treatment of influenza B
which has no known successful therapy. The efficacy of the two drugs
against influenzal pneumonia is under study.

Respiratory Syncytial Virus Infection

     Ribavirin is a broad spectrum antiviral agent approved by the FDA
for the aerosol treatment of respiratory syncytial virus (RSV) in the
neonate. A number of respiratory complications such as pneumonia and
bronchiolitis due to RSV require hospitalization of infants every year.
In a double-blind, placebo control study, 33 infants infected with RSV
were treated for 12-20 hours per day with aerosolized ribavirin or water
for 3 to 6 days (36). The infants treated with aerosolized ribavirin
responded more rapidly and with a shorter course of illness compared
with placebo. By day four, the mean severity of illness score was
significantly lower in children on ribavirin. The degree of improvement
in rales, cough, and lethargy was greater in those on ribavirin. The
treated group also showed less viral shedding in the nasal wash and
improved arterial oxygen saturation. Aerosol treatment with ribavirin
has been shown to be beneficial not only in normal infant but also in
high risk infants who either are immunocompromised or have underlying
cardiopulmonary conditions (37). More research needs to be done on the
long-term effects of ribavirin on the lungs of the neonates. Thus far,
no toxicity was noticed with aerosolized treatment. Since ribavirin has
a tendency to precipitate and clog the aerosol machine, care must be
taken to check for drug depositions regularly to assure that a proper
dosage is being administered (38).

ARENAVIRUS AND OTHER SEVERE VIRAL INFECTIONS

     Lassa fever, a serious arenavirus infection, is one of the major
causes of death in West Africa. The Mastomys rodent has been identified
as the animal reservoir for the virus. The home, rather than the field
and bush areas, has been identified as the site of transmission.
Combined clinical symptoms of fever, sore throat, and vomiting
distinguish Lassa fever from other febrile illness (39). The antiviral
screen for ribavirin demonstrated a positive effect against
arenaviruses, therefore, a study was done in Sierra Leone comparing
ribavirin with convalescent plasma in the treatment of Lassa fever. The
investigators reported striking results when patients were treated with
ribavirin within six days from onset of fever (40). Two predictors of a
fatal outcome are the level of serum aspartate aminotransferase and
serum virus levels. Patients selected who had serum aspartate
aminotransferase levels of greater than, or equal to, 150 IU per liter
normally have mortality rates of 55%. The mortality rate fell to 5%
when patients were treated with ribavirin within six days after the
onset of fever. The rate increased to 26% if the treatment began after
seven days from fever onset. Patients selected who had viremia greater
than or equal to $10^{3.6}$ $TCID_{50}$ per ml usually had mortality rates of 76%.
These patients treated within six days from onset of fever had a
mortality rate of 9% compared to 47% if treated seven or more days after
the onset of fever. Oral ribavirin was effective although the results
were not as striking. Convalescent plasma did not significantly reduce

mortality or increase the benefit of ribavirin when given together. Ribavirin not only is more effective but also is stable at room temperature and easily administered. These are two features necessary for medication used in rural, tropical areas.

A successful clinical trial using convalescent plasma was reported for another arenavirus equally sensitive to ribavirin, Junin virus, which is the etiological agent for Argentinian hemorrhagic fever (41). Patients who were given convalescent plasma within seven days from the onset of illness had a mortality rate of 1% compared to 14% for the untreated patients. These results suggest the need for a follow-up study with ribavirin, as was done with Lassa fever.

In animal and in vitro studies, ribavirin has shown potential antiviral activity for Rift Valley fever and other hemorrhagic fevers such as Crimean-Congo, Hantaan virus, Machupo, and Pichinde. Clinical trials using ribavirin for Hantaan virus, the causative factor for Korean hemorrhagic fever, are now being conducted.

CHRONIC INFECTIONS

Papilloma virus infections

Warts are generally thought of as a cosmetic problem. However, some warts can be very painful, as in genital warts; life-threatening, as in obstructive laryngeal papilloma; or a precursor to cervical cancer. Genital warts also are a source of transmission from mother to child causing laryngeal papilloma in children. Thus far, there have been a number of small studies indicating interferon is effective in reducing genital warts. It can be clearly demonstrated that intralesional administration of alpha, lymphoblastoid, or beta interferon results in the resolution of warts in approximately 50% of the interferon recipients as opposed to less than half in patients given placebo. More definitive experiments remain to be done using parenteral administration.

There is one large, ongoing trial using two doses of lymphoblastoid interferon (1 or $3x10^6$ IU/m^2 for 6 weeks) in patients with genital lesions refractory to conventional treatment. Statistical analysis showed no difference in responses between the two doses. The study using 114 patients demonstrated that the median lesion measurements were reduced significantly from baseline in both groups. There were some indications that men had a less favorable response, approximately 50% of participants had a 50% or greater reduction in lesion area compared to 75% in women. An additional four weeks of therapy increased the therapeutic effect; 22 of 70 available patients were disease-free. Of these 22 complete responders, only one exhibited disease recurrence before month 6 of the study. Analysis of graded side effects and changes in laboratory parameters showed a significantly higher frequency of side effects at the higher dose. Five dose reductions and two withdrawals occurred in the high dose group only. Additional trials are under way to evaluate combinations of interferon alfa-1 with podophyllin, laser therapy, or retinoids to enhance disease regression (Personal communication, Dr. P.K. Weck).

The usual treatment for laryngeal papilloma is repeated excision. Preliminary experiments have demonstrated that treatment with interferon following surgical removal of the lesion is effective in retarding the recurrence of new lesions. There are currently two large studies under way which may give definitive results concerning the efficacy and

toxicity of interferon in laryngeal papilloma. In one study, supported
by the National Institute of Allergy and Infectious Diseases, patients
are randomized to receive either laser surgery plus leukocyte interferon
or laser surgery alone. The interferon recipients receive $2x10^6$ IU/m^2
three times a week for one year. The study participants are evaluated
for degree of airway obstruction, frequency and duration of remission,
frequency of required surgery, and physical and developmental growth. A
total of 123 patients were enrolled in this study which is now being
completed. The analysis is complicated by the variable and
unpredictable nature of the disease. Preliminary evidence indicates
that interferon may prove to have some short term benefit early in the
treatment period but no significant difference was seen between the
controls and those treated with interferon after one year (Personal
communication, Dr. C. Laughlin).

The second study sponsored by Burroughs Wellcome uses interferon
alfa-1 in a cross-over design study. Patients are randomized into two
groups: One group receives six months of surgical management plus
interferon $5x10^6$ IU/m^2 intramuscularly for 28 days, followed by three
times weekly for an additional five months, the second group receives
surgical management alone for six months. At the end of the six months
period, the two groups are crossed-over. Patients are examined every
two months for disease severity for the duration of the study.
Statistical analysis has demonstrated a significant difference in mean
disease scores for the two groups during the first six months. In this
study, the difference was greater following cross-over (p=.001) at all
elevations. The benefits of extended therapy are currently under study.
Many questions still remain unresolved. There are some indications that
although patients respond well early during treatment, the benefits are
not long lasting. Different patients respond at different rates and to
differing degrees. It is also possible that the various subtypes of the
virus respond with varying degrees to treatment. Interferon appears to
have a beneficial effect on this disease, but the extent of the effect
remains to be shown (Personal communication, Dr. P.K. Weck).

Chronic viral hepatitis

Chronic hepatitis B with or without hepatitis delta virus has been
associated with cirrhosis and hepatocellular carcinoma. It is a serious
problem in the Far East, particularly with maternal transmission. The
seroconversion of HBe antigen to HBe antibody in chronic hepatitis B is
usually preceded by an exacerbation of the disease causing hepatic
inflammation with a rise in serum aspartate transaminase. This is then
followed by a long period of remission. Since chronic, active hepatitis
B is a disease that is dependent on continued viral replication, it
seems ideal for interferon therapy. There have been a number of pilot
studies using all types of interferon, following the initial observation
by Greenberg that interferon appears to have a beneficial effect (42).
Dosages as high as $68x10^6$ IU/day were examined. In almost all studies,
interferon was associated with a drop in the DNA polymerase activity,
disappearance of HBe antigen and Dane particles, seroconversion to HBe
antibodies and in some cases to HBs antibodies. However, there were no
cures and all beneficial responses disappeared after therapy. Studies
with interferon in chronic hepatitis consistently have revealed
different responses in different populations. It appears necessary to
define prognostic factors in order to achieve the best chances of a
favorable response. In some populations beneficial effects with
interferon can be as high as 50-60%, whereas others such as Asian
children do less well on interferon alone.

Various attempts to improve the response seen with alpha interferon have been, or are, underway. These include extending the treatment period to six or twelve months, or combining it with other antivirals such as vidarabine, vidarabine monophosphate, acyclovir, deoxyacylovir or gamma interferon. An earlier study combining interferon and vidarabine monophosphate had to be discontinued due to poor toleration of the treatment and increased neuropathic and neuromuscular toxicity. However, combination therapy with interferon and deoxyacylovir has yielded encouraging preliminary results and is currently under study in The Netherlands (43). Alpha and gamma interferon combination is under study at the NIH with preliminary results indicating some synergistic effect.

An interesting approach is to treat patients with a short course of the corticosteroid, prednisone, to induce a transient exacerbation of hepatitis prior to administering vidarabine or interferon in the hope of rendering the patient more responsive to treatment (44). Preliminary results indicate that 45-60% of patients so treated respond well to interferon therapy. A large multicenter controlled trial to compare interferon alone versus interferon following treatment with prednisone is currently under way in the U.S. In this study, patients are treated for six weeks with prednisone followed by 90 days of treatment with 5×10^6 IU recombinant interferon alpha. The immunological priming by withdrawal of prednisone followed by interferon appears to induce HBe antigen seroconversion in 80% of heterosexual males and 25% of homosexual males. Other parameters such as disappearance of viral DNA, e antigen, surface antigen and improvement of serum aminotransferase activity were also observed (45).

Similar studies have been performed in patients with chronic delta hepatitis. Preliminary results indicate that interferon inhibits the replication of the hepatitis delta agent and may improve the associated chronic liver damage. As with chronic hepatitis B, the definitive beneficial effect of interferon remains to be shown. An identical conclusion also can be reached from preliminary data on treatment of chronic non-A non-B hepatitis.

Acquired Immunodeficiency Syndrome

In the brief period since the identification of human immunodeficiency virus (HIV) as the causative agent for AIDS, a great deal of research on antiviral agents has been done. Fortunately, HIV is unique in having a reverse transcriptase enzyme not found in normal cells, therefore presenting itself as an excellent target for antiviral drugs. This target has been used to identify such agents as suramin, HPA23 (ammonium 21-tungsto-9-antimoniate), azidothymidine (AZT), and others. Suramin and HPA23 have not been selected for large scale clinical trials. Although in vitro results were promising, and in vivo viral inhibition was observed, no significant clinical benefit or immunological improvement was observed. Further these compounds do not penetrate the blood/brain barrier. Toxic reactions from suramin include lymphotoxicity, renal toxicity and adrenal toxicity. Factors that limit the efficacy of HPA23 are a short half life of approximately 20 minutes and recrudescence of infection once treatment is terminated. Thrombocytopenia was the primary side effect.

AZT is the first drug to be approved by the Food and Drug Administration for the treatment of AIDS. It has been shown to be highly specific in blocking the viral reverse transcriptase rendering it a very potent viral inhibitor at levels well below that which is cytotoxic to uninfected cells. Since the virus can reside in the brain as well as the blood, it is important for anti-HIV agents to penetrate the blood/brain barrier; AZT is able to do so. This property is becoming increasingly more important as investigators find HIV-induced neurological symptoms from dementia to peripheral neuropathy. AZT has reached the approval stage in a remarkably short period. It was first shown to be active against HIV in vitro in February 1985. In July 1985, the first patient was entered into a clinical study. A multicenter placebo control trial was initiated in early 1986 in 282 AIDS patients with pneumocystis carinii pneumonia or other AIDS-related complex (ARC). Sixteen weeks into the study an independent review committee noticed significant differences in the mortality rate. Sixteen of the 137 patients on placebo died, whereas none on AZT died. Patients taking AZT had fewer infections, improved immune function with increased circulating T cells, weight gain, and a general sense of well-being. For these reasons the independent review board prematurely terminated the study and those on placebo were given AZT. On January 16, 1987 an advisory committee at FDA recommended AZT be approved for use in AIDS patients, less than two years from the in vitro discovery. Many questions on long-term benefits and effects are still left unanswered. It is still not known how long the effects of AZT will last or whether it is effective for those who are antibody-positive in the earlier stages of the disease. More research needs to be conducted on the optimal dosage for long-term therapy. AZT's toxicity is underrated. It can cause considerable bone marrow suppression which accentuates immune suppression and leaves patients highly susceptible to bacterial infection. More than one-third of AIDS patients who receive the drug require blood transfusions or must stop using it because it impairs the body's ability to make blood. In an area with a large population of individuals with AIDS, treatment with AZT may be a potential drain on the area's blood supply. On the other hand, others appear to tolerate the drug. Some of the initial patients have continued on the drug for more than 18 months.

A further problem is that production of the drug which requires a large quantity of a scarce compound, thymidine, is expensive resulting in an annual cost of AZT treatment of over $10,000 per patient. A more effective and less toxic drug than AZT is still needed in the treatment of AIDS.

Among the more promising new agents is dideoxycytidine. The mode of action involves termination of viral DNA synthesis as with AZT. Preclinical studies have found good oral bioavailability, direct excretion by the kidney, and relative lack of toxicity in animals (46). Unlike AZT, dideoxycytidine does not reduce the intracellular pyrimidine pool, and therefore, may not result in bone marrow suppression that occurs with AZT. Clinical trials are now being conducted. Antiviral activity has also been seen with dideoxythymidine and dideoxyadenosine.

Ribavirin and foscarnet have been shown to inhibit HIV replication in vitro. Both drugs have been administered to humans for other viral infections and clinical trials are now under way to determine their efficacy in AIDS. Placebo-controlled clinical trials were conducted on homosexual males with lymphadenopathy syndrome using oral ribavirin for 24 weeks followed by no treatment for 4 weeks. Although preliminary results show some beneficial effects, HIV cultures remained positive and no immunological improvement was observed in the treatment group (47).

Before HIV was recognized as the etiological agent for AIDS, interferon was initiated in AIDS patients because of its known antiviral immunoregulatory and antiproliferating properties. Alpha interferon was one of the first antiviral drugs used against AIDS infection to treat Kaposi's sarcoma. The results of several studies using interferon were unremarkable and variable. Although some beneficial response was observed, few experienced complete remission. CMV infections were not eradicated or prevented. A beneficial clinical response appeared to be associated with patients with an intact immune system treated at the early stage of Kaposi's sarcoma. These patients usually have no history of opportunistic infection, no chronic CMV viremia, total lymphocyte count of greater than $1500/\text{mm}^3$ and T4 counts of greater than $200/\text{mm}^3$ (48). Gelmann also noted an association between patients with endogenous acid labile alpha interferon and disease progression suggesting that this may be a preclinical marker for AIDS in high risk groups.

Many other agents such as ansamycin, AL 721 (a lipid compound of neutral glycerides), and cyclosporin have been shown to have minimal or no clinical effect. Test results on gamma interferon, a known immune stimulator, showed unacceptable toxicity. Clinical trials are currently evaluating other immunomodulators such as interleukin-2, thymic humoral factors, imuthiol, etc. as conjunctive therapy to restore the damaged immune system. Newly identified compounds, 2'3'-dideoxycytidine-2'-ene and 2'3'-dideoxythymidin-2'-ene, and combinations of interferon with AZT and suramin with acyclovir show promise in vitro (49,50).

Because of the sudden urgency to find an effective anti-HIV drug, hundreds have been currently screened. Standardized procedures for screening anti-HIV drugs must be developed using a variety of cell types, including cells of human origin, and a multiplicity of infections to more accurately predict the drug response in humans.

FUTURE DIRECTIONS

Combination Drug Therapy

Many of the antiviral agents that have been tested have drawbacks. Some are only partially effective and some have excessive toxicity. One possible mechanism to improve the antiviral effect is to use these compounds in combination. A few limited applications have already been discussed. Several additional in vitro studies show clinical potential. There are at least three theoretical reasons that combination drug therapy would be beneficial. First, combination drug treatment could reduce the chances of drug resistance by attacking different sites of viral replication. A recent study on mice infected with ACV-resistant herpes simplex virus was refractory to treatment with ACV alone. However, a positive response was observed with vidarabine or a combination of vidarabine/ACV (51). If each of two antivirals affected viral replication at different sites, it would be more difficult for resistance to develop.

Second, some drug combinations result in synergy. This has a two-fold benefit leading to the third benefit--reduced toxicity. The synergy results in an improved antiviral effect and is particularly beneficial when one or both of the drugs have toxicity that limits their usefulness. The synergy allows for reduced dosage and therefore reduced toxicity. The bone marrow toxicity caused by AZT in treating HIV is a

concern when used in long-term prophylactic therapy. Studies are now
showing that, when combined with interferon, AZT is required in smaller
dosages to provide greater overall efficacy than when used alone. Drug
combinations using interferon and either ACV or triflurothymidine have
been effective clinically in treating herpes keratitis. Nonetheless,
one must exercise caution in developing drug combinations. Not all drug
combinations are effective. Sometimes combination regimens could
increase the toxicity of one or both drugs or the response of one drug
may nullify the effect of the other. Recent laboratory studies
investigating drug combinations for the treatment of AIDS found
ribavirin antagonized AZT's ability to inhibit HIV replication by
suppressing the cellular enzymes that phosphorylate AZT to its active
form. In six separate experiments using a variety of cell types, mode
of HIV inoculation, and time of drug addition, antagonism was
demonstrated (52). Caution must always be observed in clinical trials
with drug combinations. Toxicity or side effects studies with single
drugs are not enough; one must be alert to new side effects resulting
from the combination. In certain drug combinations, it may be more
effective, or less toxic to give drugs sequentially rather than together
as has been demonstrated in chronic hepatitis B.

Development of Antiviral Drugs

As stated earlier, antiviral research is now at the stage where a
rational approach can be taken to develop antiviral drugs. Many
antiviral agents being used today were originally anticancer or
antiparasitic agents later found to inhibit viruses. Although
serendipity will continue to contribute to the discovery of new drugs, a
great deal has been learned concerning the cellular and molecular
biology of viruses and the mechanism of viral replication for a targeted
approach to be feasible.

Electron microscopy, x-ray diffraction, nuclear magnetic resonance,
and computer-assisted molecular modeling technique can be used for the
analysis of the three-dimensional structure of the virus. Modeling of
the viral spikes and receptor sites in the T4 host cells will lead to a
better understanding of the attachment mechanism and has the potential
for developing blocking inhibitors. Recent crystallography using
computer-assisted molecular modeling generated a three-dimensional
structural map to study the mode of action of the antiviral agent WIN
51711 (5- 7- 4-(4,5-dihydro-2-oxazoly)phenoxy heptyl-3-methylisoxazole)
on the human rhinovirus. The mapping showed the WIN compound binding
in a canyon formed by one of the four viral coat protein of the
rhinovirus. The WIN compound fits neatly into a pocket of the viral
coat, stabilizing the virus and preventing the release of the viral
nucleic acid needed to produce progeny in the host cell (2). This
finding compliments laboratory studies that report on WIN 51711's broad
spectrum antipicornavirus activity in vitro and in poliovirus-infected
and echovirus-infected mice (53,54). Similar research is being pursued
to determine the structure of the tat III protein that triggers rapid
replication of HIV. Resulting information may lead to a drug that
inhibits this protein. Three-dimensional structural mapping of viruses
or of virus/drug interaction will greatly improve our ability to modify
structurally or design new antiviral drugs that precisely fit into
certain niches in the viral landscape. Research is in progress to
develop oligopeptide analogues, synthetic ligands, and monoclonal
antibodies that prevent viral attachment to the host cell or prevent
viral uncoating by effectively blocking active sites. Three-dimensional
studies of the virus binding-site and some of the drugs proven effective

can lead to a better understanding of their effect. This in turn can lead to the design of drugs with greater affinity and tighter binding capacity.

Similarly, a structural analysis of the viral membrane may lead to design of compounds which can alter the membrane and attachment sites. It has been demonstrated that extraction of cholesterol from the membrane of some viruses (VSV) affects its rigidity with significant reduction in their infectivity.

In addition to the tat protein, studies on the molecular biology of the virus have provided considerable information on other internal viral-specific components. Reverse transcriptase, unique to retroviruses, catalyzes several reactions that do not occur in normal cells, i.e., RNA-directed DNA polymerase activity, a closely related DNA-directed DNA polymerase activity, a ribonuclease H activity and an endonuclease. An understanding of their chemistry, particularly the first three which are essential for generating double-stranded DNA copies of the viral RNA, can lead to the design of compounds which can inhibit their activity. Ideally, such an inhibitor would be a compound which is inert until activated by the target enzyme.

As logical as the targeted approach appears, it is a difficult procedure to pursue. In some instances, such as crystallography large amounts of highly purified proteins are required. In others, it requires the combined talents of scientists in several disciplines. This venture is costly in time and money with no certain outcome. History has shown that many compounds which proved to be highly effective antiviral agents in vitro were disappointing in vivo. The targeted approach, however, must be attempted. If it does not yield an effective antiviral agent it will without question yield invaluable information which can lead eventually to antivirals against other viruses.

Another area which must be pursued to increase the success of antiviral therapy is improved drug delivery to the targeted site. Ideally, the drug should be available at a continuous concentration for the required period of time at the targeted site. Experiments have shown that liposome-encapsulated drugs are more effective than the unencapsulated drug at the same dosage. The liposome prevents the rapid destruction of the drug, allowing for a slow release of the drug into the bloodstream. Liposomes can also be made to be taken up selectively by the reticuloendothelial system. In a murine model, concentrations of ribavirin were 25 times greater in the spleen, 10 times greater in the lungs, and 4 times greater in the liver when administered encapsulated in liposome rather than in a free state (Personal communication, Dr. E. Kurstak). To increase target specificity further, liposomes may be incorporated with monoclonal antibodies that are directed to specific target sites. The unfortunate drawback of liposomes is their fragile nature. They have served to prove the value of the system but more stable encapsulating materials are needed. Polymeric encapsulates may serve as a possible alternative. Research to improve drug delivery, such as in retrograde axonal flow of agents attached to protein, albumin microspheres, and drug delivery devices is in progress.

Challenges

Although it appears that there has been an explosion of research and progress in the area of antiviral drug development, much still remains to be done. It is true that the AIDS epidemic has attracted much attention and financial support and that much of the information obtained from research will benefit work on other viral infections. However, viruses listed on Table 1 warrant intensive research in their own right. For many of these viruses such as papilloma and hepatitis B, research is hindered by the fact that a cell culture has still not been found to study the virus effectively and others lack suitable animal models. In another area of importance, viruses such as influenza with rapidly changing surface antigens and viruses with animal reservoirs have characteristics that make control of the virus more difficult. Recently, it was thought that viral infections were no longer of prime importance since vaccines were available to control the major problems. Not only has the introduction of HIV changed this view but there are now indications of viral involvement in the etiology of some chronic and degenerative diseases such as diabetes, cancer, arthritis, multiple sclerosis, and heart disease.

Table 1. Viruses warranting increased efforts for
         antiviral agent development

Human immunodeficiency	Parainfluenza
Herpes	Hepatitis
Papilloma	Enteroviruses
Influenza	Calici
Rhino	Arena
Corona	Arbo
Respiratory Syncytial	Rabies

Latency is one of the major obstacles preventing antiviral treatment from being a "cure" for viral infections such as hepatitis, papilloma, herpes, and AIDS. These viruses as well as some others, have the ability to integrate their genome into the nucleic acid of the host cell. They then can replicate in the host cell indefinitely without destroying it. Eventually, due to a stimulus which is not fully understood, the virus can again replicate causing activation of the disease. Several unconventional viruses are known to cause a slow persistent viral disease such as kuru and Creutzfeldt-Jakob spongiform encephalopathies. In addition, the more familiar conventional viruses that cause persistent infections are adenovirus which persists in tonsils causing periodic disease, congenital rubella which causes a variety of malformations, subacute sclerosing panencephalitis (SSPE) which is associated with measles, polyomavirus such as JC virus which can cause progressive multifocal leukoencephalopathy, herpes simplex virus which causes recurrent genital and oral lesions, varicella-zoster virus which causes shingles, and retrovirus which causes AIDS or leukemia. Many questions on the mechanism of latency remain to be answered. It is not certain whether a specific protein is needed to activate the latent viral genome. If so, an inhibiting agent can be synthesized to prevent this activation. More research is needed to investigate if many latent infections are actually suppressed by the host immune system and why the immune system fails during a recrudescence. Once the viral genome is incorporated, a persistent infection may also evolve with chronic illness. Until an agent is developed to suppress recrudescence or destroy all latently infected cells, individuals with latent or persistent viral infections face long-term and possibly lifetime treatment. Any new drug developed for such infections, therefore, must be easily administered, either orally or topically, and must have a low incidence of long-term toxic side effects.

This in turn will lead to potential problems of antiviral drug resistance as has been observed with antibiotics. As antiviral drugs make their debut as acceptable therapy for viral infections, we must be alert to the development of resistant strains. A dichotomy exists between the development of a less toxic antiviral agent and one that is less likely to cause resistance. As antiviral drugs are developed which target specific viral chemical events to make them less toxic, this, in turn, provides a greater opportunity for viral resistance. Since immunocompromised patients undergoing cancer treatment or organ transplantation are the patients in greatest need of available antiviral agent therapy, it is understandable that this is the population where drug resistant strains of viruses have been reported (55). Although drug resistance has been observed, it has not been a serious problem to date. In most instances reported to date, the resistant strain has been less virulent than the parent strain. Further, when co-cultivated, the parent strain overcomes the resistant strain. Nonetheless, it may be only a matter of time until a serious drug-resistant strain will be reported and a workable solution, such as an alternative antiviral agent which targets another site or a combination drug treatment which targets several sites, should be available.

If antivirals are to be effective, they must be used during the period of viral replication which in many instances occurs very early in the disease. Early diagnosis of the viral infection, therefore, is imperative. Although major efforts and funding are concentrated on antiviral drug development, the effectiveness of most drugs depends on a simple rapid technique to identify early a specific virus as the etiological agent, so that immediate treatment with the proper drug can begin. As mentioned previously, the mortality rate for herpes encephalitis is dramatically reduced with fewer debilitating sequelae when ACV is administered early in the course of the infection. Although a more rapid and accurate assay using antigen detection is desirable, brain biopsy is the only accurate assay for this infection. Even when antiviral agents are not available for treatment, rapid viral identification is beneficial to the patient since it can properly direct treatment. The use of antibiotics or other non-effective therapies can be avoided, providing a more cost-effective and therapeutically sound treatment for the patient. Such instances include pneumonia (respiratory viruses), aseptic meningitis, acute gastroenteritis (rotaviruses), congenital illness (CMV, HIV, rubella, enteroviruses) conjunctivitis (adenovirus, HSV), etc.

Currently virus isolation and identification can be reported within 2-6 days from receipt of specimen. It can be relatively rapid when the cytopathic effect is readily recognized as in the case of RSV, influenza or enterovirus or when there is a characteristic hemadsorption. Much emphasis currently is being placed on immunological techniques for rapid identification. Whenever possible, culture antigen amplification should be performed prior to indirect fluorescent antibody techniques, preferably using monoclonal antibody. In some instances direct fluorescent antibody techniques are used to identify virus in the specimen such as tissue scraping. Direct detection by electron microscope examination of feces and other secretions can be accomplished in 15 minutes for rotavirus identification.

Among the newer diagnostic techniques are nucleic acid hybridization technique for detecting viral nucleic acids in specimens. Solid-phase immunoassay (SPIA) is used for antigen or antibody detection. The indicator for SPIA's is either a radioactive label for RIA or the action of an enzyme label on an added substrate as in the

ELISA. Commercial kits are now available for hepatitis, rotavirus, and HIV identification.

Finally and most importantly, the greatest challenge we face is proving efficacy once an agent is identified. The inclination is to rush into clinical trial without sufficient thought to the natural history of the disease, show benefit in a few patients without the proper controls and rush to publication. Then pressure is brought on authorities to release the drug for general use. This results in a great disservice to the field and to the patient.
The efficacy of a drug can be determined more rapidly if results of all clinical trials can readily be accepted and validly compared. This can only come about if efforts are made to insure that clinical trials are double-blind and placebo-controlled. Care must be taken to select techniques that identify the virus properly and to develop objective measures to gauge viral and clinical responses with a proper understanding of the normal progression of the disease. Only then will our resources in time, money, and volunteers be properly and fully used. Without a placebo control, an identified beneficial clinical response may actually be a part of the natural course of the disease. Even when the results are negative, a properly controlled trial is useful in providing more insight into the pathogenesis of the disease.

Based on the number of agents approved for antiviral therapy since the last meeting, progress in this field has been slow. Only three agents--ribavirin for respiratory syncytial virus infections, oral ACV for genital herpes, and azidothymidine for AIDS have been made available. However, the field has been greatly expanded and much excitement has been generated. There are several new investigators entering the field, particularly in the area of targeted antiviral development. There are also several compounds against HIV entering clinical studies. I am most optimistic that the progress to take place between this and the next meeting in this series will be very gratifying.

References

1. G. J. Galasso, Antiviral agents: Why not a "penicillin" for antiviral infections?, in: "Targets for the Design of Antiviral Agents?," E. DeClercq and R. T. Walker, eds., Plenum Press, New York (1984).
2. T. J. Smith, M. J. Kremer, M. Luo, G. Vriend, E. Arnold, G. Kamer, M. G. Rossmann, M. A. McKinlay, G. D. Diana and M. J. Otto, The site of attachment in human rhinovirus 14 for antiviral agents that inhibit uncoating, Science 233:1286-1293 (1986).
3. D. H. Shepp, P. S. Dandliker and J. D. Meyers, Treatment of varicella-zoster virus infection in severely immunocompromised patients: A randomized comparison of acyclovir and vidarabine, NEJM 314:208-212 (1986).
4. J. L. Vilde, F. Bricaire, C. Leport, M. Renaudie and F. Brun-vezinet, Comparative trial of acyclovir and vidarabine in disseminated varicella-zoster virus infections in immunocompromised patients, J. Med. Virol. 20:127-134 (1986).
5. S. Feldman, P. K. Robertson, L. Lott and D. Thornton, Neurotoxicity due to adenine arabinoside therapy during varicella-zoster virus infections in immunocompromised children, JID 154:889-893 (1986).
6. B. Leyland-Jones, H. Donnelly, S. Goshen, P. Myskowski, A. L. Donner, M. Fanucchi, J. Fox and the Memorial Sloan-Kettering Antiviral Working Group, 2'-Fluoro-5-iodoarabinosylcytosine, a

new potent antiviral agent: Efficacy in immunosuppressed
individuals with herpes zoster, JID 154:430-436 (1986).

7. R. J. Whitley, C. A. Alford, Jr., M S. Hirsch, R. T. Schooley, J.
P. Luby, F. Y. Aoki, D. Hanley, A. J. Nahmias and S-J Soong,
Vidarabine vs. acyclovir therapy in herpes simplex
encephalitis, NEJM 314:144-149 (1986).

8. R. J. Whitley, A. J. Nahmias, S-J Soong, G. J. Galasso, C. L.
Fleming and C. A. Alford, Jr., Vidarabine therapy of neonatal
herpes simplex virus infection, J. Ped. 66:495-501 (1980).

9. R. J. Whitley, A. Arvin, L. Corey, D. Powell, S. Plotkin, S. Starr,
C. A. Alford, Jr., J. Connor, A. J. Nahmias, S-J Soong and the
NIAID Collaborative Study Group, Vidarabine versus acyclovir
therapy of neonatal herpes simplex virus infection, Society
for Pediatric Research, Abstract #987, Washington, D.C.
(1986).

10. S. E. Straus, H. E. Takiff, M. Seidlin, S. Bachrach, L. Linninger,
J. J. DiGiovanna, K. A. Western, H. A. Smith, S.
Nusinoff-Lehrman, T. Creagh-Kirk and D. W. Alling, Suppression
of frequently recurring genital herpes: A placebo controlled
double-blind trial of oral acyclovir, NEJM 310:1545-1550
(1984).

11. J. M. Douglas, C. Critchlow, J. Benedetti, G. J. Mertz, J. D.
Conner, M. A. Hintz, A. Fahnlander, M. Remington, C. Winter
and L. Corey, A double-blind study of oral acyclovir for
suppression of recurrences of genital herpes simplex virus
infection, NEJM 310:1551-1556 (1984).

12. S. E. Straus, M. Seidlin, H. E. Takiff, J. F. Rooney, S.
Nusinoff-Lehrman, S. Bachrach, J. M. Felser, J. J. DiGiovanni,
G. J. Grimes, Jr., H. Krakauer, C. Hallahan and D. Alling,
Double-blind comparison of weekend and daily regimens of oral
acyclovir for suppression of recurrent genital herpes,
Antiviral Res. 6:151-159 (1986).

13. J. L. Drucker, L. G. Davies, and the Herpes Collaborative Group,
Abstr., 2nd World Congress Sex. Transm. Dis., (1986).

14. D. H. Shepp, P. S. Dandliker, N. Flournoy and J. D. Meyers,
Once-daily intravenous acyclovir for prophylaxis of herpes
simplex virus reactivation after marrow transplantation, J.
Antimicrob. Chemother. 16:389-395 (1985).

15. J. M. Douglas, Jr., F. N. Judson, M. J. Levin, J. A. Bosso, S. L.
Spruance, J. M. Johnston, L. Corey, J. A. McMillan, L. B.
Weiner and J. A. Frank, Jr., A placebo-controlled trial of
topical 8% arildone cream early in recurrent genital herpes,
Antimicrob. Ag. Chemother. 29:464-467 (1986).

16. S. L. Sacks, J. Portnoy, D. Lawee, W. Schlech III, F. Y. Aoki, D.
L. Tyrrell, M. Poisson, C. Bright, J. Kaluski and the Canadian
Cooperative Study Group, Clinical course of recurrent genital
herpes and treatment with foscarnet cream: Results of a
Canadian multicenter trial, JID 155:178-185 (1987).

17. Collaborative DHPG Treatment Study Group, Treatment of serious
cytomegalovirus infections with
9-(1,3-dihydroxy-2-propoxymethyl)guanine in patients with AIDS
and other immunodeficiencies, NEJM 314:801-805 (1986).

18. O. L. Laskin, C. M. Stahl-Bayliss, C. M. Kalman and L. R. Rosecan,
Use of ganciclovir to treat serious cytomegalovirus infections
in patients with AIDS, JID 155:323-327 (1987).

19. O. Ringden, H. Wilczek, B. Lonnqvist, G. Gahrton, B. Wahren and J.
O. Lernestedt, Foscarnet for cytomegalovirus infections,
Lancet 1:1503-1504 (1985).

20. G. Klintmalm, B. Lonnqvist, B. Oberg, G. Gahrton, J. O. Lernestedt,
G. Lundgren, O. Ringden, K. H. Robert, B. Wahren and C.G.

Groth, Intravenous foscarnet for the treatment of severe cytomegalovirus infection in allograft recipients, <u>Scand. J. Inf. Dis.</u> 17:157-163 (1985).

21. H. A. Blacklock, P. Griffiths, P. Stirk and H. G. Prentice, Specific hyper-immune globulin for cytomegalovirus pneumonitis, <u>Lancet</u> 2:152-153 (1985).

22. J. Rush and J. Mills, Effects of combinations of difluromethylornithine and 9[(1,3-dihydroxy-2-propoxy)methyl]-guanine (DHPG) on human cytomegalovirus, <u>J. Med. Virol.</u> 21:269-276 (1987).

23. S. R. Turk, C. Shipman, Jr., R. Nassiri, G. Genzlinger, S. H. Krawczyk, L. B. Townsend and J. C. Drach, Pyrrolo[2,3-d]pyrimidine nucleosides as inhibitors of human cytomegalovirus, <u>Antimicrob. Ag. Chemother.</u> 31:544-550 (1987).

24. J. Andersson, S. Britton, I. Ernberg, U. Andersson, W. Henle, B. Skoldenberg and A. Tisell, Effect of acyclovir on infectious mononucleosis: A double-blind placebo-controlled study, <u>JID</u> 153:283-290 (1986).

25. R. M. Douglas, B. W. Moore, H. B. Miles, L. M. Davies, N. M. H. Graham, P. Ryan, D. A. Worswick and J. K. Albrecht, Prophylactic efficacy of intranasal $alpha_2$-interferon against rhinovirus infections in the family setting, <u>NEJM</u> 314:65-70 (1986).

26. F. G. Hayden, J. K. Albrecht, D. L. Kaiser and J. M. Gwaltney, Jr., Prevention of natural colds by contact prophylaxis with intranasal $alpha_2$-interferon, <u>NEJM</u> 314:71-75 (1986).

27. L. C. Jennings and E. C. Dick, Transmission and control of rhinovirus colds, <u>Eur. J. Epidemiol.</u> (in press).

28. E. C. Dick, S. U. Hossain, K. A. Mink, C. K. Meschievitz, S. B. Schultz, W. J. Raynor and S. L. Inhorn, Interrruption of transmission of rhinovirus colds among human volunteers using virucidal paper handkerchiefs, <u>JID</u> 153:352-356 (1986).

29. G. A. Eby, D. R. Davis and W. W. Halcomb, Reduction in duration of common colds by zinc gluconate lozenges in a double-blind study, <u>Antimicrob. Ag. Chemother.</u> 25:20-24 (1984).

30. F. C. Geist, J. A. Bateman and F. G. Hayden, <u>In vitro</u> activity of zinc salts against human rhinoviruses, <u>Antimicrob. Ag. Chemother.</u> 31:622-624 (1987).

31. R. Dolin, R. C. Reichman, H. P. Madore, R. Maynard, P. N. Linton and J. Webber-Jones, A controlled trial of amantadine and rimantadine in the prophylaxis of influenza A infection, <u>NEJM</u> 307:580-584 (1982).

32. F. G. Hayden and A. S. Monto, Oral rimantadine hydrochloride therapy of influenza A virus H_3N_2 subtype infection in adults, <u>Antimicrob. Ag. Chemother.</u> 29:339-341 (1986).

33. J. Thompson, W. Fleet, E. Lawrence, E. Pierce, L. Morris and P. Wright, A comparison of acetaminophen and rimantadine in the treatment of influenza and infection in children, <u>J. Med. Virol.</u> 21:249-255 (1987).

34. S. Z. Wilson, B. E. Gilbert, J. M. Quarles, V. Knight, H. W. McClung, R. V. Moore and R. B. Couch, Treatment of influenza A (H_1N_1) virus infection with ribavirin aerosol, <u>Antimicrob. Ag. Chemother.</u> 26:200-203 (1984).

35. H. W. McClung, V. Knight, B. E. Gilbert, S. Z. Wilson, J. M. Quarles and G. W. Divine, Ribavirin aerosol treatment of influenza B virus infection, <u>JAMA</u> 249:2671-2674 (1983).

36. C. B. Hall, J. T. McBride, E. E. Walsh, D. M. Bell, C. L. Gala, S. Hildreth, L. G. TenEyck and W. J. Hall, Aerosolized ribavirin treatment of infants with respiratory syncytial virus

infection: A randomized double-blind study, NEJM 308:1443-1447 (1983).

37. C. B. Hall, J. T. McBride, C. L. Gala, S. W. Hildreth and K. C. Schnabel, Ribavirin treatment of respiratory syncytial viral infection in infants with underlying cardiopulmonary disease, JAMA 254:3047-3051 (1985).

38. C. B. Hall and J. T. McBride, Vapors, viruses and views, AJDC 140:331-332 (1986).

39. J. B. McCormick, I. J. King, P. A. Webb, K. M. Johnson, R. O'Sullivan, E. S. Smith, S. Trippel and T. C. Tong, A case control study of the clinical diagnosis and course of Lassa fever, JID 155:445-455 (1987).

40. J. B. McCormick, I. J. King, P. A. Webb, C. L. Scribner, R. B. Craven, K. M. Johnson, L. H. Elliott and R. Belmont-Williams, Lassa Fever: Effective therapy with ribavirin, NEJM 314:20-26 (1986).

41. J. I. Maiztegui, N. J. Fernandez and A. J. de Damilano, Efficacy of immune plasma in treatment of Argentine haemorrhagic fever and association between treatment and late neurological syndrome, Lancet 2:1216-1217 (1979).

42. H. B. Greenberg, R. B. Pollard, L. I. Lutwick, P. B. Gregory, W. S. Robinson and T. C. Merigan, Effect of human leucocyte interferon on hepatitis B virus infection in patients with chronic active hepatitis, NEJM 295:517-522 (1976).

43. R. A. deMan, S. W. Schalm, R. A. Heytink, R. Charmuleau, H. Reesink, J. den Ouden, R. Grym, M. deJong, J. T. M. v.d.Heyden and F. J. W. ten Kate, Interferon plus descyclovir in chronic hepatitis B: Incidence of virus marker elimination and reactivation, J. Med. Virol. 21:120A (1987).

44. O. Yokosuka, M. Omata, F. Imazeki, K. Hirota, J. Mori, K. Uchiuma, Y. Ito and K. Okuda, Combination of short-term prednisolone and adenine arabinoside in the treatment of chronic hepatitis B, Gastroenterology 89:246-251 (1985).

45. R. Perillo, F. Regenstein, M. Peters, C. Bodicky and C. Campbell, A randomized, controlled trial of prednisone withdrawal followed by recombinant alpha interferon (rIFN-∝) in the treatment of chronic hepatitis B, J. Med. Virol. 21:124A-125A (1987).

46. H. Mitsuya and S. Broder, Strategies for antiviral therapy in AIDS, Nature 325:773-778 (1987).

47. R.B. Roberts, P.N.R. Haseltine, P.W.A. Mansell, G.M. Dickinson, J.M. Leedom and K.M. Johnson, Ribavirin delays progression of the lymphadenopathy syndrome to the acquired immune deficiency syndrome, Clin. Res. 35:616A (1987).

48. E. P. Gelman, O. T. Preble, R. Steis, H. C. Lane, A. H. Rook, M. Wesley, J. Jacob, A. Fauci, H. Masur and D. Longo, Human lymphoblastoid interferon treatment of Kaposi's sarcoma in the acquired immune deficiency syndrome: Clinical response and prognosis parameters, Am. J. Med. 78:737-741 (1985).

49. K.L. Hartshorn, M.W. Vogt, T-C Chou, R.S. Blumberg, R. Byington, R.T. Schooley and M.S. Hirsch, Synergistic inhibition of human immunodeficiency virus in vitro by azidothymidine and recombinant alpha A interferon, Antimicrob. Ag. Chemother. 31:168-172 (1987).

50. L. Resnick, P. D. Markham, K. Veren, S. Z. Salahuddin and R. C. Gallo, In vitro suppression of HTLV-III/LAV infectivity by a combination of acyclovir and suramin, JID 154:1027-1030 (1986).

51. R.F. Schinazi, Drug combination for treatment of mice infected with acyclovir-resistant herpes simplex virus, Antimicrob. Ag. Chemother. 31:477-479 (1987).

52. M.W. Vogt, K.L. Hartshorn, P.A. Furman, T-C Chou, J.A. Fyfe, L.A. Coleman, C. Crumpacker, R.T. Schooley, S. Hirsch, Ribavirin antagonizes the effect of azidothymidine on HIV replication, Science 235:1376-1379 (1987).

53. M.A. McKinlay and B.A. Steinberg, Oral efficacy of WIN 51711 in mice infected with human poliovirus, Antimicrob. Ag. Chemother. 29:30-32 (1986).

54. M.A. McKinlay, J.A. Frank, Jr., D.P. Benziger and B.A. Steinberg, Use of WIN 51711 to prevent echovirus type 9-induced paralysis in suckling mice, JID 154:676-681 (1986).

55. C.S. Crumpacker, L.E. Schnipper, S.I. Marlowe, P.N. Kowalsky, B.J. Hershey and M.S. Levin, Resistance to antiviral drugs of herpes simplex virus isolated from a patient treated with acyclovir, NEJM 306:343-346 (1982).

PARTICIPANTS

AL-NAKIB, W., MRC Common Cold Unit, Harvard Hospital, Coombe Road, Salisbury, Wilts SP2 8BW, Great Britain

AMEYE, C., Universitäts-Augenklinik, Moorenstrasse 5, D-4000 Düsseldorf 1, W.-Germany

ANDRIES, K., Department of Life Sciences, Janssen Pharmaceutica, Turnhoutseweg, 30, B-2340 Beerse, Belgium

AUER, M., Abteilung Biophysik, Max-Planck-Institut für Medizinische Forschung, Jahnstrasse 29, D-6900 Heidelberg, W.-Germany

BABA, M., Department of Bacteriology, Fukushima Medical College, Fukushima 960, Japan

BALZARINI, J., Rega Institute for Medical Research, Katholieke Universiteit Leuven, Minderbroedersstraat 10, B-3000 Leuven, Belgium

BARTH, W., Chemistry Department, Bayer AG, Aprather Weg, D-5600 Wuppertal 1, W.-Germany

BARTSCH, D., Department of Experimental Virology, Institute of Sera and Vaccines, W. Pieck Street 108, Praha 10, Czechoslovakia

BATTISTINI, C., R. & D. - Antiinfectives, Farmitalia Carlo Erba, Via dei Gracchi 35, 20146 Milano, Italy

BEATON, G., Department of Chemistry, University of Birmingham, P.O.Box 363, Birmingham B15 2TT, Great Britain

BENTRUDE, W.G., Department of Chemistry, University of Utah, Henry Eyring Building, Salt Lake City, Utah 84112, U.S.A.

BERNAERTS, R., Rega Institute for Medical Research, Katholieke Universiteit Leuven, Minderbroedersstraat 10, B-3000 Leuven, Belgium

BOBEK, M., Grace Cancer Drug Center, Roswell Park Memorial Institute, 666 Elm Street, Buffalo, New York 14263, U.S.A.

BORCHARDT, R.T., Department of Pharmaceutical Chemistry, School of Pharmacy, The University of Kansas, Malott Hall, Lawrence, Kansas 66045-2504, U.S.A.

BREUER, E., Department of Pharmaceutical Chemistry, School of Pharmacy, Faculty of Medicine, The Hebrew University of Jerusalem, P.O.Box 12065, Jerusalem 91120, Israel

BROWN, D.M., MRC Lab of Molecular Biology, Hills Road, Cambridge CB2 2QH, Great Britain

CAMERON, J., Microbial Biochemistry Department, Glaxo Group Research Ltd., Greenford Road, Greenford, Middlesex UB6 OHE, Great Britain

CANONICO, P.G., Department of Antiviral Studies, U.S. Army Research Institute of Infectious Diseases, Fort Detrick, Frederick, Maryland 21701, U.S.A.

CHEN, M.S., Institute of Agriscience, Antimicrobial Research, Syntex Research, 3401 Hillview Avenue, Palo Alto, California 94303, U.S.A.

CHEN, Y.-Q., Shanghai Institute of Organic Chemistry, Academia Sinica, 345 Lingling Lu, Shanghai, China

CHRISTODOULOU, C., MRC Lab of Molecular Biology, Hills Road, Cambridge CB2 2QH, Great Britain

CHRISTOPHERS, J., Department of Virology, Withington Hospital, West Didsbury, Manchester M20 8LR, Great Britain

CHU, C.K., College of Pharmacy, University of Georgia, Athens, Georgia 30602, U.S.A.

COE, P.L., Department of Chemistry, University of Birmingham, P.O.Box 363, Birmingham B15 2TT, Great Britain

COOLS, M., Rega Institute for Medical Research, Katholieke Universiteit Leuven, Minderbroedersstraat 10, B-3000 Leuven, Belgium

COSSTICK, R., Department of Organic Chemistry, The Robert Robinson Laboratories, P.O.Box 147, Liverpool L69 3BX, Great Britain

COSTI, M.P., Dipartimento di Scienze Farmaceutiche, Universita degli Studi di Modena, Via S. Eufemia 19, 41100 Modena, Italy

CROSS, S.S., Antiviral Program, SeaPharm, R.R. ≠, Box 196-A, Fort Pierce, Florida 33450, U.S.A.

DE CLERCQ, E., Rega Institute for Medical Research, Katholieke Universiteit Leuven, Minderbroedersstraat 10, B-3000 Leuven, Belgium

DE FAZIO, G.M.M., Seçao de Virologia Fitopatologica e Fitopatologia, Instituto Biologico, Avenida Cons. Rodrigues Alves 1252, Caixa Postal 7119, Sao Paulo 01000, Brasil

DIANA, G.D., Virology, Sterling-Winthrop Research Institute, Columbia Turnpike, Rensselaer, New York 12144, U.S.A.

DO ROSARIO VASCONCELOS, M., Av. de Saboia no 41 2°A, Edificio Saboia, 2765 Monte Estoril, Portugal

DRACH, J.C., Department of Oral Biology, School of Dentistry, The University of Michigan, Ann Arbor, Michigan 48109, U.S.A.

EGGERS, H.J., Institut für Virologie der Universität zu Köln, Fürst-Puckler Strasse 56, 5000 Köln 41, W.-Germany

ESBER, E., Office of Biological Research and Review Center for Drugs and Biologies, Food and Drug Administration, Room 13-B-43, Parklawn Building, Rockville, Maryland 20857, U.S.A.

FARROW, S., Department of Chemistry, University of Birmingham, P.O.Box 363, Birmingham B15 2TT, Great Britain

FIELD, H.J., Department of Clinical Veterinary Medicine, University of Cambridge, Madingley Road, Cambridge CB3 OES, Great Britain

FIGLEROWICZ, M., Institute of Bioorganic Chemistry, Polish Academy of Sciences, Noskowskiego 12/14, 61-704 Poznan, Poland

FOX, J.J., Donald S. Walker Laboratory, Sloan-Kettering Institute for Cancer Research, 145 Boston Post Road, Rye, New York 10580, U.S.A.

GABRIELSEN, B., Department of Antiviral Studies, U.S. Army Research Institute of Infectious Diseases, Fort Detrick, Frederick, Maryland 21701, U.S.A.

GALASSO, G.J., Extramural Affairs, National Institutes of Health, Building Shannon, Room 111, Bethesda, Maryland 20892, U.S.A.

GALLO, R.C., National Cancer Institute, National Institutes of Health, Building 37, Room 6A09, Bethesda, Maryland 20892, U.S.A.

GHAZZOULI, I., Pharmaceutical Research and Development Division, Bristol-Myers Company, 5 Research Parkway, P.O.Box 5100, Wallingford, Connecticut 06492-7660, U.S.A.

GOSSELIN, G., Laboratoire de Chimie Bio-Organique, Université des Sciences et Techniques du Languedoc, Place Eugène Bataillon, 34060 Montpellier, France

GUPTA, V.S., Department of Veterinary Physiological Sciences, College of Veterinary Medicine, University of Saskatchewan, Saskatoon, Saskatchewan S7N OWO, Canada

HANSEN, A.J., Research Station, Summerland, British Columbia VOH 1ZO, Canada

HARNDEN, M.R., Research Division, Beecham Pharmaceuticals, Biosciences Research Centre, Great Burgh, Yew Tree Bottom Road, Epsom, Surrey KT18 5XQ, Great Britain

HERDEWIJN, P., Rega Institute for Medical Research, Katholieke Universiteit Leuven, Minderbroedersstraat 10, B-3000 Leuven, Belgium

HINCAL, A.A., Pharmaceutical Technology Department, Faculty of Pharmacy, Hacettepe University, Hacettepe, Ankara, Turkey

HOLY, A., Institute of Organic Chemistry and Biochemistry, Czechoslovak Academy of Sciences, Flemingovo namesti 2, 16610 Praha 6, Czechoslovakia

HUMMEL, K., Merz & Co., GmbH & Co., Eckenheimer Landstr. 100-104, D-6000 Frankfurt am Main, W.-Germany

HUTCHINSON, D.W., Department of Chemistry, University of Warwick, Coventry CV4 7AL, Great Britain

JÄHNE, G., Antiviral Drug Research Group, Hoechst Aktiengesellschaft, Postfach 80 03 20, D-6230 Frankfurt am Main 80, W.-Germany

KENDE, M., Department of Antiviral Studies, U.S. Army Medical Research Institute of Infectious Diseases, Fort Detrick, Frederick, Maryland 21701, U.S.A.

KERN, E.R., Division of Infectious Diseases, Department of Pediatrics, School of Medicine, The University of Utah, Salt Lake City, Utah 84132, U.S.A.

KHWAJA, T.A., School of Medicine, University of Southern California, 1303 North Mission Road, Los Angeles, California 90033, U.S.A.

KOOMEN, G.J., Laboratory of Organic Chemistry, University of Amsterdam, Nieuwe Achtergracht 129, 1018 WS Amsterdam, The Netherlands

KRISTOFFERSON, A., Antiviral Chemotherapy, Research & Development Laboratories, Astra Alab AB, S-151 85 Södertälje, Sweden

LA COLLA, P., Istituto di Microbiologia, Universita' Degli Studi di Cagliari, Via Porcell N. 4, 09124 Cagliari, Italy

LANGEN, P., Zentralinstitut für Molekularbiologie, Akademie der Wissenschaften der DDR, Robert-Rössle-Strasse 10, D-1115 Berlin-Buch, DDR

LAUGHLIN, C.A., National Institute of Allergy and Infectious Diseases, National Institutes of Health, Building Westwood, Room 750, Bethesda, Maryland 20892, U.S.A.

LINDBORG, B., Antiviral Chemotherapy, Research & Development Laboratories, Astra Alab AB, S-151 85 Södertälje, Sweden

MAAG, H., Institute of Bio-Organic Chemistry, Syntex Research, 3401 Hillview Avenue, Palo Alto, California 94303, U.S.A.

MANFREDINI, S., Dipartimento di Scienze Farmaceutiche, Universita di Ferrara, Via Scandiana 21, 44100 Ferrara, Italy

MARTIN, J.A., Roche Products Ltd., P.O.Box 8, Welwyn Garden City, Hertfordshire AL7 3AY, Great Britain

MARTIN, J.C., Anti-Infective Chemistry, Bristol-Myers Company, 5 Research Parkway, P.O.Box 5100, Wallingford, Connecticut 06492-7660, U.S.A.

MELCHERS, W., S.S.D.Z., Postbus 5010, 2600 GA Delft, The Netherlands

NEENAN, J.P., Department of Chemistry, College of Science, Rochester Institute of Technology, One Lomb Memorial Drive, Rochester, New York 14623-0887, U.S.A.

NOBLE, S.A., Microbiological Chemistry Department, Glaxo Group Research Ltd., Greenford Road, Greenford, Middlesex UB6 OHE, Great Britain

OTTO, M.J., Virology, Sterling-Winthrop Research Institute, Columbia Turnpike, Rensselaer, New York 12144, U.S.A.

OTVOS, L., Central Research Institute for Chemistry, The Hungarian Academy of Sciences, P.O.Box 17, H-1525 Budapest, Hungary

PAGANO, J.S., Lineberger Cancer Research Center, 237H, Division of Health Affairs, The University of North Carolina at Chapel Hill, Chapel Hill, North Carolina 27514, U.S.A.

PAUWELS, R., Rega Institute for Medical Research, Katholieke Universiteit Leuven, Minderbroedersstraat 10, B-3000 Leuven, Belgium

PFLEIDERER, W., Fakultät für Chemie, Universität Konstanz, Postfach 5560, D-7750 Konstanz 1, W.-Germany

PIFAT, D.Y., Department of Antiviral Studies, U.S. Army Research Institute of Infectious Diseases, Fort Detrick, Frederick, Maryland 21701, U.S.A.

PRUSOFF, W.H., Department of Pharmacology, Yale University School of Medicine, Sterling Hall of Medicine, 333 Cedar Street, P.O.Box 3333, New Haven, Connecticut 06510, U.S.A.

RENIS, H.E., Cancer & Viral Diseases Research, The Upjohn Company, Kalamazoo, Michigan 49001, U.S.A.

REVANKAR, G.R., Medicinal Chemistry, Nucleic Acid Research Institute, ICN Plaza, 3300 Hyland Avenue, Costa Mesa, California 92626, U.S.A.

ROBERTS, S.M., Department of Chemistry, University of Exeter, Stocker Road, Exeter EX4 4QD, Great Britain

ROBINS, M.J., Cancer Research Center, Brigham Young University, Provo, Utah 84602, U.S.A.

ROBINS, R.K., Molecular Research, Nucleic Acid Research Institute, ICN Plaza, 3300 Hyland Avenue, Costa Mesa, California 92626, U.S.A.

ROSENTHAL, H.A., Bereich Medizin (Charité) der Humboldt-Universität zu Berlin, Lehrstuhl für Virologie, Schumannstrasse 20/21, Postfach 140, D-1040 Berlin, DDR

ROSENTHAL, S., Zentralinstitut für Molekularbiologie, Akademie der Wissenschaften der DDR, Lindenberger Weg 70, D-1115 Berlin-Buch, DDR

ROSENWIRTH, B., Department of Virology, Sandoz Forschungsinstitut, Brunnerstrasse, 59, A-1235 Wien, Austria

SACKS, S.L., Department of Medicine, Health Sciences Centre Hospital, The University of British Columbia, 2211 Wesbrook Mall, Vancouver, British Columbia V6T 1W5, Canada

SÁGI, G., Central Research Institute for Chemistry, Hungarian Academy of Sciences, P.O.Box 17, H-1525 Budapest, Hungary

SÁGI, J., Central Research Institute for Chemistry, Hungarian Academy of Sciences, P.O.Box 17, H-1525 Budapest, Hungary

SAKUMA, T., Department of Microbiology, Asahikawa Medical College, 4-5 Nishikagura, Asahikawa 078-11, Japan

SCHELLEKENS, H., Primate Center TNO, Lange Kleiweg, 151, Postbus 5815, 2280 HV Rijswijk, The Netherlands

SCHROEDER, C., Bereich Medizin (Charité) der Humboldt-Universität zu Berlin, Institut für Virologie, Schumannstrasse 20/21, Postfach 140, D-1040 Berlin, DDR

SCHWARTZ, J., Department of Antiviral Chemotherapy, Schering Corporation, 60 Orange Street, Bloomfield, New Jersey 07003, U.S.A.

SCZAKIEL, G., Institut für Virologie, Deutsches Krebsforschungszentrum, Im Neuenheimerfeld 280, D-6900 Heidelberg, W.-Germany

SECRIST III, J.A., Organic Chemistry Research Department, Southern Research Institute, 2000 Ninth Avenue South, P.O.Box 55305, Birmingham, Alabama 35255-5305, U.S.A.

SEELA, F., Laboratorium für Organische und Bioorganische Chemie, Fachbereich Biologie/Chemie, Universität Osnabrück, Barbarastrasse 7, D-4500 Osnabrück, W.-Germany

SELWAY, J., Department of Biochemical Virology, Wellcome Research Laboratories, Langley Court, Beckenham, Kent BR3 3BS, Great Britain

SERAFINOWSKI, P., Cancer Research Campaign Laboratory, Clifton Avenue, Sutton, Surrey SM2 5PX, Great Britain

SHANNON, W.M., Microbiology and Virology Division, Southern Research Institute, 2000 Ninth Avenue South, P.O.Box 3307-A, Birmingham, Alabama 35255-5305, U.S.A.

SHIPMAN, Jr., C., Department of Oral Biology, School of Dentistry, The University of Michigan, Ann Arbor, Michigan 48109, U.S.A.

SIDWELL, R.W., Department of Animal, Dairy and Veterinary Sciences, College of Agriculture, Utah State University, Logan, Utah 84322-5600, U.S.A.

SIM, I.S., Department of Experimental Oncology and Virology, Hoffmann-La Roche Inc., Nutley, New Jersey 07110, U.S.A.

SMEE, D.F., Antiviral Testing Program, Nucleic Acid Research Institute, ICN Plaza, 3300 Hyland Avenue, Costa Mesa, California 92626, U.S.A.

STERNBERG, M.G., Department of Crystallography, Birbeck College, University of London, Mallet Street, London WC1 E7HX, Great Britain

STOLLAR, V., Department of Molecular Genetics and Microbiology, University of Medicine & Dentistry of New Jersey, 675 Hoes Lane, Piscataway, New Jersey 08854-5635, U.S.A.

UEDA, T., Faculty of Pharmaceutical Sciences, University of Hokkaido, Sapporo 060, Japan

UNGHERI, D., R. & D. - Antiinfectives, Farmitalia Carlo Erba, Via Papa Giovanni XXIII 23, 20014 Nerviano (Milano), Italy

VONKA, V., Department of Experimental Virology, Institute of Sera and Vaccines, W. Pieck Street 108, Praha 10, Czechoslovakia

WALKER, R.T., Department of Chemistry, University of Birmingham, P.O.Box 363, Birmingham B15 2TT, Great Britain

WATAYA, Y., Faculty of Pharmaceutical Sciences, Okayama University, Tsushima, Okayama 700, Japan

WIEWIOROWSKA, D., Institute of Bioorganic Chemistry, Polish Academy of Sciences, Noskowskiego 12/14, 61-704 Poznan, Poland

WIEWIOROWSKI, M., Institute of Bioorganic Chemistry, Polish Academy of Sciences, Noskowskiego 12/14, 61-704 Poznan, Poland

WIGDAHL, B., Department of Microbiology, The Milton S. Hershey Medical Center, The Pennsylvania State University, Hershey, Pennsylvania 17033, U.S.A.

WINKLER, I., Abteilung für Chemotherapie, Hoechst Aktiengesellschaft, Postfach 80 03 20, D-6230 Frankfurt am Main 80, W.-Germany

WONG-STAAL, F., National Cancer Institute, National Institutes of Health, Building 37, Bethesda, Maryland 20892, U.S.A.

ZAHER, H., Department of Pharmaceutical Chemistry, School of Pharmacy, Faculty of Medicine, The Hebrew University of Jerusalem, P.O.Box 12065, Jerusalem 91120, Israel

INDEX